NONMAMMALIAN GENOMIC ANALYSIS
A Practical Guide

NONMAMMALIAN GENOMIC ANALYSIS

ANALYSIS

A Practical Guide

BRUCE BIRREN

Whitehead Institute/MIT Center for Genome Research

Cambridge, Massachusetts

ERIC LAI

Glaxo Wellcome

Research Triangle Park

North Carolina

ACADEMIC PRESS

San Diego   New York   Boston   London
Sydney   Tokyo   Toronto

Academic Press, Inc.
525 B Street, Suite 1900, San Diego, California 92101-4495, USA
http://www.apnet.com

Academic Press Limited
24-28 Oval Road, London NW1 7DX, UK
http://www.hbuk.co.uk/ap/

Library of Congress Cataloging-in-Publication Data

Nonmammalian genomic analysis : a practical guide / edited by Bruce
     Birren, Eric Lai.
          p.   cm.
     Includes bibliographical references (p.   ) and index.
     ISBN 0-12-101285-9 (alk. paper)
     1. Gene mapping--Laboratory manuals.   2. Microbial genetics-
-Laboratory manuals.   3. Plant genetics--Laboratory manuals.
I. Birren, Bruce W.   II. Lai, Eric Hon-Cheong, date.
QH445.2.N663   1996
574.87'3382--dc20   96-33958
                                                             CIP

PRINTED IN THE UNITED STATES OF AMERICA
96 97 98 99 00 01 EB 9 8 7 6 5 4 3 2 1

Contents

3

Isolation and Analysis of High-Molecular-Weight DNA from Plants   61

Martin Ganal

4

Generating and Using DNA Markers in Plants   75

J. Antoni Rafalski, Julie M. Vogel, Michele Morgante, Wayne Powell, Chaz Andre, and Scott V. Tingey

5

Genome Mapping of Protozoan Parasites by Linking Clones   135

S. P. Morzaria

6

*Macrorestriction Mapping and Analysis
of Bacterial Genomes*   165

Ute Römling, Rainer Fislage, and Burkhard Tümmler

7

Cosmid Cloning with Small Genomes   197

Rainer Wenzel and Richard Herrmann

8

*Construction of P1 Artificial Chromosome (PAC) Libraries
from Lower Vertebrates*   223

Chris T. Amemiya, Tatsuya Ota, and Gary W. Litman

9

The Selection of Chromosome-Specific DNA Clones from African Trypanosome Genomic Libraries   **257**

Sara E. Melville, Nancy S. Shepherd, Caroline S. Gerrard, and Richard W. F. Le Page

10

Analysis of the Dictyostelium discoideum Genome   **293**

Adam Kuspa and William F. Loomis

11

Integrated Genome Mapping by Hybridization Techniques   **319**

Jörg D. Hoheisel, Elmar Maier, Richard Mott, and Hans Lehrach

Contributors

Numbers in parentheses indicate the pages on which the authors' contributions begin.

Christopher T. Amemiya (223), Center for Human Genetics, Boston University School of Medicine, Boston, Massachusetts 02118

Chaz Andre (75), Applied Biosystems Division, Perkin Elmer Corporation, Foster City, California 94404

Louis Bernier (25), Centre de Recherche en Biologie Forestière, Faculté de Foresterie et de Géomatique, Université Laval, Ste-Foy, Québec G1K 7P4, Canada

Bruce Birren (1), Whitehead Institute/MIT Center for Genome Research, Cambridge, Massachusetts 02139

Ken Dewar (25), Department of Biology, University of Pennsylvania, Philadelphia, Pennsylvania 19104

Rainer Fislage (165), Klinische Forschergruppe, Medizinische Hochschule Hannover, D-30623 Hannover, Germany

Martin Ganal (61), Institute for Plant Genetics and Crop Plant Research, D-06466 Gatersleben, Germany

Caroline S. Gerrad (257), Department of Pathology, Division of Microbiology and Parasitology, University of Cambridge, Cambridge CB2 1QP, United Kingdom

Richard Herrmann (197), ZMBH Mikrobiologie, Universitüt Heidelberg, D-69120 Heidelberg, Germany

Jörg D. Hoheisel (319), German Cancer Research Center, D-69120 Heidelberg, Germany

Adam Kuspa (293), Department of Biochemistry, Baylor College of Medicine, Houston, Texas 77030

Eric Lai (1), Glaxo Wellcome, Research Triangle Park, North Carolina 27709

Richard W. F. Le Page (257), Department of Pathology, Division of Microbiology and Parasitology, University of Cambridge, Cambridge CB2 1QP, United Kingdom

Jennifer S. Lee (1), Whitehead Institute, Massachusetts Institute of Technology Center for Genome Research, Cambridge, Massachusetts 02139

Hans Lehrach (319), Imperial Cancer Research Fund, Lincoln's Inn Fields, London WC2A 3PX, United Kingdom

Roger C. Levesque (25), Microbiologie Moléculaire et Génie des Protéines, Faculté de Médecine et Pavillon Charles-Eugène Marchand, Université Laval, Ste-Foy, Québec G1K 7P4, Canada

Gary W. Litman (223), All Children's Hospital and University of South Florida, St. Petersburg, Florida 33701

William F. Loomis (293), Department of Biology, University of California, San Diego, La Jolla, California 92093

Elmar Maier (319), Imperial Cancer Research Fund, Lincoln's Inn Fields, London WC2A 3PX, United Kingdom

Sara E. Melville (257), Department of Pathology, Division of Microbiology and Parasitology, University of Cambridge, Cambridge CB2 1QP, United Kingdom

Michele Morgante (75), Biotechnology Research, Agricultural Products Department, E. I. du Pont de Nemours and Co. (Inc.), Wilmington, Delaware 19880

S. P. Morzaria (135), International Livestock Research Institute, Nairobi, Kenya

Richard Mott (319), Imperial Cancer Research Fund, Lincoln's Inn Fields, London WC2A 3PX, United Kingdom

Tatsuya Ota (223), Center for Human Genetics, Boston University School of Medicine, Boston, Massachusetts 02118

Wayne Powell (75), Biotechnology Research, Agricultural Products Department, E. I. du Pont de Nemours and Co. (Inc.), Wilmington, Delaware 19880

J. Antoni Rafalski (75), Biotechnology Research, Agricultural Products Department, E. I. du Pont de Nemours and Co. (Inc.), Wilmington, Delaware 19880

Ute Römling (165), Klinische Forschergruppe, Medizinische Hochschule Hannover, D-30623 Hannover, Germany

Nancy S. Shepherd (257), Glaxo Wellcome Inc., Research Triangle Park, North Carolina 27709

Scott V. Tingey (75), Biotechnology Research, Agricultural Products Department, E. I. du Pont de Nemours and Co. (Inc.), Wilmington, Delaware 19880

Burkhard Tümmler (165), Zentrum Biochemie II, Medizinische Hochschule Hannover, D-30623 Hannover, Germany

Julie M. Vogel (75), Biotechnology Research, Agricultural Products Department, E. I. du Pont de Nemours and Co. (Inc.), Wilmington, Delaware 19880

Rainer Wenzel (197), Landeskriminalamt, 55118 Mainz, Germany

Preface

The past decade has seen an explosion in our ability to generate genome maps and isolate genes, with much of this progress being attributable to the concerted effort dubbed the Human Genome Project. However valuable the specific information gathered about the human genome proves to be, the greatest legacy of the Human Genome Project may be that it focused effort on the development of techniques for genome analysis. The power of these techniques is often most striking when they are applied to problems outside human or mammalian genetics. For example, by permitting electrophoretic karyotyping and purification, these techniques have revolutionized studies of chromosome structure, gene mapping, and population biology for species with chromosomes small enough to be resolved by pulsed-field gel electrophoresis. In some cases, the smaller size of nonmammalian genomes means that complete maps may be developed with very little effort or commitment of resources. In the case of organisms with large genomes, these new and efficient methods become essential when genetic maps and mapping resources are lacking. The genomes of some of the world's economically most important species of animals, plants, and pathogens are still very poorly understood, and limited funding restricts the options of a research laboratory. This book is intended to promote the spread of the new techniques of genome analysis. The procedures selected are those that will be useful to investigators working with a wide variety of organisms, from microorganisms and parasites to complex eucaryotes. Because the examples are all drawn from studies of nonmammalian organisms, there is little need to sift through nonrelevant information, such as approaches specific to human or mouse genetics.

Our goal is to present protocols for procedures exactly as they are carried out in the labs where they are successfully used. Since many of the users of this work will have limited prior exposure to molecular biology, let alone genomic analysis, the procedures are presented in a thorough and detailed fashion. This is important, since failure to successfully transfer many procedures from one lab to another often lies in the details that at first seem "routine." Each chapter contains background material to allow novices to understand why each step is used and information on troubleshooting to help identify both when the procedures are working correctly and when they are not, and if so, what corrective measures should be taken. The uniform format permits researchers to rapidly determine both the reagents needed for the work and the steps that must be performed. We expect the book to be used directly at the lab bench by graduate students, postdoctoral scientists, and other researchers carrying out genome analysis, as well as by advanced undergraduates and those directing research efforts who are interested in the strategies and different approaches available for making maps and cloning genes. Although each procedure is illustrated with work involving a single species, all the methods presented will be valuable for study of virtually any genome, and thus we expect each chapter to be of interest to those interested in an overview of the methodology. The book begins with techniques that do not involve cloning, with the first chapter on pulsed-field gels describing methods that underlie virtually all methods described in the rest of the book. The remaining chapters of the book consider different cloning strategies, concluding with an approach to genome mapping that integrates all the kinds of mapping information described in previous chapters.

Bruce Birren
Eric Lai

<hr>

Introduction to Pulsed-Field Gels and Preparation and Analysis of Large DNA

Jennifer S. Lee, Bruce Birren, and Eric Lai

I. Pulsed-Field Gel Electrophoresis (PFGE)

A. Introduction

In 1983 Schwartz and his co-workers demonstrated that yeast chromosomes can be separated in agarose gels by using two alternating electric fields of different orientation (Schwartz et al, 1982). This technology is now known as pulsed-field gel electrophoresis (PFGE). PFGE can resolve DNA from a few kilobases (kb) to more than 10 megabases (Mb) long (Orbach *et al.*, 1988), extending the size range of resolution for DNA molecules orders of magnitude beyond conventional agarose electrophoresis. The ability to separate intact chromosomes from microorganisms has revolutionized gene and genome mapping in these species (Carle and Olson, 1985; Fan *et al.*, 1989). Similarly, pulsed-field gel (PFG) techniques have been the basis for many of the advances in large fragment cloning and physical mapping (Shizuya *et al.*, 1992; Dausset *et al.*, 1992). PFGE has become an essential technique for the characterization and analysis of chromosomes and genomic DNA and has recently been reviewed in detail (Birren and Lai, 1993). As both a preparative and an analytical tool, PFGE is central to all aspects of genome analysis, and its use is fundamental to nearly every chapter in this book. In this chapter we review the parameters that govern the migration of DNA in PFGE and provide guidelines and protocols for preparing and manipulating high-molecular-weight DNA.

NONMAMMALIAN GENOMIC ANALYSIS: A PRACTICAL GUIDE

B. Size-Dependent Separation in PFGE

When DNA in an agarose gel is placed in an electric field, the DNA molecules become elongated and oriented with the electric field as they move through the gel (Smith *et al.,* 1989; Schwartz and Koval, 1989). The leading end represents the "head" and the following end the "tail." In conventional electrophoresis employing a constant electric field, all DNA molecules larger than 15–20 kb travel unidirectionally through the gel with the same mobility. PFGE utilizes two alternately pulsing electric fields of different orientation (Schwartz and Cantor, 1984; Carle and Olson, 1984). With each electric field switch, the DNA reorients to align itself with the new field direction prior to beginning to migrate (Smith *et al.,* 1990). Therefore, the DNA actually migrates in a series of brief steps of alternating direction, with the electric fields regulated to ensure that the net direction of DNA migration is a straight line. The time required for DNA molecules to reorient with each change in field direction is a function of the molecular weight. Thus, by periodically forcing the DNA to reorient, PFGE establishes size-dependent separation for large DNA molecules.

The angle between the two alternating fields is known as the reorientation angle and represents the angle that the DNA must turn through to reorient with each change in field direction. When the two alternating fields are oriented in opposite directions (i.e., a reorientation angle of 180°), the process is referred to as field inversion gel electrophoresis (FIGE) (Carle *et al.,* 1986). In most other cases, the angle between the two fields ranges from 106° to 120°.

C. Switch Time Governs PFG Resolution

PFGs can be run with very little effort or specialized knowledge. However, the researcher must have an adequate understanding of the factors affecting DNA mobility in PFGs to generate reproducible and successful cloning and mapping experiments. Choosing optimal gel conditions in PFGE depends mainly upon the size range of the DNA fragments to be resolved. The goal is to achieve the highest band resolution with minimum run time. Separation of the desired size range of DNA fragments depends on the time required for the molecules to fully reorient from one field direction to the other. The duration of each electric field pulse is referred to as the switch time, or switch interval. For any given switch time, a specific size range of DNA molecules will have had sufficient time to completely reorient and begin to migrate under the influence of the new field. For each switch time, all molecules larger than a certain limit will not have had adequate time to fully reorient with each field switch; these molecules will not resolve from all the DNA in the sample that is equal to or greater than its size. Therefore, the switch time is the single most important parameter in determining which molecules are resolved in a PFG.

Each set of PFGE conditions will separate a specific size range of molecules, but the upper size limit of resolution is primarily determined by the switch time. As the switch time is lengthened, the size range of molecules that can be resolved increases. Figure 1.1 shows that, with all other gel parameters held constant, increasing the switch interval from 45 to 75 sec increases the size range that is separated from 677 to 960 kb, and a further increase to 105 sec permits separation of DNA up to over 1100 kb. Chapter 2 illustrates the importance of careful control of switch time: minor changes in switch time can permit resolution of molecules that are otherwise not separated. However, as the switch time is lengthened, although a larger size range of molecules is being resolved, the resolution between the smaller molecules run on each gel is diminished. This is also illustrated in Fig. 1.1. The separation between the 50- and 100-kb bands of the lambda ladder in the left lane of each panel decreases as the switch time is increased. Therefore, obtaining optimal resolution on a PFG requires use of the minimal switch time necessary to effectively separate the molecules of interest. Using much longer switch times than needed will reduce resolution of the sample bands as well as the molecular size markers, reducing the ability to determine accurate sizes.

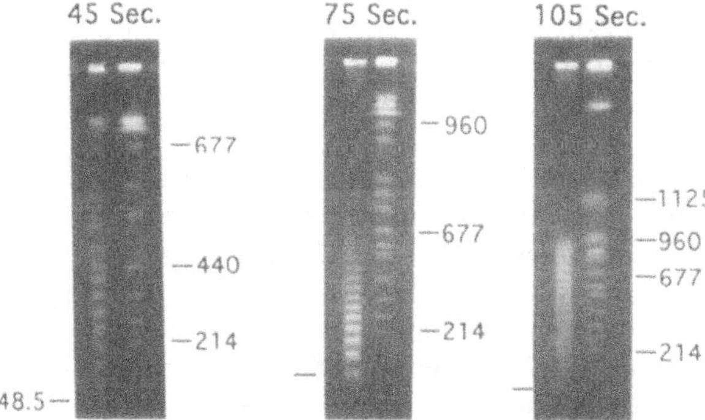

Figure 1.1 Switch time determines the resolution and size range separated by a pulsed-field gel. With all other gel parameters held constant, Lambda ladders (left lane) and yeast chromosomes (right lane) were separated with switch intervals of 45, 75, or 105 sec. Each pulsed-field gel can separate a certain size range of DNA molecules, and lengthening the switch time increases this size range. However, as the switch time is lengthened, the separation between the fragments of different sizes (the resolution) diminishes. Gel conditions used were 14 = 9A C, 1% SeaKem LE agarose, 0.5× TBE, 120 = 9A reorientation angle at 6 V/cm.

In all PFGs, regardless of the particular switch time used, there will be a portion of the gel in which DNA molecules migrate in an order that does not reflect increasing size. Some molecules will migrate faster than molecules which are actually smaller, a phenomenon known as band inversion. In most PFGE band inversion occurs only in a small region of the gel, normally near the sample well, and therefore does not interfere with determination of sizes. In a FIGE gel, this reversal of the relationship between mobility and size can occur at either the bottom or the top of the gel and can greatly reduce the amount of gel in which useful size information can be determined. The impact of this phenomenon is minimized by (i) using size markers that extend beyond the range of the molecules being studied, (ii) using size markers that have unique sizes and are easily distinguished, such as yeast chromosomes, in addition to markers with regular spacing, such as "ladders," and (iii) analysis of the fragments on gels of several different switch times.

The decreased resolution accompanying longer switch times and the separation of greater size ranges of DNA produce two important practical results. First, obtaining high-resolution separations for molecules of very different sizes will require different PFG runs. For example, the switch time necessary to clearly resolve a 2-Mb DNA molecule will poorly resolve a 200-kb fragment. In this case, two separate PFGs must be run with switch times appropriate for each set of size ranges. Second, accurately determining the size of an uncharacterized molecule usually requires multiple PFGs. The preliminary PFGs need a broad switch time that ensures the fragment will be resolved in order to estimate the size. Subsequent PFGs can use a switch time specific to the observed size range of the fragments which would then provide more accurate size information relative to the size markers.

In Fig. 1.2, DNA mobility in PFGs run with different switch times is shown for DNAs of three different size ranges. The gel conditions that give rise to this migration are shown in Table 1.1, and for Fig. 1.2 a single switch time is used throughout the run duration. Just as the PFGE conditions for DNAs of these different size ranges are very different, the time scales of the three curves are notably different. Switch times can be chosen from Fig. 1.1 by noting the shortest time that permits resolution of the desired size range. For example, Fig. 1.2A shows that a switch time of approximately 8 sec would be appropriate to resolve molecules of 35 kb, provided that the gel is run according to the conditions shown in Table 1.1. Molecules of 35 kb are also resolved using longer switch times, such as those shown in Fig. 1.2B and 1.2C, but the conditions separating larger DNA offer poor resolution for the 35-kb DNA fragments.

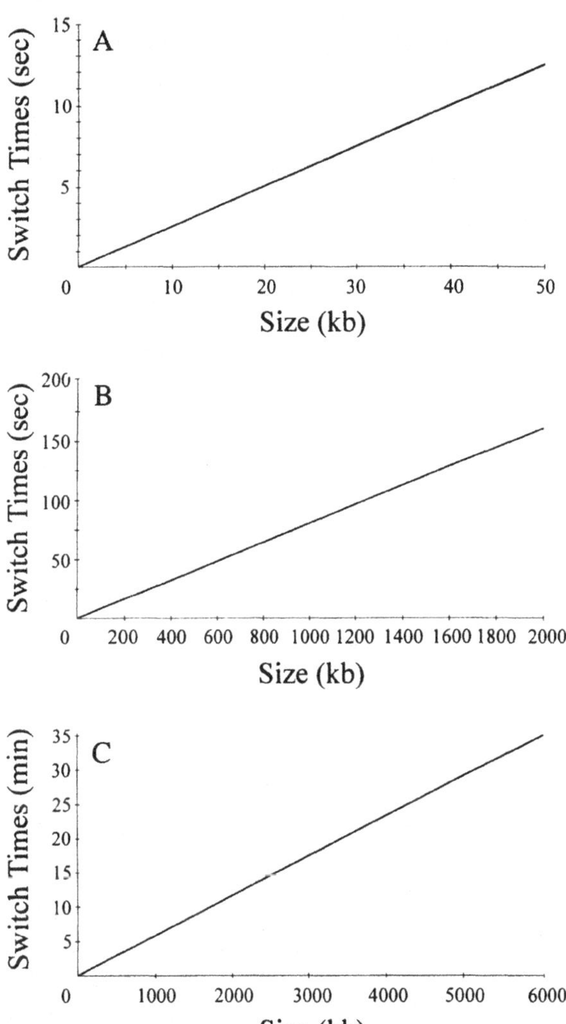

Figure 1.2 Determining the optimum switch time for separating different sized DNA. The sizes of the largest molecules that could be resolved in pulsed-field gels using a constant switch interval are shown for DNAs of three different size ranges. Reproducing the resolution depicted requires use of the gel conditions used to generate each of the three curves, given in Table 1.2. To select an appropriate switch time, i.e., one providing optimal resolution of the desired size range of molecules, one should choose from these curves the shortest time that permits resolution of the desired size range. These curves can be used to predict the separation obtained when using ramped switch times by calculating the average switch time.

Table 1.1
Separation Conditions of Different Size DNA Fragments[a]

Parameter	Small DNA (1–50 kb)	Medium DNA (50–2000 kb)	Big DNA (>2000 kb)
Voltage gradient (V/cm)	9	6	2[b]
Reorientation angle	120°	120°	106°
Temperature	14°C	14°C	14°C
Buffer	0.5× TBE	0.5× TBE	1× TAE
Agarose concentration	1% Low EEO[c]	1% Low EEO	0.8% Low EEO

[a]These are "standard" conditions for pulsed-field gels, and are chosen to offer the optimal separation in terms of resolution and run time for each of the different size ranges of DNA. Many variations on each of these parameters will also be effective.

[b]Lower voltage gradients (1.5 V/cm) may be necessary for resolving fragments over 4 Mb. Further discussion of separation of Mb DNA can be found in Chapter 2.

[c]Low EEO agarose refers to EEO values of less than approximately 0.13.

Often, the switch time is progressively changed during the run, so that the fields alternate more frequently at the beginning of the run than at the end. This is termed switch time ramping. PFGs that are run with a constant switch time over the entire run duration normally have a specific area of the gel where the fragment mobility is linear with respect to fragment size. Switch time ramping increases the proportion of the gel in which the fragment mobility is linear to its size. This is especially true for FIGE, where failure to ramp the switch time leads to a large portion of the gel that cannot be used due to band inversion. With switch time ramping, the DNA molecules migrate with a mobility that reflects the average of all the switch times used. Thus, the actual size range of molecules separated on a gel with ramped switch times can be predicted by esimating the average switch time between the initial and final value. For example, a gel run using an initial switch time of 20 sec and a final switch time of 120 sec would separate the same approximate size range as a gel run using a constant switch time of 70 sec. While Fig. 1.2 shows the mobility of DNA run with a constant switch time, these curves can also be used to predict the separation obtained when using ramped switch times by calculating the average switch time.

D. Establishing Effective Separation Conditions

While the switch time is the primary determinant of the migration of DNA in PFGE, all of the other conditions under which the gel is run also influence the speed and resolution of the separation (Birren *et al.*, 1988). The greatest changes in DNA migration and resolution occur with switch time variation, followed by changes in the voltage gradient. Most of the parameters that must be selected for a PFG are similar to those involved in conventional electrophoresis: the temperature, the duration of the gel run, the agarose concentration and electroendosmotic (EEO) value, and the buffer type and concentration. The impact of variations in these parameters is much more pronounced in PFGE than in conventional electrophoresises, though these conditions do not greatly change the outcome of a PFG. PFGs also involve the additional electric field parameters, namely the duration and direction of the alternating fields.

Table 1.1 presents conditions that are effective for routine separation of DNA molecules from 1 kb to 10 Mb. Specific applications may benefit from variation of these conditions, for example, to achieve higher resolution over small defined size range or more rapid separations of lower resolution. Depending on the specific parameters changed and/or the combined effect of several parameter changes, some cases will produce minor changes in the migration of the DNA (on the order of 20%), while others can cause a complete failure to resolve the fragments of interest. Therefore, understanding the relationship of these factors to the migration of the DNA allows the researcher to optimize the conditions for any given separation and understand the possible ways to obtain satisfactory separation. When contemplating changes in PFGE conditions, it is important to remember that any change is likely to affect not only the speed of the separation but the resolution as well.

The largest difference between the conditions used for the different DNA size ranges in Table 1.1 is in the voltage gradient. For molecules smaller than 50 kb, voltage gradients as high as 20 V/cm may be used for PFGs (Wagner and Lai, 1994), though commercially available PFGE boxes are limited to gradients of 9 V/cm. These high-voltage gradients produce rapid DNA separations by dramatically increasing the rate of DNA migration. In contrast, with increasing size, DNA molecules require a reduction in the voltage gradient. As described in Chapter 2, separation of *Schizosaccharomyces pombe* chromosomes (3.5 to 5.7 Mb) can occur only with gradients of 2 V/cm or less (Smith *et al.*, 1987; Vollrath and Davis, 1987). This reduction in voltage gradient necessitates such long gel runs for large DNA that changes in other gel conditions, such as the reorientation angle, become especially important to minimize run durations.

Although a range of temperatures from 4 to 30°C may be used for PFGs, the temperature must be carefully and consistently regulated throughout the PFGE. A pump and usually a heat exchanging system ensures uniform migration and resolution across the lanes of the gel and over the duration of the gel run. PFGE buffers are usually either Tris Borate (TBE) or Tris Acetate (TAE) buffers. TBE is preferred for routine use, since it requires changing less frequently than TAE. For separation of very large DNA, the increased separation speed of TAE buffers is valuable. At the lower voltage gradients used for large DNA PFGE, the buffer will not break down as rapidly as at the higher voltages used for separation of smaller DNA. A reorientation angle of 120° is effective for all PFG separations, though reductions in this angle will allow more rapid separations of DNAs of all sizes.

Table 1.2 summarizes the role of each PFGE parameter and lists the effects of its increase or decrease on the rate of migration of the DNA

Table 1.2
Parameters That Affect Migration of DNA in Pulsed-Field Gels

Parameters	Function/ description	Effect of increase	Effect of decrease
Voltage gradient	Potential difference between electrodes in gel box, force driving DNA "down" gel.	Increased DNA mobility—shorter run times; decreases upper size limit of DNA separation. Will require more cooling to maintain temperature.	Decreased mobility—longer run times; increases upper size limit of separation.
Switch time, fixed	Duration of each electric field pulse. Dictates size range of fragments that will be resolved. Produces distinct zones of differing resolution in the gel.	Increases upper size limit of separation; resolution decreased.	Decreases upper size limit of separation; resolution increased.
Switch time, ramped	Progressive change in the switch time during the duration of the run. DNA migration is more linear with respect to size.	Using a broader range of times for the ramp will allow a large size range of fragments to be separated.	Using a narrower range of times for the ramp reduces the size range of fragments to be separated.
Reorientation angle	Angle between electric field directions, dictates degree of horizontal and vertical movement of fragments.	Decreases rate of migration, larger horizontal component of electric field vector.	Increases rate of migration, especially important for DNA >2Mb.

continues

and the resolution obtained in a PFG. This table can be used to predict the affect of changes in each of these parameters on the PFG. For example, if the gel will be run at temperatures lower than the recommended 14°C, a longer run time is necessary to produce the same degree of separation. More importantly, each factor that influences the rate at which the DNA migrates also affects the size range of molecules that is resolved: each change in a run parameter will alter the size range of molecules that are separated and their resolution. As listed in Table 1.3, otherwise un-

Table 1.2   continued

Parameters	Function/ description	Effect of increase	Effect of decrease
Temperature	Affects both migration rate and resolution, must be constant across the gel to achieve uniform migration.	Increases migration rate of DNA and upper limit of size range separated. Decreases resolution.	Decreases migration rate of DNA and upper limit of size range separated. Increases resolution.
Duration of gel run	Dictates spacing between separated fragments, but not which fragments will be resolved. Obtain sharp bands by running for minimum duration necessary. Only increases in switch time will increase resolution.	Increases spacing between bands, bands less sharp.	Decreases spacing between bands, bands sharper.
Agarose concentration	Density of matrix dictates speed of DNA migration.	Sharper bands, slower migration.	Less sharp bands, faster migration.
Buffer concentration	Buffer ions carry current and buffer pH. Ions are depleted with successive runs.	Slows DNA migration. Needs changing less frequently.	Increases DNA migration. Needs changing more often.
Agarose EEO value	Electroendosmosis (EEO) reflects the amount of residual charge in agarose which retards migration of DNA.	Slows rate of DNA migration.	Increases rate of DNA migration. "Pulsed-field gel" agaroses sold for separation of Mb DNA have lowest EEO values.

Table 1.3
Interaction of Switch Time with Other PFGE Parameters

Effect of change in gel conditions	Examples	To maintain resolution
Increased rate of DNA migration	Increased voltage gradient Increased temperature Decreased agarose concentration or EEO value Decreased buffer concentration	Shorten switch time
Decreased rate of DNA migration	Decreased voltage gradient Decreased temperature Increased agarose concentration or EEO value Increased buffer concentration	Lengthen switch time

desirable changes in resolution can be compensated for by concomitant changes in the switch time. In some instances the relationship is easy to predict. When the voltage gradient is reduced by half, the switch time must be approximately doubled to maintain comparable resolution. In most cases, achieving optimum separations requires some amount of trial and error, varying the switch time with most other parameters held constant (for further details see Chapter 2 and Birren and Lai, 1993).

E. Selection of a PFGE Instrument

Initially, PFGE boxes were home-made instruments involving a variety of designs and names, for which ease of use was not a primary concern (reviewed in Lai *et al.*, 1989). Today, commercial PFGE boxes that are simple to use and offer a range of features are widely available. Nearly all commercially available PFGE instruments are now based on the CHEF (Contour-clamped Homogeneous Electric Field) (Chu *et al.*, 1986) or FIGE (Carle *et al.*, 1986) designs, both of which produce DNA migration in a straight line. The more advanced CHEF systems incorporate the additional features and flexibility of the PACE (Programmable Autonomously Controlled Electrodes) gel box (Clark *et al.*, 1988). Selection of a PFGE instrument depends on the needs of the researcher and the type of separations to be performed. FIGE will adequately separate "small" DNA fragments (under 200 kb) and has limited use with large DNA, such as generating a large-fragment restriction map around a particular gene or preparing a few blots of separated yeast chromosomes for mapping. FIGE has the advantage that, aside from the switching unit, it requires only standard components such as a gel box and power supply that are usually already present in the lab and can be used after PFGE is no longer needed. FIGE systems are available that use a constant voltage for the forward and reverse fields (such as the Hoefer Switchback). For fragments under 100 kb, superior resolution can be obtained by using a FIGE system that varies the voltage instead of the time (Birren *et al.*, 1989), such as the FIGE Mapper from Bio-Rad. However, for projects involving whole genome mapping, electrophoretic karyotyping, or large fragment cloning, a reliable and flexible PFGE device is essential. The Bio-Rad CHEF DRII and the Pharmacia Gene Navigator are simple fixed-angle (120°) CHEF systems, adequate for separating DNA from 50 kb to 2 Mb if long run times are not a concern. However, features of more advanced systems can noticeably improve PFGE separations and reduce run durations. For example, the Bio-Rad CHEF DRIII has a variable reorientation angle from 106° to 120°. This allows reduction of the reorientation angle which can save days of electrophoresis time for separations of Mb DNA (see Chapter 2). When separating P1, BAC, or PAC digests, the combined use of reorientation angles less than 120° and higher voltage gradients can reduce run durations from 16 to 4 hr (Birren and Lai, 1995). The Bio-Rad CHEF Mapper has the most advanced features, permitting any number of fields to be used with any orientation and duration. In addition, it contains an algorithm that chooses optimal separation conditions based on the size of the DNA molecules of interest. This permits implementation of the most effective separation conditions, specific to each application.

II. Materials

A. Electrophoresis Buffers

PFGE buffers are usually either TBE or TAE. TBE is preferred for routine use, due to its higher buffering capacity; 0.5× TBE buffer can be used for several runs of 30 hr at 6 V/cm without requiring changing or replenishment of the buffer. The buffer should always be changed and the gel box rinsed prior to any preparative electrophoresis. TAE will provide faster migration of the DNA than TBE, and is therefore recommended for larger, slow-moving DNA fragments, though the buffer must be changed with each gel run. Do not include ethidium bromide in the buffer or the gel, because ethidium bromide will slow down the migration of the DNA and alter the size range of fragments separated.

1. 1× TBE

Component	Final concentration (m*M*)	Amount needed to prepare 1 liter of a 10× concentrated solution
Tris	89	108 g Tris base
Borate	89	55 g boric acid
EDTA	25	9.3 g disodium EDTA·2 H_2O
H_2O	—	Add to bring final volume to 1 liter

The pH of this mixture will be 8.3.

2. 1× TAE

Component	Final concentration (m*M*)	Amount needed to prepare 1 liter of a 50× concentrated solution
Tris	40	242 g Tris base
Acetate	40	57.1 ml glacial acetic acid
EDTA	25	100 ml 0.5 *M* EDTA, pH 8.0
H_2O	—	add to bring final volume to 1 L

B. Solutions

1. Bacterial Lysis Solution

Final concentration	Stock solutions	Amount needed to prepare 50 ml
10 m*M* Tris, pH 7.5	1 *M*	0.5 ml
50 m*M* NaCl	4 *M*	0.625 ml
100 m*M* EDTA	0.5 *M*, pH 8.0	10.0 ml
0.2% Na deoxycholate		0.1 g
0.5% sarcosyl, Na salt, H_2O		0.25 g in 38.75 ml

This solution may by prepared ahead of time. Add egg white lysozyme to 1 mg/ml final concentration immediately prior to use.

2. Digestion Buffer

Component	Final concentration, DB 0.5
EDTA	0.5 *M*
Lauroyl sarcosine, Na salt	1%
Proteinase K	0.5 mg/ml

Prepare fresh buffer by dissolving sarcosyl in EDTA by shaking. Add proteinase K as a 20 mg/ml stock solution.

3. YPD

YPD is a complete medium for yeast composed of yeast extract, peptone, and dextrose. To prepare 1 liter of YPD mix:

10 g Bacto–yeast extract
20 g Bacto–peptone
Add H_2O to bring final volume to 1 liter.
Sterilize by autoclaving. Add 40 ml of 50% glucose.

4. LB

LB is a complete medium for bacteria. To prepare 1 liter of LB mix:

10 g Bacto–tryptone
5 g Bacto–yeast extract
10 g NaCl
Add H_2O to bring final volume to 1 liter.
Sterilize by autoclaving.

C. Choice of Agarose

Most PFGs are cast at a 1% concentration using standard agarose sold for
DNA electrophoresis. The agarose should be certified for use in molecular
biology, because contaminants in impure agarose can degrade DNA or
inhibit subsequent enzymatic reactions. Faster DNA migration is obtained
with agaroses of low electroendosmosis (EEO), which reflects the internal
charge of the agarose. For routine use, low EEO agarose (such as SeaKem
LE FMC BioProducts) is effective and inexpensive. The use of medium
EEO agarose will reduce the speed of DNA migration by approximately
10–15%. For separating DNA molecules larger than 2 Mb, the very long
run times needed can be reduced by using agarose of even lower EEO
values (often sold as Pulsed-Field Gel agarose) and/or at a concentration
of 0.7%—see Chapter 2 for discussion. For preparation of DNA samples
in solid agarose, highly purified and quality tested low-melting agarose
(e.g., InCert agarose, FMC BioProducts) is of value only when the samples
will be used subsequently for restriction digestion. For preparation of intact chromosomes for separation by electrophoresis, conventional low-
melting agarose is effective.

III. Preparation of Nonmammalian Chromosomes

A. General Principles

Traditional techniques for purification of DNA involve organic extraction
of proteins and alcohol precipitation. These procedures involve shear
forces that will break large DNA fragments to an average size of no more
than a few hundred kilobases. At the same time that they developed the
electrophoretic techniques to separate large DNA, Schwartz and Cantor

(1984) developed methods for purifying megabase-sized DNA in solid agarose to protect the DNA during preparation. Intact cells are mixed with low-percentage low-melting agarose which is then allowed to harden in molds. These solid samples are then treated with enzymes and detergents that will digest the cell wall, membranes, proteins, and other cellular debris, allowing them to diffuse out of the agarose, leaving only the nucleic acid. At its simplest, the treatment can require only proteinase K (or another protease), detergent, and EDTA (used to inhibit endogenous nucleases that could degrade the DNA during the incubations). This is usually the only treatment necessary for organisms that lack a cell wall. Digesting a cell wall usually requires an additional step, to allow access of the cell to the reagents used for DNA isolation. The methods, most often enzymatic, are specific to each organism and vary with differences in the nature of the cell wall.

Detailed protocols for preparing high-molecular-weight DNA from a variety of organisms may be found within the other chapters of this book and elsewhere (Birren and Lai, 1993). While the exact protocol for isolation of large or chromosomal DNA will vary from organism to organism, there are general principles that apply in most cases:

(1) The cells should be from a healthy, actively growing source. Cultures in which growth has ceased often will have undergone thickening of the cell wall, making rapid lysis more difficult, and some amount of cell death, resulting in DNA degradation. When a choice of tissues exists, tissues containing low levels of nucleases should be used for DNA preparation.

(2) Most enzymes are less active in the presence of agarose, and hence cell walls are more efficiently digested with the cells in solution rather than after embedding the cells in agarose. Resulting spheroplasts must be osmotically stabilized during any washing steps to prevent lysis prior to embedding in agarose.

(3) Add EDTA to all solutions as early in the process as possible to prevent nucleolytic DNA degradation. Low-melting agarose should be used because it will remain liquid when cooled to temperatures that will not damage the cells. If the DNA will be digested with restriction enzymes after preparation, the low-melting agarose must be free of compounds that inhibit enzyme activity.

(4) Once cell walls have been digested, rapidly perform any necessary washes and embed the resulting spheroplasts in agarose.

(5) Sufficient time should be allowed to fully digest the embedded cells since residual cellular components can degrade the DNA on storage or interfere with subsequent enzymatic treatment.

(6) Dialysis of the samples after DNA isolation should be extensive enough to remove small DNA fragments as well as the reagents used for DNA preparation, which can severely inhibit enzyme activity.

B. Procedures

Additional procedures for preparation of high-molecular-weight DNA are found in many chapters of this book.

1. Preparation of Yeast Chromosomes by Embedding Intact Cells in Agarose

Preparation of spheroplasts is more efficient in solution than after embedding cells in agarose. Therefore, a higher yield of chromosomes is obtained by first preparing spheroplasts in solution and then embedding these in agarose for digestion with proteinase K, as described in Chapter 2. However, the following is a simple method that is effective for preparing chromosomes from *Saccharomyces cerevisiae*.

(1) Inoculate 100 ml of YPD medium with 0.1 ml of a saturated culture and grow for approximately 16 hr to achieve cultures in late log or early stationary phase of growth.

(2) Harvest cells in the centrifuge by spinning for 5 min at 5000 rpm in a Sorvall GSA rotor.

(3) Discard the supernatant and resuspend the cell pellet in 20 ml of 50 mM EDTA, pH 8.0, to wash the cells. Collect cells by spinning again for 5 min as before and decant supernatant.

(4) Resuspend cells in 6 ml 50 mM EDTA, pH 8.0, and mix in 160 ml 10 mg/ml Zymolyzase 20T.

(5) Briefly warm cells to 37°C and mix in 9 ml 1.2% low-melting agarose. The final concentration of cells in agarose should be around 2 × 10^9 cells per milliliter.

(6) Rapidly pipette the mixture into molds to avoid solidification of the sample before it is in the molds.

(7) Allow to harden for 5–15 min at 4°C or on ice.

(8) Transfer the samples into a screw cap centrifuge tube and cover the samples in 0.25 M EDTA, pH 8.0, and 5% β-mercaptoethanol. Cap the tube tightly and seal with Parafilm. Incubate overnight at 37°C to generate spheroplasts.

(9) Discard solution in a ventilated fume hood and rinse the samples several times with 0.25 M EDTA, pH 8.0. Discard the rinse solution in the fume hood.

(10) Add 1–2 vol of digestion buffer (enough to cover) and incubate 24–48 hr with gentle agitation at 50°C.

(11) Store the samples in digestion buffer at 4°C, where they will be stable for many months. This protocol will yield sufficient material for hundreds of lanes in PFGs.

(12) Prior to electrophoresis, wash samples in 50 mM EDTA, pH 8.0, or electrophoresis buffer using several changes of at least 5 ml.

2. Preparation of Chromosomal DNA from Bacteria

High-molecular-weight DNA can be prepared from most bacteria by embedding the cells in agarose and then performing a two-step process that first generates spheroplasts and then lyses and digests them with detergent and proteases. The following protocol is effective for bacteria that are sensitive to lysozyme. Alternative treatments may be required to generate spheroplasts from other bacterial strains (Birren and Lai, 1993).

(1) Grow cells in 10 ml rich medium to mid log phase. Overnight (saturated) cultures of *Escherichia coli* will provide DNA of sufficient quality for most applications. Determine the cell density by reading an OD_{600} prior to harvest; 10 ml of *E. coli* harvested at 4–5×10^{8} cells per milliliter will yield sufficient DNA for approximately 40 gel lanes containing 0.5 µg DNA per lane.

(2) Arrest cell growth either by chilling cultures by swirling flasks in ice water or by adding chloramphenicol to a final concentration of 0.2 mg/ml and continuing incubation for 1 hr. Chloramphenicol will synchronize cultures with respect to chromosome replication to give equal representation of all sequences.

(3) Harvest cells by centrifuging for 10 min at 4000*g*.

(4) Discard supernatant and wash the cell pellet by resuspending in 2 ml of 200 mM NaCl, 10 mM Tris, pH 7.2, 100 mM EDTA.

(5) Collect the cells by centrifuging at 4000*g* for 10 min.

(6) Discard supernatant and resuspend cells in 0.5 ml of 200 mM NaCl, 10 mM Tris, pH 7.2, 100 mM EDTA by pipetting. Briefly warm to 37°C.

(7) Add an equal volume of 1% low-melting agarose, prepared in water and cooled to 40°C. Mix cells with agarose by pipetting and transfer mixture to sample molds.

(8) Cool molds on ice or at 4°C for 5–15 minutes until solid.

(9) Transfer samples from molds to a screw cap plastic centrifuge tube and 4-ml lysis solution tube. Incubate at 37°C for 2–16 hr to allow spheroplasts to form.

(10) Discard the lysis solution, taking care to retain the samples, and add 4 ml digestion buffer. Incubate at 50°C for 12–36 hr with gentle agitation.

(11) Store samples in this solution at 4°C, where they will be stable for weeks or months. Wash samples by placing in several changes of 50 mM EDTA prior to use.

C. Controlling and Determining DNA Concentration

The final concentration of DNA in the agarose is determined by the initial concentration of cells in the liquid agarose. However, once the agarose has solidified and the DNA has been prepared, the concentration of the DNA cannot be altered, except by procedures likely to break the large DNA. Therefore, it is essential to determine the cell number as accurately as possible prior to embedding. DNA samples that are prepared at too dilute a concentration may not allow visualization of the samples after electrophoresis. DNA samples that are prepared at too high a concentration can also fail to produce clear bands and will not migrate at a rate that accurately reflects the size of the DNA fragments. Even when embedding a constant cell number, variations in the efficiency of cell wall digestion will cause fluctuation in the final concentration of DNA seen after electrophoresis.

The yield of high-molecular-weight DNA varies from preparation to preparation. Preparing samples at several different concentrations at and around the expected value is valuable, especially for the first few attempts at preparing DNA from a new species or strain. The optimal amount of DNA to prepare in the agarose samples will vary with the size of the genome being studied, as well as the intended use of the samples. For chromosomal DNA where the only use will be PFGE separations and blotting, the proper amount is that which gives clear, well-resolved bands (see Chapter 2). In general, the "best-looking" PFG results from DNA samples that contain 20–50 ng per band, enough for ethidium bromide visualization but not enough to overload the gel. A simple rule is to calculate the cell density that will provide approximately 50 ng DNA per band and prepare three concentrations of cell–agarose plugs at fivefold the calculated density, at the calculated density, and at one-fifth the calculated density. This will ensure that one of the sample preparations will be within the desirable range. Note that for very large DNA, molecules greater than 2 Mb, increasing the concentration of the DNA can be counterproductive, since the chromosomal bands become smeared and less distinct at too high a concentration. For high-molecular-weight DNA that will be used for cloning, the optimum DNA concentration often can be derived only through pilot cloning experiments.

The concentrations of samples in agarose may be determined in one of two ways. In the first, a small portion of the sample can be melted and diluted and the amount of DNA determined by either optical density or fluorimetry. The drawback of this approach is that it gives a reading of the total nucleic acid present. It does not distinguish between intact chromosomal DNA and smaller broken fragments (or RNA). Also, intact cells that have remained in the agarose without having been lysed can obscure the results. These difficulties are overcome by the second approach, which after electrophoresis compares the fluorescence of a small amount of solid or liquid sample to a known amount of standards. This need not involve PFGE and can be accomplished with a conventional "mini-gel" in which the samples are run briefly at high voltage using different dilutions of a known amount of large DNA, such as intact lambda DNA or a *Hin*dIII digest of lambda DNA.

IV. Enzymatic Reactions Using DNA–Agarose Plugs

A. General Principles

In cases where DNA samples are to be separated as intact chromosomes, no treatment of the DNA beyond purification is required. In other cases, such as long-range restriction mapping or preparative uses for cloning, high-molecular-weight DNA in agarose is intended as a substrate for subsequent enzymatic treatments. Choosing enzymes that will be effective under these conditions, as well as controlling their activity, requires consideration of factors not often encountered in traditional molecular biology.

There are three approaches to enzymatic treatment of DNA samples in agarose plugs. First, the agarose can be removed by digestion with agarase before performing other enzymatic treatments. The second method does not require removing the agarose but simply melting the agarose to enhance the mixing of buffers and enzymes. This is done by melting the DNA–agarose plugs at 65°C for 10 min and then placing the sample at 37°C, where it will remain liquid. In each of these methods, working with large DNA in solution instead of in solid agarose will lead to breakage of very long DNA molecules, and extreme care must be taken to minimize this damage. However, many enzymatic reactions are more efficient in solution than in solid agarose, and these methods are used widely for restriction enzyme digestions where the resulting fragments are expected to be less than 500 kb. Whenever DNA in agarose is melted, care should be taken to use the lowest possible temperature (65°C) and to maintain a sufficient salt concentration to prevent denaturation of the DNA (e.g., 50 m*M* NaCl). In the third method, used to produce very large restriction fragments, enzymes and buffers diffuse into the agarose plugs which remain solid throughout the procedure.

Generating partial digests of DNA in agarose, e.g., for preparing material for cloning, requires a slightly different strategy. Because the diffusion of the enzymes into the plugs is limited by the agarose, a gradient of enzyme activity is established, with maximal activity at the outside of the agarose block and no activity near the center. To allow equal access of the enzyme to the entire DNA sample, the enzyme is allowed to diffuse into the agarose at a low temperature (e.g., 4°C) in the absence of magnesium. Under these conditions the enzyme will penetrate the entire block but will not be active. Upon raising the temperature and adding magnesium (which will diffuse quickly) the reaction can be initiated. The extent of digestion can be controlled by limiting the amount of enzyme added, the amount of magnesium added, or the amount of time allowed after addition of the magnesium. Detailed procedures related to enzymatic treatments of DNA from various organisms can be found in other chapters in this book.

B. Selection of Restriction Enzymes

Choosing the appropriate restriction enzymes for cloning, fingerprinting, or long-range mapping will depend primarily on the desired cutting frequency, the degree of methylation, and the ratio of G/C to A/T content in the genome being studied. In the idealized case of a random sequence that has an equal proportion of A/T:G/C, the frequency of any restriction site is given by 4^n, where n equals the number of bases in the recognition sequence of the enzyme. From this relationship we obtain the generalized estimate that 6 base recognition enzymes will cut on average every 4^6, or about every 4 kb. However, the genomes of different organisms show such a wide variation in base composition that the actual frequency with which enzymes will cleave DNA is drastically different depending on the ratio of G/C to A/T in the recognition site. Therefore, choosing restriction enzymes that will cut DNA infrequently and generate large DNA fragments must take into account the base composition of the genome in question. Enzymes that have proven useful in large fragment cloning and PFG mapping are listed in Table 1.4.

V. Southern Blotting of Pulsed-Field Gels

Many applications of PFGE call for blotting and hybridization of the separated DNA. These molecules are too large to transfer from the gel effi-

Table 1.4
Useful Restriction Enzymes for Large Fragment Cloning and PFG Mapping

Intron-encoded nucleases

I-*Ceu* I	5′ TAACTATAACGGTCCTAA$^\nabla$GGTAGCGA 3′
I-*Sce* I	5′ TAGGGATAA$^\nabla$CAGGGTAAT 3′
I-*Ppo* I	5′ ATGACTCTCTTAA$^\nabla$GGTAGCCAAA
PI-*Tli* I	5′ GGTTCTTTATGCGGACAC$^\nabla$TGACGGCTTTATG 3′
PI-*Psp* I	5′ TGGCAAACAGCTATTAT$^\nabla$GGGTATTATGGGT 3′
PI-*Sce* I	5′ ATCTATGTCGGGTGC$^\nabla$GGAGAAAGAGGTAATGAAATGGCA 3′

Restriction enzymes with a >6-bp recognition site

*Asc*I	GG$^\nabla$CGCGCC
*Csp*I (*Rsr*II)	CG$^\nabla$G(A,T)CCG
*Fse*I	GGCCGG$^\nabla$CC
*Not*I	GC$^\nabla$GGCCGC
*Pac*I	TTAAT$^\nabla$TAA
*Pme*I	GTTT$^\nabla$AAAC
*Sfi*I	GGCCNNNN$^\nabla$NGGCC
*Sgr*AI	C(A,G)$^\nabla$CCGG(T,C)G
*Srf*I	GCCC$^\nabla$GGGC
*Sse*8388871	CCTGCA$^\nabla$GG
*Swa*I	ATTT$^\nabla$AAAT

Restriction endonucleases useful for mapping of bacterial genomes (compiled by Römling *et al.*, Chapter 6)

	Recognition sequence	Application (example)
Enzymes with 4 and 5 base recognition sites		
*Dpn*I	G^{m6}A$^\nabla$TC	*Pseudomonas aeruginosa* PAO
*Nci*I	CC$^\nabla$(G,C)GG	*Campylobacter jejuni*
Enzymes with 6 base recognition sites containing exclusively A/T or G/C		
*Apa*I	GGGCC$^\nabla$C	*Campylobacter jejuni*
*Asn*I (*Ase*I)	AT$^\nabla$TAAT	*Rhodobacter sphaeroides* 1.2.4.
*Bgl*I	GCCNNNN$^\nabla$NGGC	Mollicutes
*Bss*HII	G$^\nabla$CGCGC	*Campylobacter jejuni*
*Dra*I	TTT$^\nabla$AAA	*Streptomyces coelicolor* M145
*Eag*I	C$^\nabla$GGCCG	*Sulfolobus acidocaldarius*
*Nae*I	GCC$^\nabla$GGC	*Haemophilus influenzae* Rd
*Sac*II (*Sst*II)	CCGC$^\nabla$GG	*Clostridium perfringens*
*Sma*I	CCC$^\nabla$GGG	Mollicutes
*Ssp*I	AAT$^\nabla$ATT	*Thermus thermophilus*

continues

ciently without some prior treatment to reduce their length; fragmentation of the DNA must precede transfer. Two methods of fragmenting the DNA can be used: either depurination with acid or exposure of the ethidium bromide-stained gel to a known dose of UV light. Generally, the procedures that produce successful Southern transfer and hybridization for genomic DNA in conventional agarose gels may be used for DNA in

Table 1.4 *continued*

Enzymes found to cut less frequently than expected based on G/C content:

For bacteria with G/C content below 35%:

*Asp*7181	G$^\nabla$GTACC	
*Bam*HI	G$^\nabla$GATCC	Mollicutes
*Fsp*I	TGC$^\nabla$GCA	*C. perfringens*
*Kpn*I	GGTAC$^\nabla$C	Mollicutes
*Mlu*I	A$^\nabla$CGCGT	*Borrelia burgdorferi*
*Nru*I	TCG$^\nabla$CGA	*Helicobacter pylori* VA802
*Sal*I	G$^\nabla$TCGAC	*C. jejuni*
*Xho*I	C$^\nabla$TCGAG	Mollicutes

For bacteria with G/C content between 45 and 70%:

*Avr*II	C$^\nabla$CTAGG	*Myxococcus xanthus*
		Anabaena sp. strain PCC 7120
*Nhe*I	G$^\nabla$CTAGC	*Neisseria gonorrhoeae*
		Methanobacterium thermoautotorphicum
*Spe*I	A$^\nabla$CTAGT	*R. sphaeroides* 2.4.1
*Xba*I	T$^\nabla$CTAGA	*Thermococcus celer* Va13

For bacteria with high G/C content (above 70%):

*Bfr*I	C$^\nabla$TTAAG	*Streptomyces lividans* 66
*Bst*BI	TT$^\nabla$CGAA	*T. thermophilus*
*Eco*RI	G$^\nabla$AATTC	*T. thermophilus*
*Eco*RV	GAT$^\nabla$ATC	*T. thermophilus*
*Hpa*I	GTT$^\nabla$AAC	*T. thermophilus*
*Mfe*I	C$^\nabla$AATTG	*T. thermophilus*
*Nde*I	CA$^\nabla$TATG	*T. thermophilus*
*Sna*BI	TAC$^\nabla$GTA	*R. sphaeroides* 2.4.1.

For aerchaebacteria (regardless of G/C content):

*Bam*HI	G$^\nabla$GATCC	*Haloferax mediterranei*
*Bcl*I	T$^\nabla$GATCA	*Methanococcus voltae*
*Bgl*II	A$^\nabla$GATCT	*M. voltae*
*Pvu*I	CGAT$^\nabla$CG	*M. voltae*

PFGs, without modification. Although the extent of depurination is highly temperature dependent, acid treatment is simple, effective, and the most commonly used method. Nicking with UV light has the advantage that it is highly reproducible since UV ovens available for cross linking nucleic acids to filters can be used to deliver a specific dose of energy. As long as gels are stained to the same extent by using the same concentration of ethidium bromide for the same length of time, the amount of nicking will be uniform across the gel and between different gels. The output of most UV transilluminators falls dramatically with the age of the bulbs and UV filter. Therefore, conditions for one light box may not be effective when used on another, and this requires calibration by trial Southern transfer using different UV exposure times. Overtreatment with UV light should be avoided, because thymine dimers generated by UV will interfere with the DNA's ability to hybridize a probe.

VI. Troubleshooting Pulsed-Field Gels

Problems with pulsed-field gels most typically arise from three aspects of the process: gel conditions, sample preparation, and restriction digestion. Most difficulties from gel conditions arise either from inattention to the set up and maintenance of the equipment or from failure to control each of the gel parameters. For example, if the gel box is not level or if the buffer flow rate is not appropriate, DNA migration will not be straight. Failure to clean out gel boxes frequently, especially before a preparative run, can lead to degradation of the DNA due to nucleases that accumulate with fungal growth in the buffer. If gel boxes are not used every few days, the buffer should be drained after each run. With more frequent use, buffer may be left in the box but should be kept chilled to reduce microbial growth. Failure to reproduce published separations may result from either failure to use the exact conditions reported or failure of the original investigators to fully report the conditions they used. Comparison of size marker migration to the published figure should provide confirmation that the size range of interest is being separated to the desired extent. However, otherwise reliable size markers may not produce clear bands when run under inappropriate conditions. Therefore, it is important to confirm that the conditions being used mimic those desired to the greatest extent possible.

Problems with sample preparation most often come from degradation of the genomic DNA, during the process of embedding the cells and digesting them, or from embedding improper concentrations of cells. Degradation of the DNA can be minimized by using healthy growing cultures, proceeding through the process as quickly as possible, finding conditions that most quickly digest cell walls, and maintaining high EDTA concentrations (at least 100 mM) to inhibit endogenous nucleases. Failure to completely remove protein from the DNA can allow nucleases that will degrade the samples over time to remain active, even when stored in the presence of EDTA. In some cases, using different strains will produce cleaner DNA samples due to variations in the endogenous levels of nuclease.

Difficulties are often encountered when digesting high-molecular-weight DNA, because residual amounts of the reagents used to prepare the DNA (such as detergents, proteases, and EDTA) will strongly inhibit most enzymes. For this reason, extensive washing of the samples is needed prior to enzymatic treatment. Intact chromosomal DNA in agarose can be maintained for months or years and will not diffuse out of the agarose in a few hours or days of washing. Failure to completely remove the protein from the samples can also inhibit enzymatic digestion or lead to degradation of the DNA when the samples are incubated with magnesium. For samples that are being used for restriction digests, it is important always to perform a mock restriction digestion to assay for the presence of residual nuclease activity. The sample should be incubated in the presence of restriction enzyme buffer (including magnesium) without adding restriction enzyme and compared by PFGE to an untreated sample. Degradation of the incubated sample may indicate residual nuclease in the sample or contamination of the reagents used for the digestion, such as the restriction buffer or the BSA. Further digestion of the samples with protease and more extensive washing may be necessary to remove incompletely digested protein or residual cellular debris.

References

Birren, B., and Lai, E. (1993). "Pulsed Field Gel Electrophoresis: A Practical Guide." Academic Press, San Diego, CA.

Birren, B., and Lai, E. (1995). Rapid pulsed field separation of DNA molecules up to 250 kb. *Nucleic Acids Res.* **22,** 5366–5370.

Birren, B. W., Lai, E., Clark, S. M., Hood, L., and Simon, M. I. (1988). Optimized conditions for pulsed-field-gel electrophoretic separations of DNA. *Nucleic Acids Res.* **16,** 7563–7582.

Birren, B. W., Lai, E., Hood, L., and Simon, M. I. (1989). Pulsed field gel electrophoresis techniques for separating 1 to 50 kilobase DNA fragments. *Anal. Biochem.* **177,** 282–286.

Carle, G. F., and Olson, M. V. (1984). Separation of chromosomal DNA molecules from yeast by orthogonal-field-alternation gel electrophoresis. *Nucleic Acids Res.* **12,** 5647–5664.

Carle, G. F., and Olson, M. V. (1985). An electrophoretic karyotype for yeast. *Proc. Natl. Acad. Sci. U.S.A.* **82,** 3756–3760.

Carle, G. F., Frank, M., and Olson, M. V. (1986). Electrophoretic separation of large DNA molecules by periodic inversion of the electric field. *Science* **232,** 65–68.

Chu, G., Vollrath, D., and Davis, R. W. (1986). Separation of large DNA molecules by contour-clamped homogeneous electric fields. *Science* **234,** 1582–1585.

Clark, S. M., Lai, E., Birren, B. W., and Hood, L. (1988). A novel instrument for separating large DNA molecules with pulsed homogeneous electric fields. *Science* **241,** 1203–1205.

Dausset, J., Ougen, P., Abderrahim, H., Billault, A., Sambucy, J.-L., Cohen, D., and Le Paslier, D. (1992). The CEPH YAC Library. *Behring Inst. Mitt.* **91,** 13–20.

Fan, J. B., Chikashige, Y., Smith, C. L., Niwa, O., Yanagida, M., and Cantor, C. R. (1989). Construction of a *Not*I restriction map of the fission yeast *Schizosaccharomyces pombe* genome. *Nucleic Acids Res.* **17,** 2801–2818.

Lai, E., Birren, B. W., Clark, S. M., Simon, M. I., and Hood, L. (1989). Pulsed field gel electrophoresis. *BioTechniques* **7,** 34–42.

Orbach, M. J., Vollrath, D., Davis, R. W., and Yanofsky, C. (1988). An electrophoretic karyotype of *Neurospora crassa. Mol. Cell. Biol.* **8,** 1469–1473.

Schwartz, D. C., and Cantor, C. R. (1984). Separation of yeast chromosome-sized DNAs by pulsed field gradient gel electrophoresis. *Cell (Cambridge, Mass.)* **37,** 67–75.

Schwartz, D. C., and Koval, M. (1989). Conformational dynamics of individual DNA molecules during gel electrophoresis. *Nature (London)* **338,** 520–522.

Schwartz, D. C., Saffran, W., Welsh, J., Haas, R., Goldenberg, M., and Cantor, C. R. (1982). New techniques for purifying large DNAs and studying their properties and packaging. *Cold Spring Harbor Symp. Quant. Biol.* **47,** 189–195.

Shizuya, H., Birren, B., Mancino, V., Slepak, T., Tachiiri, Y., Kim, U.-J., and Simon, M. (1992). Cloning and stable maintenance of 300 kb fragments of human DNA in Escherichia coli using an F-factor based vector. *Proc. Nat. Acad. Sci. U.S.A.* **89,** 8794–8797.

Smith, C. L., Matsumoto, T., Niwa, O., Klco, S., Fan, J. B., Yanagida, M., and Cantor, C. R. (1987). An electrophoretic karyotype for *Schizosaccharomyces pombe* by pulsed field gel electrophoresis. *Nucleic Acids Res.* **15,** 4581–4488.

Smith, S. B., Aldridge, P. K., and Callis, J. B. (1989). Observation of individual DNA molecules undergoing gel electrophoresis. *Science* **243,** 203–206.

Smith, S. B., Gurrieri, S., and Bustamante, C. (1990). Fluorescence microscopy and computer simulations of DNA molecules in conventional and pulsed field gel electrophoresis. *In* "Electrophoresis of Large DNA Molecules: Theory and Applications" (E. Lai and B. Birren, eds.), pp. 55–80. Cold Spring Harbor Lab. Press, Cold Spring Harbor, NY.

Vollrath, D., and Davis, R. W. (1987). Resolution of DNA molecules greater than 5 megabases by contour-clamped homogeneous electric fields. *Nucleic Acids Res.* **15,** 7865–7876.

Wagner, L., and Lai, E. (1994). Separation of large DNA molecules with high voltage pulsed field gel electrophoresis. *Electrophoresis* **15,** 1078–1083.

2

Electrophoretic Karyotyping in Fungi

Ken Dewar, Louis Bernier, and Roger C. Levesque

I. Introduction

Following studies of the viscoelastic properties of DNA, Schwartz *et al.* (1982) devised a technique whereby a periodic alternation between two electric fields in different orientation could be used to separate sizes of DNA molecules much beyond the capacity of conventional agarose gel electrophoresis. Improvements in the understanding of the technique, and the demonstration that chromosome-sized DNAs could be isolated from cells embedded in agarose, allowed separations to >700 kb (Schwartz and Cantor, 1984; Carle and Olson, 1984). Carle and Olson (1985), using PFGE and chromosome specific probes, were able to correlate the PFGE-separated DNAs with the *Saccharomyces cerevisiae* genetic map to show that these molecules were intact chromosomes. PFGE techniques were subsequently adapted to separate the chromosomal DNAs of many fungi (Table 2.1) and other micro-organisms (examples being Van der Ploeg *et al.*, 1984; Kemp *et al.*, 1985; Higashiyama and Yamada, 1991). At the same time, other applications of PFGE were being developed: fingerprinting and long-range restriction enzyme-based mapping (Smith *et al.*, 1987; Ganal *et al.*, 1989), long-range chromosome walking (Kenwrick *et al.*, 1987; Rommens *et al.*, 1989), large insert DNA cloning systems (Burke *et al.*, 1987; Sternberg, 1990; Shizuya *et al.*, 1992; Iaonnou *et al.*, 1994), and chromosome fragmentation techniques (Vollrath *et al.*, 1988; Ferrin and Camerini-Otero, 1991; Strobel and Dervan, 1991). This repertoire of techniques has now made PFGE an indispensable tool for the

NONMAMMALIAN GENOMIC ANALYSIS: A PRACTICAL GUIDE

Table 2.1
Electrophoretically Karyotyped Fungi, and the Method Used To Digest Their Cell Walls

Species	Cell wall treatment	Reference[a]
Absidia glauca	*Streptomyces* No. 6	Kayser and Wöstemeyer (1991)
Acremonium chrysogenum	Novozym	Walz and Kück (1991)
Agaricus bisporus	Novozym	Royer *et al.* (1992)
Aspergillus nidulans	Novozym	Brody and Carbon (1989)
A. *niger*	Novozym	Debets *et al.* (1990)
A. Section *Flavi*	Novozym + Driselase	Keller *et al.* (1992)
Beauveria bassiana	Pronase	Pfiefer and Khachatourians (1993)
B. *nivea*	Novozym	Stimberg *et al.* (1992)
Candida albicans	Zymolyase	Snell and Wilkins (1986)
C. *boidinii*	Zymolyase	Kobori *et al.* (1991)
C. *glabrata*	Zymolyase	Magee and Magee (1987)
C. *guillieermondii*	Zymolyase	Magee and Magee (1987)
C. *lusitaniae*	Zymolyase	Merz *et al.* (1992)
C. *parapsilosis*	Zymolyase	Magee and Magee (1987)
C. *shehatae*	Zymolyase	Passoth *et al.* (1992)
C. *stellatoidea*	Zymolyase	Magee and Magee (1987)
C. *tropicalis*	Zymolyase	Magee and Magee (1987)
	Lyticase	Doebbeling *et al.* (1993)
C. *utilis*	Zymolyase	Stoltenburg *et al.* (1992)
Cephalosporium acremonium	Novozym	Skatrud and Queener (1989)
Cercospora kikuchii	Novozym + β-glucuronidase	Hightower *et al.* (1995); Upchurch *et al.* (1991)[b]
Cladosporium fulvum	Novozym	Talbot *et al.* (1991)
Cochliobolus heterostrophus	Novozym (or β-glucuronidase) + Driselase + Chitinase	Tzeng *et al.* (1992); Yoder (1988)[b]
Colletotrichum gloeosporioides	Novozym	Masel *et al.* (1990)
Coprinus cinereus	Novozym + Chitinase	Zolan *et al.* (1992)
Curvularia lunata	Novozym	Osiewacz and Ridder (1991); Osiewacz and Weber (1989)[b]
Endomyces fibuliger	Zymolyase	Naumova *et al.* (1993)
Endomycopsella vini	Zymolyase	Naumova *et al.* (1993)
E. *cratagensis*	Zymolyase	Naumova *et al.* (1993)
Entyloma spp.	Novozym	Boekhout *et al.* (1992)
Erysiphe graminis	Physical disruption	Borbye *et al.* (1992)
Filobasidiella neoformans	Novozym	de Jonge *et al.* (1986)
Fusarium oxysporum	Novozym	Boehm *et al.* (1994)
Hansenula spp.	Novozym	de Jonge *et al.* (1986)
Itersonilia spp.	Novozym	Boekhout *et al.* (1991)
Kluyveromyces spp.	Novozym	de Jonge *et al.* (1986)
K. *marxianus* var. *marxianus*	Zymolyase	Lehmann *et al.* (1992)
Leptosphaeria maculans	*T. harzianum* lytic enzyme	Taylor *et al.* (1991)
	T. harz lyticenzyme + Driselase + Chitinase + β-glucuronidase	Morales *et al.* (1993)
	Pronase E	Plummer and Howlett (1993)
Magnaporthe grisea	Novozym	Talbot *et al.* (1993)
Melanotaenium spp.	Novozym	Boekhout *et al.* (1992)
Metarhizium anisopliae	Novozym	Boekhout *et al.* (1992)

continues

Table 2.1   continued

Species	Cell wall treatment	Reference[a]
Mucor circinelloides	Protoplast-forming enzyme	Nagy *et al.* (1994)
Nectria haematococca	Novozym	Miao *et al.* (1991b)
Neurospora crassa	Novozym	Orbach *et al.* (1988)
Ophiostoma ulmi s.l.	Novozym	Royer *et al.* (1991)
	Driselase	This study
	Aspergillus lytic enzyme	This study
	Cytophaga lytic enzyme	This study
	R. solani lytic enzyme	This study
	T. harzianum lytic enzyme	This study
Parasitella parasitica	*Streptomyces* No. 6	Burmester and Wöstemeyer (1994)
Penicillium janthinellum	Novozym	Kayser and Schulz (1991)
Phoma tracheiphila	Funcelase	Rollo *et al.* (1989)
Phytophthora megasperma	Novozym	Howlett (1989)
Pichia stipitis	Zymolyase	Passoth *et al.* (1992)
Podospora anserina	Novozym	Osiewacz *et al.* (1990)
Pythium sylvaticum	Driselase + Pronase	Martin (1995a,b)[b]
Rhodosporium toruloides	Novozym	de Jonge *et al.* (1986)
Rhodotorula mucilaginosa	Novozym	de Jonge *et al.* (1986)
Saccharomyces cerevisiae	Zymolyase	Schwartz and Cantor (1984)
	Lyticase	Pasero and Marilley (1993)
	Physical disruption	Kwan *et al.* (1991)
	No treatment	McCluskey *et al.* (1990)
S. bayanus	Zymolyase	Naumov *et al.* (1992)
S. capsularis	Zymolyase	Naumov *et al.* (1992)
S. paradoxis	Zymolyase	Naumova *et al.* (1993)
Saccharomycopsis malanga	Zymolyase	Naumova *et al.* (1993)
Schizophyllum commune	Novozym	Horton and Raper (1991)
Schizosaccharomyces pombe	Novozym	Vollrath and Davis (1987)
	Lyticase	Birren and Lai (1993)
	Zymolyase	Birren and Lai (1993)
Schwanniomyces spp.	Zymolyase	Janderova and Sanca (1992)
Septoria nodorum	Novozym	Cooley and Caten (1991); Cooley *et al.* (1988)[b]
S. tritici	Novozym	McDonald and Martinez (1991)
Tilletia spp.	Novozym + μ-glucoronidase	McCluskey *et al.* (1990)
	No treatment	McCluskey *et al.* (1990)
Tilletiopsis spp.	Novozym	Boekhout *et al.* (1992)
Tolypocladium spp.	Novozym	Stimberg *et al.* (1992)
Trichoderma harzianum	Novozym	Herrera-Estrella *et al.* (1993)
T. reesei	Novozym	Carter *et al.* (1992)
T. viride	Novozym	Herrera-Estrella *et al.* (1993)
Ustilago hordei	Novozym	McCluskey and Mills (1990)
	Novozym + β-glucoronidase	McCluskey *et al.* (1990)
	No treatment	McCluskey *et al.* (1990)
U. maydis	Novozym	Kinscherf and Leong (1988)
Yarrowia lipolytica	Zymolyase	Naumova *et al.* (1993)

[a]Only the earliest electrophoretic karyotyping reference is given.
[b]Sphaeroplasting protocol used, if not given in the original karyotyping reference.

analysis of genomes ranging from bacteria to animals to higher order plants.

PFGE has become increasingly important in fungal genome analysis. For many fungi, it is possible to separate some of all of their chromosomes. The ability to construct electrophoretic karyotypes, assign markers to their chromosomes, and create chromosome-enriched genomic or cDNA libraries is an asset to any study of genome structure. Since many fungi do not have mating systems which can be controlled within a laboratory environment, the inability to perform crosses and follow the inheritance of genetic/phenotypic characters has precluded an understanding of their genetics. Fungal ultrastructural karyotypes have also been difficult to obtain, generally because pachytene bivalents are small and diffuse, and hence are difficult to view by light microscopy. Electrophoretic karyotyping and related techniques thus provide a set of tools that complement existing genetic approaches, or can be used to study genome organization when no other tools are available.

This chapter provides the procedures and principles useful in the construction of fungal electrophoretic karyotypes. It is oriented toward fungi for which PFGE techniques have not been used, and is intended to help answer the following questions: (1) are there culture conditions which provide a better starting material? (2) How can chromosome-sized DNAs be prepared? (3) How can sample quality be judged? and (4) How should PFGE be used to develop useful separations of fungal chromosomes?

Because fungal electrophoretic karyotyping covers a wide range of systems and applications—from ascomycetes to zygomycetes, from animal to insect to plant pathogens, and from genetically well-characterized to genetically unknown—the different and sometimes conflicting uses of terminology can be a problem. This work has taken a general approach. We refer to *strain* as a genetically pure culture, generally arising from a single cell. Our usage of *strain* is synonymous with *isolate, culture,* or *clone* as used elsewhere. Similarly, we use *population* as a regroupment of strains on the basis of a genetic, geographical, or physiological distinction, synonymous with *race, subgroup, pathotype,* or *biotype.* We refer to *PFGE* as the range of applications whereby alternations between differently oriented electric fields are used to separate DNA molecules in an agarose matrix. We use *chDNAs* (after Miao *et al.,* 1991b) to refer to chromosome-sized DNAs; *chDNA preparations* should not be confused with *DNA extractions,* which we use to describe the purification of DNA by conventional molecular biology methods. We use *separation* and *resolution* to describe the ability to clearly distinguish between distinct chDNA bands following PFGE.

II. Choice of Sample Material

The choice of sample material has two components: (1) the decision of which strain or strains to study; and (2) the decision of what culturing conditions to use to obtain the best possible chDNAs. The choice of strains can range from a single strain of particular importance (Smith *et al.*, 1991) to a population-wide survey (Boehm *et al.*, 1994). However, due to the surprisingly high level of genome plasticity in fungi (Kistler and Miao, 1992), it cannot be concluded that the karyotype of a single strain represents the species karyotype. Thus, even if the focus is on one strain, a sampling approach is necessary to determine the relationship between the genome structure of the strain of interest and the species at large. The range of strains in the population to be sampled must also be included during the testing and development of the protocols for electrophoretic karyotyping. Just as culturing conditions can vary among different groups of strains, so may the more particular conditions required for spheroplasting and the PFGE separation of chromosomes.

Once the strains to be analyzed have been selected, a knowledge and control of growth conditions can often be exploited to obtain improved chDNAs. The overall goal is to obtain the number of genomes (ploidy level × nuclei/cell × number of cells), reproducibly embedded in agarose, that gives the sharpest, cleanest possible chDNA bands after PFGE. Whether a fungus is vegetatively haploid *(Schizosaccharomyces pombe)*, vegetatively diploid *(Candida albicans)*, culturable as a haploid or a diploid *(S. cerevisiae)*, or culturable as a heterokaryon *(Sordaria macrospora)* will have an effect on chDNA preparations, since ploidy level and the number of nuclei per cell changes the chDNA concentration/cell.

The types of cells used for chDNA preparation greatly affect chDNA quality. Fungi can be cultured as budded cells (yeast-like fungi), as mycelia (filamentous fungi), or both (dimorphic fungi). In yeast-like and dimorphic fungi, budded cells are used: they are easier to manipulate (pipetting of cell suspensions is possible) and quantitate (chDNA concentration can be inferred from cell concentration). ChDNA preparations from filamentous fungi are more difficult. Since the ratio of nuclei:cell volume decreases as hyphal strands elongate, chDNA concentration tends to decrease with culture age. Thus, to maintain a high enough chDNA concentration, mycelial chDNAs are usually prepared from freshly germinated conidia.

Although successful electrophoretic karyotypes have been produced from fungal material obtained from a wide variety of growth conditions, an attention to media formulation and culturing conditions can greatly aid the facility and efficiency by which chDNAs are isolated. This is principally due to the effects of growth factors and nutrient availability on cell wall formation, and hence spheroplasting efficiency. Electrophoretic karyotypes have often been generated using cells grown in rich media, and it has been reported that the culturing of *S. cerevisiae* in starvation medium leads to cell walls more resistant to enzymatic digestion (Birren and Lai, 1993). In other cases, particular media and incubation conditions have been used to induce budding cell cultures in dimorphic fungi (Dewar and Bernier, 1993) or to increase conidiation for filamentous fungi (Orbach *et al.*, 1988).

III. Sample Preparation

Once the growth conditions have been established such that a sufficient quantity of cells susceptible to enzymatic cell wall digestion can be reproducibly obtained, the preparation of chDNAs is easy. While chDNAs approaching 750 kb can be isolated from liquid lysates when care is taken (Carle and Olson, 1984; Borbye *et al.*, 1992), current chDNA preparation procedures are almost universally performed by embedding cells in agarose and continuing subsequent treatments by diffusing various solutions and enzymes into and out of the agarose blocks (Schwartz and Cantor, 1984). There are two main variations of this technique; one is to remove the cell walls before embedding the spheroplasts, and the other is to remove the cell walls after the whole cells have been embedded within the agarose matrix. There are also other variations, including physically disrupting the cell walls (Kwan *et al.*, 1991; Borbye *et al.*, 1992) or isolating chDNAs without treating the cell walls at all (McCluskey *et al.*, 1990).

Fungal chDNA preparations are composed of three steps: (1) the growth and harvesting of suitable cells, (2) spheroplasting and embedding (in either order), and (3) lysis and deproteination. There are only two fundamental differences between chDNA protocols: the first is the manner in which the cells are obtained, and the second is the manner by which the cell walls are removed. As shown in Table 2.1, the enzymatic digestion of cell walls has most often been conducted with the enzymes Zymolyase or Novozym (Novozym 234), both of which are commercially available from a variety of suppliers. The protocols given below have been adapted from the original Zymolyase procedure developed for *S. cerevisiae* (Schwartz and Cantor, 1984) and the Novozym procedure developed for *Neurospora crassa* (Orbach *et al.*, 1988). Each protocol has proven to be applicable across a wide range of fungi (Table 2.1). In the cases where the enzymes have not worked well, similar protocols have been developed using different enzymes or combinations of enzymes.

A. Materials

1. Growth Media

a. YPD

Final concentration	To prepare 500 ml
1% (w/v) yeast extract	Dissolve 5 g yeast extract, 10 g bacto-peptone,
2% (w/v) bacto-peptone	and 10 g dextrose in dH_2O to 500 ml
2% (w/v) dextrose	Dispense into 125-ml flasks at 10 ml/flask
	Autoclave 15 min, 121°C

b. *Ophiostoma ulmi s.l.* Minimal Medium

Final concentration	To prepare 1 liter
10 mM L-asparagine	Dissolve 1.5 g L-asparagine, 1.0 g
7.5 mM KH_2PO_4	KH_2PO_4, 0.5 g $MgSO_4\cdot\&H_2O$, 0.1 g
2 mM $MgSO_4$	$CaCl_2\cdot2H_2O$, 500 µg H_3BO_3, 400 µg
700 µM $CaCl_2$	$MnSO_4\cdot7H_2O$, 400 µg $ZnSO_4\cdot7H_2O$, 200
8 µM H_3BO_3	µg $Na_2MoO_4\cdot2H_2O$, 200 µg $FeCl_3\cdot6H_2O$,
1.5 µM $MnSO_4$	100 µg pyridoxine-HCl, 40 µg
1.5 µM $ZnSO_4$	$CuSO_4\cdot5H_2O$, and 20 g sucrose in dH_2O
0.8 µM Na_2MoO_4	to 1 liter
0.7 µM $FeCl_3$	Dispense into 125-ml flasks at 50 ml/flask
0.5 µM pyridoxine-HCl	Autoclave 15 min, 121°C
0.2 µM $CuSO_4$	
2% (w/v) sucrose	

c. *O. ulmi s.l.* Complete Medium

Final concentration	To prepare 1 liter
15 mM ammonium sulfate	Prepare as described above,
7.5 mM KH_2PO_4	substituting 2.0 g ammonium sulfate
2 mM $MgSO_4$	for 1.5 g ʟ-asparagine, and adding 5 g
700 µM $CaCl_2$	yeast extract and 5 g malt extract
8 µM H_3BO_3	Dispense into 125-ml flasks at 50 ml/
1.5 µM $MnSO_4$	flask
1.5 µM $ZnSO_4$	Autoclave 15 min, 121°C
0.8 µM Na_2MoO_4	
0.7 µM $FeCl_3$	
0.5 µM pyridoxine-HCl	
0.2 µM $CuSO_4$	
2% (w/v) sucrose	
0.5% (w/v) yeast extract	
0.5% (w/v) malt extract	

2. Stock Solutions

a. 0.5 M EDTA (pH 8)
b. 1 M Tris–HCl (pH 7.8)
c. 5 M NaCl
d. 10% (w/v) SDS
e. 2 M Sorbitol (pH 5.8)
f. 14.4 M β-mercaptoethanol
g. Proteinase K

Final concentration	To prepare 10 ml
20 mg/ml Proteinase K	Dissolve 200 mg proteinase K in 10 ml 50 mM Tris-HCl (pH 7.8)

h. Sodium deoxycholate

Final concentration	To prepare 10 ml
10% (w/v) Sodium deoxycholate	Dissolve 1.0 g sodium deoxycholate in 10 ml dH$_2$O Aliquot and store at −20°C

i. DNase-free RNase A

Final concentration	To prepare 10 ml
20 mg/ml RNase A	Dissolve 200 mg RNase A in 10 ml TE Immerse in boiling water 20 min Aliquot and store at −20°C

3. Sphaeroplasting and Plug Treatment Solutions

a. Sorbitol–EDTA

Final concentration	To prepare 200 ml
1 M Sorbitol (pH 5.8) 50 mM EDTA (pH 8)	Mix 100 ml 2 M Sorbitol (pH 5.8), 20 ml 0.5 M EDTA (pH 8), and 80 ml dH$_2$O

b. Sorbitol–EDTA-β-mercaptoethanol

Final concentration	To prepare 50 ml (prepared fresh and used in a fume hood)
1 M Sorbitol (pH 5.8) 50 mM EDTA (pH 8) 150 mM β-mercaptoethanol	Add 500 μl 14.4 M β-mercaptoethanol to 50 ml Sorbitol–EDTA

c. EDTA–β-Mercaptoethanol

Final concentration	To prepare 1 liter (prepared fresh and used in a fume hood)
5 mM EDTA (pH 8) 300 mM β-mercaptoethanol	Mix 10 ml 0.5 M EDTA (pH 8), 970 ml dH$_2$O, and 20 ml 14.4 M β-mercaptoethanol

d. Zymolyase Solution

Final concentration	To prepare 10 ml
(i) 2.5 mg/ml Zymolyase 100T (ii) 10 mg/ml Zymolyase 25T	(i) Dissolve 25 mg Zymolyase 100T in 10 ml Sorbitol–EDTA, aliquot, and store at −20°C (ii) Dissolve 100 mg Zymolyase 20T in 10 ml Sorbitol–EDTA, aliquot, and store at −20°C

e. Novozym Solution

Final concentration	To prepare 100 ml (prepared fresh)
1 M Sorbitol (pH 5.8) 10 mg/ml Novozym	Dissolve 4 g Novozym 234 in 400 ml 1 M Sorbitol (pH 5.8)

f. Low-Melting-Point Agarose Solution

Final concentration	To prepare 100 ml
2.45 LMP agarose 100 mM EDTA (pH 8)	Mix 2.4 g LMP agarose, 20 ml 0.5 M EDTA (pH 8), and 80 ml dH$_2$O Heat gently to dissolve agarose

g. Cell Lysis Solution

Final concentration	To prepare 100 ml
1 M NaCl 100 mM EDTA (pH 8) 10 mM Tris-HCl (pH 7.8) 0.5% (w/v) Sarkosyl 0.2% (w/v) Sodium deoxycholate 20 μg/ml RNase A	Dissolve 0.5 g Sarkosyl in 56.9 ml dH$_2$O, 20 ml 5 M NaCl, 20 ml 0.5 M EDTA (pH 8), 1 ml 1 M Tris-HCl (pH 7.8), 2 ml 10% (w/v) sodium deoxycholate, 100 μl 20 mg/ml DNase-free RNase A

h. Digestion Solution

Final concentration	To prepare 100 ml
250 mM EDTA (pH 8) 50 mM Tris-HCl (pH 7.8) 1% (w/v) Sarkosyl 0.5 mg/ml proteinase K	Dissolve 1.0 g Sarkosyl in 50 ml 0.5 M EDTA (pH 8), 42.5 ml dH$_2$O, 5 ml 1 M Tris-HCl (pH 7.8), 2.5 ml 20 mg/ml Proteinase K

i. Plug Storage Solution

Final concentration	To prepare 100 ml
100 mM EDTA (pH 8) 1% (w/v) Sarkosyl	Dissolve 1.0 g Sarkosyl in 20 ml 0.5 M EDTA (pH 8), 80 ml dH$_2$O

B. Representative Protocols

1. *Saccharomyces cerevisiae* and Other Fungi

We recommend the preparation of *S. cerevisiae* chromosomes for several reasons. (1) As shown by the number of fungi susceptible to Zymolyase treatment (Table 2.1), it is a protocol that works well for many fungi. (2) The preparation of *S. cerevisiae* chromosomes is a method of becoming familiar with the types of manipulations used in other chDNA preparations. (3) Preparing *S. cerevisiae* chromosomes from well-characterized strains like YPH80 or YNN295 provides a set of precisely defined chromosomal size standards, for little effort or cost. (4) *Saccharomyces cerevisiae* chDNA preparations are useful in other applications, including the analysis of yeast artificial chromosomes.

This procedure is based on spheroplasting before embedding. A similar protocol where spheroplasting is performed after embedding can be found in Birren and Lai (1993). A 10-ml late exponential phase ($\approx 10^8$ cells/ml) YPD culture of the diploid strain YNN295 will give 2 ml of plugs—sufficient for hundreds of 5-mm-wide gel lanes. This protocol can be scaled up or down as need be, although instead of scaling down, it is just as easy to form fewer plugs at the end.

(1) Grow YNN295 in 10 ml of YPD in a 125-ml flask at 30°C with agitation (200–300 rpm) to obtain a late exponential phase culture ($\approx 10^8$ cells/ml). This can be done overnight if the culture is inoculated with 20–50 μl of a saturated culture or may take 1–2 days if inoculated with colonies.

(2) Transfer the culture to a 50-ml centrifuge tube and collect the cells by a 5-min centrifugation at 3000 g at 4°C. Discard the supernatant.

(3) Resuspend the cells by pipetting in 2 ml of Sorbitol–EDTA. Collect the cells by a 5-min centrifugation at 3000g at 4°C. Discard the supernatant.

(4) Resuspend the cells by pipetting in 1 ml of Sorbitol–EDTA–β-mercaptoethanol, then incubate at room temperature 5 min.

(5) Add 40 μl of Zymolyase 100T solution (or 50 μl of Zymolyase 20T solution), then incubate at 37°C for 30 min.

(6) Calculate the spheroplasting efficiency by withdrawing an aliquot (5–10 μl) of the cell solution to a microscope slide, then (at 20–40× magnification) determine the percentage of cells that lyse following the addition of an equal volume of 10% SDS.[1]

(7) When sufficient spheroplasting has occurred, warm the cells to 42°C and gently mix by pipetting with an equal volume of molten 42°C 2.4% LMP agarose. Form plugs and let them harden on ice at least 10 min.

(8) Transfer the plugs to a 50-ml centrifuge tube and submerge them in digestion solution. Incubate at 50°C for 24 h or longer.

(9) The plugs can be stored at 4°C (months to years) or at room temperature (hours to months) in digestion solution or plug storage buffer.

2. *Neurospora crassa* and Other Fungi

This is an adaptation of the *N. crassa* procedure of Orbach *et al.* (1988), and is a representative protocol for embedding yeast-like cells after digestion with Novozym 234. For *O. ulmi s.l.*, cultures are grown at room temperature in minimal or complete medium (Bernier and Hubbes, 1990) with agitation ($\approx$100 rpm). Generally, 50 ml media in 125-ml flasks, after inoculation with fresh mycelial plugs, will generate uninucleate haploid cell cultures at 10^8–10^9 cells/ml after 5–7 days. Typically, 50 ml of cells will provide 5 ml of plugs, sufficient for hundreds of gel lanes. Using this protocol, *O. ulmi s.l.* chDNAs can be prepared after treatment with Novozym 234, or after treatment with several other lytic enzyme preparations (Fig. 2.1).

(1) Grow *O. ulmi s.l.* in 50 ml medium in a 125-ml flask with agitation ($\approx$100 rpm) for 5–7 days to obtain a late exponential phase culture (10^8–10^9 cells/ml).

(2) Transfer the culture to a 50-ml centrifuge tube and collect the cells by a 10-min centrifugation at 3000*g* at 4°C. Discard the supernatant.

(3) Gently resuspend the cells by pipetting in 50 ml of EDTA–β-mercaptoethanol. Incubate for 20 min at room temperature with agitation ($\approx$100 rpm). Collect the cells by a 10-min centrifugation at 3000*g* at 4°C. Discard the supernatant in a fumehood.

---

[1]We aim for $\approx$80% spheroplasting efficiency. If this has not occurred after the 30-min incubation period, the incubation can be extended and/or fresh enzyme can be added. If spheroplasting efficiency continues to be low, verify enzyme activity, cell concentration, and culturing conditions.

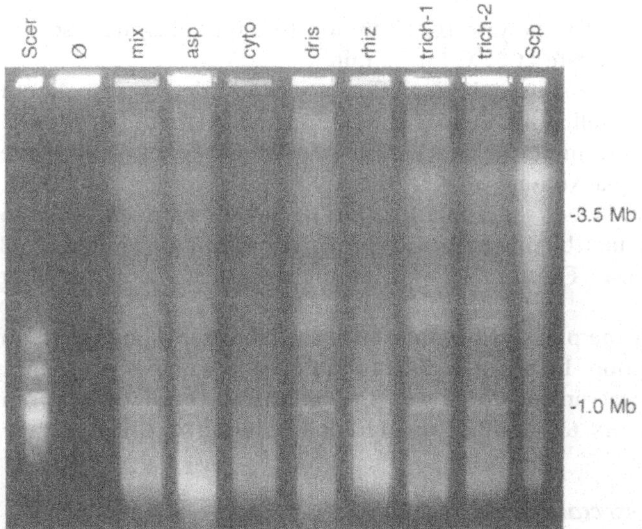

Figure 2.1 Chromosomal DNAs of *O. ulmi s.l.* strain CESS16K after treatment with different lytic enzymes. All *O. ulmi s.l.* samples were prepared following the accompanying *N. crassa* procedure (Orbach *et al.,* 1988), with only the type and quantity of lytic enzyme being altered. **Scer** and **Scp:** chromosomal size standards *S. cerevisiae* strain YNN295 and *S. pombe* strain 972. Ø: Treatment with no lytic enzymes. **mix:** Treatment with a cocktail of lytic enzymes (Sigma) from *Aspergillus* spp. (400 µg/ml), Cytophaga spp. (1 mg/ml), Driselase (1 mg/ml), *Rhizoctonia solani* (1 mg/ml), and *Trichoderma harzianum* (1 mg/ml). **asp:** Treatment with lytic enzyme from *Aspergillus* spp. (5 mg/ml). **cyto:** Treatment with lytic enzyme from *Cytophaga* spp. (2 mg/ml). **dris:** Treatment with Driselase (5 mg/ml). **rhiz:** Treatment with two separate lots of lytic enzyme from *Trichoderma harzianum* (5 mg/ml). **PFGE conditions:** a 1% SeaKem GTG agarose (FMC BioProducts) gel in 0.5× TBE (Birren and Lai, 1993) chilled to 14°C was run at 2 V/cm with an included field angle of 106° for 96 hr using a linearly increasing switch time ramp of 15 to 30 min.

(4) Resuspend the cells by pipetting in 20 ml of Sorbitol–EDTA. Collect the cells by a 10-min centrifugation at 3000*g* at 4°C. Discard the supernatant in a fumehood.

(5) Resuspend the cells by pipetting in 20 ml of Novozym solution. Incubate for 1 hr with agitation (≈100 rpm). Calculate spheroplasting efficiency as already described.

(6) Collect the cells by a 10-min centrifugation at 3000*g* at 4°C. Carefully remove the supernatant, as the cell pellet can be very loose.[2]

[2]This protocol collects and embeds the spheroplasts, nonspheroplasts, and cellular debris. More detailed procedures for purifying sphaeroplasts can be found in Orbach *et al.* (1988) or Royer *et al.* (1991).

(7) Gently resuspend the cells by pipetting in 2.5 ml of 2 *M* Sorbitol (pH 5.8). Warm the cell suspension to 42°C and mix by gentle pipetting with an equal volume of molten 42°C 2.4% LMP agarose. Form plugs and let them harden on ice at least 10 min.

(8) Transfer the plugs to 50-ml tubes and submerge them in cell lysis solution. Incubate at 37°C for 24 hr or more.

(9) Discard the cell lysis solution and submerge the plugs in digestion solution. Incubate at 50°C for 24 hr or more.[3]

(10) The plugs can be stored as already described.

3. Chopped Inserts and Agarose Bead Encapsulation

DNA concentration profoundly affects DNA migration during PFGE, so when electrophoretic karyotypes are being compared, it is essential to load equivalent quantities of chDNA in each well. Since chDNA band shape corresponds to plug size and shape, a combination of thick and thin plugs on the same gel will confuse interpretations even if chDNA concentration per lane is standardized. There are three ways to overcome this problem. The most obvious is to make plugs of equivalent chDNA concentration, which may or may not be feasible. The other two methods rely on making the plugs small enough that they can be loaded into the wells of the gel using large bore pipette tips. An additional advantage of these techniques is that the small plug size permits a more efficient diffusion, thereby reducing the quantity of enzymes and incubation times required for subsequent treatments.

In the chopped insert technique, chDNA plugs are formed in typical plug molds, then chopped with a razor blade into sizes small enough for pipetting. Using *C. albicans*, Wang and Schwartz (1993) have shown that chDNAs of at least 2 Mb do not show appreciable damage after chopping. In the agarose bead encapsulation technique, The DNA–LMP agarose solution is mixed vigourously with mineral oil before the agarose is allowed to set. Forcing the agarose to set quickly under agitation causes the formation of small agarose spheres containing the chDNAs. Agarose bead encapsulation has been used for *S. cerevisiae*, *Hansenula wingei*, and *S. pombe* (all commercially available from BRL), demonstrating that the technique can be used to prepare chDNAs of at least 5.7 Mb. Agarose bead encapsulation protocols can be found in Overhauser and Radic (1987), Bakalinsky (1990), and Mills *et al.* (1995).

---

[3]Both the cell lysis and digestion treatments usually require no longer than 24 hr; however, extended incubations do not harm the plugs and may give more complete lysis or deproteination.

IV. Constructing Electrophoretic Karyotypes

A. Chromosomal Size Standards

During every stage in the construction of an electrophoretic karyotype, it is important to have the appropriate chromosomal size standards. PFGE size standards serve four purposes, of which sizing may be the least important. Size markers serve as a means for verifying that the gel worked properly—this can be important if there are atypic migrations being caused by nonuniform electric fields, power shortages, buffer temperature changes, etc., or if there is sample degradation occurring due the presence of contaminants in the plug, gel, or buffer. Markers aid in identifying the size range of DNA molecules being separated in the gel. Markers also aid in cross-referencing between gels, so that the effects of changes in PFGE parameters can be monitored. Having confidence that the gels function as planned, being able to identify the regions of separation, and being able to correlate changes in PFGE conditions with differences in chDNA separations are all essential when trying to derive the appropriate PFGE conditions for unknown samples.

The three most widely used chromosomal size standards are concatamers of λ bacteriophage, chromosomal DNAs of well-characterized *S. cerevisiae* strains (YPH80 and YNN295), and chromosomal DNAs of *S. pombe*, λ ladders range in size from 50 kb to ≈1 Mb, *S. cerevisiae* chromosomes range from 200 kb to ≈2 Mb, and *S. pombe* contains three chromosomes estimated to be 3.5, 4.6, and 5.7 Mb (Fan *et al.*, 1988). To fill in the gap between the largest *S. cerevisiae* chromosomes and the chromosomes of *S. pombe*, two other commercially available chromosome standards have been used, *H. wingei* and *C. albicans* (both with chromosomes from 1 to 3 Mb). Other chDNA standards, although not commercially available, include the fungal strains for which the procedures for chDNA preparations and PFGE separations have already been developed. Still other markers include nondigested very-high-molecular-weight DNA (>10 Mb) and partially digested very-high-molecular-weight DNAs.

B. Constructing Electrophoretic Karyotypes

There is a rational approach by which PFGE can be used to determine the electrophoretic karyotype of a fungus. The three major steps, represented by the boxes in Fig. 2.2, are (1) the verification of chDNA quality, (2) the testing of different PFGE conditions to determine the chromosome size range and distribution for the strains of interest, and (3) the

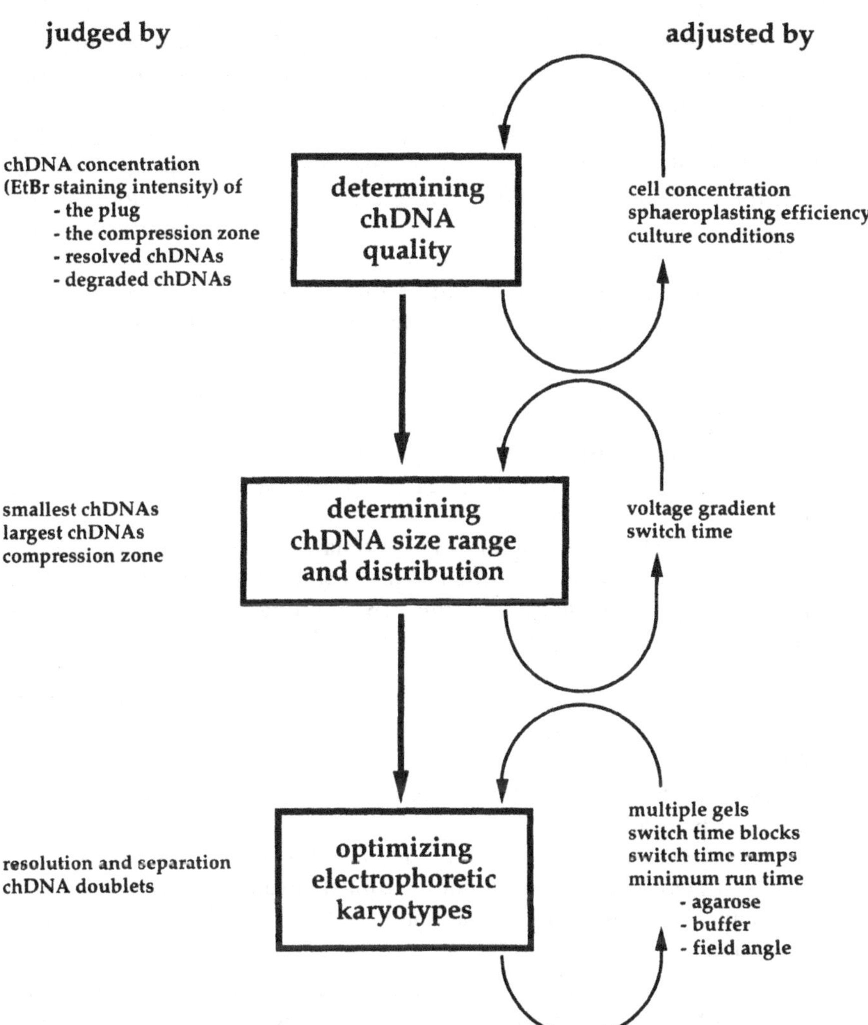

Figure 2.2 Constructing electrophoretic karyotypes.

development of PFGE protocols for the most efficient separations of the chromosomes of interest.

1. Determining ChDNA Quality

Before investing substantial effort in determining the appropriate PFGE conditions for good chDNA separations, it is necessary to verify the quality of the chDNAs. Good quality chDNA plugs satisfy the following three criteria: (1) the concentration of chDNAs available for migration is high enough to permit chDNA band visualization following PFGE, but (2) not so high as to inhibit separations; and (3) the ratio of intact to degraded chDNAs is high enough that interpretations after EtBr staining are facile and unambiguous. Depending on the research goals, two additional criteria can also be important: (4) equivalent chDNA concentrations among different preparations, and (5) restriction enzyme digestibility.

Tests of chDNA quality should be rapid, allow high throughput, and be as informative as possible. One very simple test is to run the newly prepared plugs alongside of *S. cerevisiae* size standards under conditions separating the *S. cerevisiae* chromosomes (Fig. 2.3). This test offers several advantages:

(1) It is rapid. PFGE separations of DNA molecules ranging from 200 kb to 2 Mb can be achieved in 24 hr or less using very simple PFGE programs.

(2) ChDNA concentration can be estimated by comparing the sample chDNAs to the *S. cerevisiae* chromosomes. Even if the sample chDNAs remain in the compression zone, the EtBr staining intensity can indicate whether there is too little or too much chDNA for useful PFGE separations.

(3) Spheroplasting efficiency can be evaluated by comparing the EtBr staining intensities of the chDNAs and the material remaining in the plug.[4]

(4) Variation in chDNA concentration between different preparations can be evaluated.

(5) The ratio of intact to degraded chDNAs can be estimated.

(6) Depending on the sizes of the fungal chromosomes, some chDNA bands may be resolved, further aiding in judging chDNA quality and concentration.

If the PFGE conditions have already been derived for resolving the chDNAs of the strain of interest, another effective quality check is to run a gel using the optimal PFGE conditions, but for a shorter duration than normally used. While the chDNA bands will not migrate as far, or separate as much, they will still be resolved. As traditional chromosome size standards are less useful in this type of test, a plug from a previous batch of confirmed good quality provides the best reference.

These quality checks are very important, but caution is required, since there are two situations where the results could lead to false interpreta-

---

[4]An example of insufficient spheroplasting is shown in lane Ø of Fig. 1.

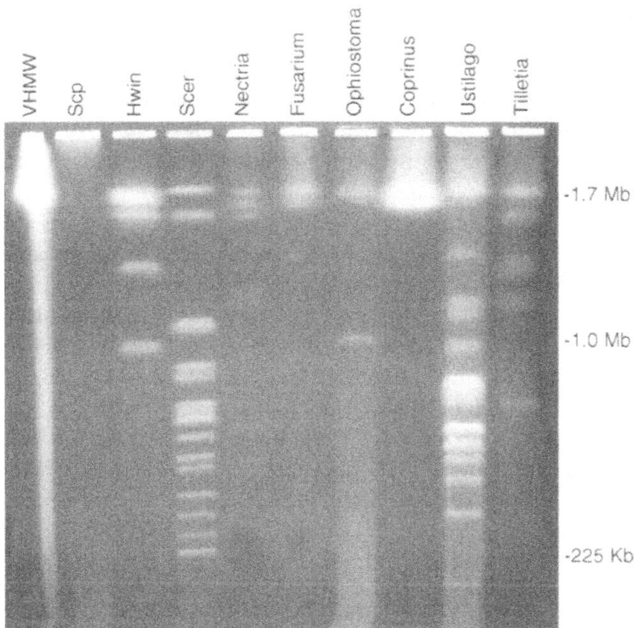

Figure 2.3 Determining chDNA quality. **VHMW:** undigested very-high-molecular-weight DNA of *Meleagris gallopavo* (courtesy of Ed Smith), prepared following Birren and Lai (1993). **Scp, Hwin, Scer:** chromosomal size standards *S. pombe*, *H. wingei*, and *S. cerevisiae* (Bio-Rad). **Nectria:** chDNAs of *Nectria haematococca* MP VI (courtesy of Corby Kistler and Ulla Benny), prepared following Powell and Kistler (1990). **Fusarium:** chDNAs of *Fusarium oxysporum* f. sp. *cubense* VCG 1214 (courtesy of Corby Kistler and Ulla Benny), prepared following Boehm *et al.* (1994). **Ophiostoma:** chDNAs of *Ophiostoma ulmi s.l.* CESS16K, prepared following Dewar and Bernier (1993). **Coprinus:** chDNAs of *Coprinus cinereus* Java-6 (courtesy of Mimi Zolan), prepared following Zolan *et al.* (1992). **Ustilago:** chDNAs of *Ustilago hordei* 2.1 (courtesy of Dallice Mills and Brian Russell), prepared following McCluskey and Mills (1990). **Tilletia:** chDNAs of *Tilletia caries* M7-1 (courtesy of Dallice Mills and Brian Russell), prepared following Russell and Mills (1993). **PFGE conditions:** a 1% SeaKem GTG agarose gel in 0.5× TBE chilled to 14°C was run at 6 V/cm with an included field angle of 120° for 24 hr using a linearly increasing switch time ramp of 60 to 120 sec.

tions. In PFGE separations to 1–2 Mb, larger chDNAs will comigrate in the compression zone, but some very large chDNAs may never leave the plug.[5] Similarly, some circular DNAs can remain trapped in the plug. In both these cases, the higher EtBr staining in the plug could be misconstrued as evidence of suboptimal spheroplasting.

[5]An example is shown with *S. pombe* in Fig. 3.

2. Determining Chromosome Size Range and Distribution

A pulsed-field gel, run under the simplest conditions (one constant switch time throughout the duration of the run), contains at least five zones of separation. From lowest to highest molecular weight, these are (1) a migration front containing the molecules too small to be resolved under the PFGE conditions used; (2) a zone where a size range of molecules is resolved; (3) an inflection point; (4) a second zone where a different size range of molecules is resolved; (5) a compression zone containing molecules small enough to migrate yet too large to be resolved; and possibly (6) a set of molecules that are unable to leave the plug. The two zones of resolution, the inflection point, and the compression zone are based on regions I to IV as defined by Vollrath and Davis (1987). Whereas the occurrence of these zones is a function of switch time, their limits and ranges are affected by the other PFGE parameters: agarose and buffer type and concentration, field strength, included field angle, temperature, and DNA concentration (see Chapter 1).

In many cases, however, pulsed-field gels are not run using the simplest conditions. Changing switch times during a run, either progressively (ramping) in a linear or nonlinear fashion or in discrete steps (blocks), has often been used to improve chDNA separations. Still other techniques, including the use of secondary pulses (brief intervals of migration in a different direction between the major pulses) and interrupts (pauses between pulses), may also be useful. Yet as the programs used for PFGE separations become more complex, so does the identification of the various zones of resolution in the gel. Thus, while these programs may aid in improving separations when chromosome size and distribution are already known, their use in "first tries" is probably unwise.

At its most basic, the determination of appropriate PFGE conditions entails identifying which chDNAs occur in which zones of the gel under which conditions, then using this knowledge to develop efficient pulsed-field gel separation regimens. There are two approaches by which this can be accomplished. One is strictly a marker-based approach, whereby the separations of the unknown chDNAs are monitored under gel conditions known to separate the reference chDNAs. The hope is that the PFGE conditions optimal for the standards will lead to good separations of the sample chDNAs. As an example, the unknown chDNAs could be run alongside of *S. cerevisiae*, then *H. wingei*, then *S. pombe*, under the conditions best for each of those markers. In this manner, the presence of "small," "medium," and "large" chDNAs could be identified, and the conditions for improved separations of the unknown chDNAs could be deduced. The advantage of this approach is that the markers serve as internal standards for each gel: because a particular pattern of the marker chDNAs is expected, it is immediately evident whether the desired PFGE conditions were duplicated and hence whether the gel worked as expected or not.

There are, however, several disadvantages to relying exclusively upon this approach. Effectively reproducing a set of PFGE conditions can be a major problem: not only must the electrophoresis parameters be duplicated exactly (voltage gradient, included field angle, temperature, switch time, run time, the types and concentrations of agarose and buffer), but the same marker must be used at an equivalent concentration. Duplicating PFGE protocols can range from being annoying (preparing different stocks of buffer) to costly (purchasing different brands of agarose) to impossible (due to differences between pulsed-field gel systems). In addition, commercial suppliers provide the PFGE conditions optimal for their markers. The more complicated the protocol, the more specific it becomes in what it can and cannot separate, thus the less valuable it becomes in general utility. *Saccharomyces cerevisiae* chromosomes, for example, do not differ by constant lengths, hence programs designed to separate those chDNAs may do a poor job of separating similarly sized chDNAs with a different size distribution (i.e., *S. cerevisiae* vs *Ustilago* in Fig. 2.3). The final, overriding problem of this approach is that all the effort is dedicated to separating the marker chDNAs and not the chDNAs of the sample of interest.

The second approach to constructing electrophoretic karyotypes relies upon the principles of PFGE. In this approach, a series of gels is run to develop a broad outline of the numbers and sizes of the unknown chDNAs. This preliminary knowledge of which chDNAs separate under which conditions is then used to derive conditions more appropriate to the sample of interest. While this approach also employs chromosomal size standards, the focus is reversed—instead of using gels to separate the chromosomes of the marker strains, the markers are used to indicate the size ranges being separated on the gel. A major advantage of this approach is that it is "self-contained," the electrophoretic karyotype is constructed with the materials at hand (the PFGE system available, the agarose and buffer normally used, etc.), without regard to the PFGE conditions of others. Another important advantage is that the attention remains firmly upon the strain of interest—while markers are used to quantitate the regions of separation on the gel, it is how these regions affect the separation of the unknown chDNAs that is used to deduce improved separations.

3. Optimizing Electrophoretic Karyotypes

To illustrate the three stages involved in constructing and optimizing electrophoretic karyotypes (Fig. 2.2), we performed the series of gels presented in Fig. 2.3, 2.4, and 2.5. Each of the gels contained the same 10 samples. The first 4, used as size references, consisted of very-high-molecular-weight avian DNA, and the chromosome size standards *S. pombe*, *H. wingei*, and *S. cerevisiae* (Bio-Rad). The remaining 6 "unknown" samples were chDNAs prepared from *Nectria haematococca*, *Fusarium oxysporum*, *O. ulmi s.l.*, *Coprinus cinereus*, *Ustilago hordei*, and *Tilletia caries*.

ChDNA quality was assessed by running the samples under conditions known to separate most of the *S. cerevisiae* chromosomes (Fig. 2.3). In this gel, chDNAs from 200 kb to ≈1.7 Mb were resolved, with larger-sized chDNAs comigrating in the compression zone. The intensity of fluorescence of the resolved chromosomes and compression zone chDNAs in-

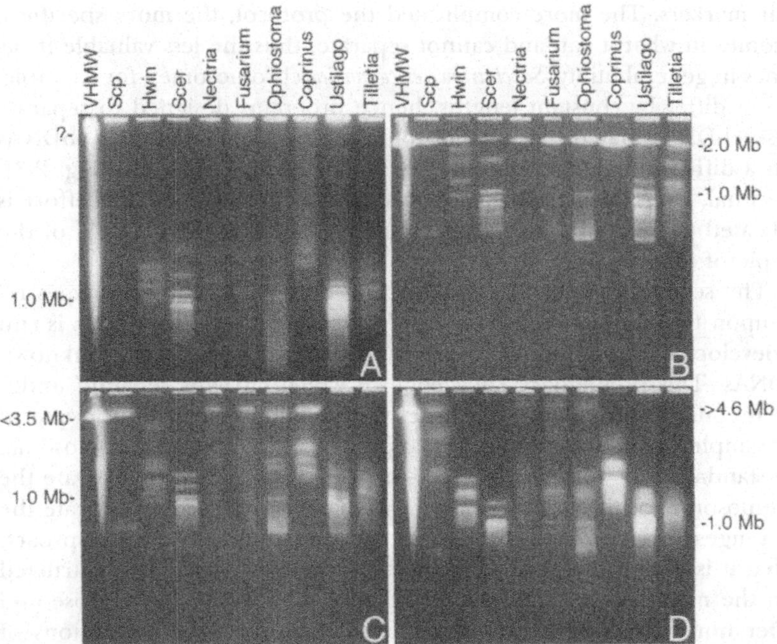

Figure 2.4 Determining chDNA size range and distribution. Samples as in Fig. 3. **PFGE conditions: A,** a 1% SeaKem GTG agarose gel in 0.5× TBE chilled to 14°C was run at 3 V/cm with an included field angle of 120° for 60 hr using a constant switch time of 10 min. **B,** a 1% SeaKem GTG agarose gel in 0.5× TBE chilled to 14°C was run at 2 V/cm with an included field angle of 106° for 72 hr using a constant switch time of 10 min. **C,** a 1% SeaKem GTG agarose gel in 0.5× TBE chilled to 14°C was run at 2 V/cm with an included field angle of 106° for 72 hr using a constant switch time of 20 min. **D,** a 1% SeaKem GTG agarose gel in 1× TAE (Birren and Lai, 1993) chilled to 14°C was run at 2 V/cm with an included field angle of 120° for hr using a constant switch time 30 min.

dicated that each of the six "unknowns" contained intact chromosomes at concentrations sufficient for EtBr visualization after PFGE.

Having verified that the chDNAs were of good quality, we were prepared to invest further effort in determining chromosome size range and distribution. We next ran a series of gels intended to resolve progressively larger chDNAs (Fig. 2.4). As the resolution of larger DNA molecules requires a reduction in the voltage gradient used, we began by reducing the voltage gradient to 3 V/cm and extending the switch time to 10 min (Fig. 2.4A). Although chDNAs ranging from 1.0 to 2.3 Mb were resolved in the gel, the lack of a defined compression zone indicated that this combination of PFGE parameters was not appropriate. After a further reduction of the voltage gradient to 2 V/cm, and a change of the included field angle to 106°, discrete compression zones were visible (Fig. 2.4B, 2.4C, 2.4D). Whereas a constant switch time of 10 min now resolved chDNAs from 1.0 to ≈2.0 Mb (Fig. 2.4B), lengthening the switch time to 20 min resulted in the resolution of chDNAs to almost 3.5 Mb (Fig.

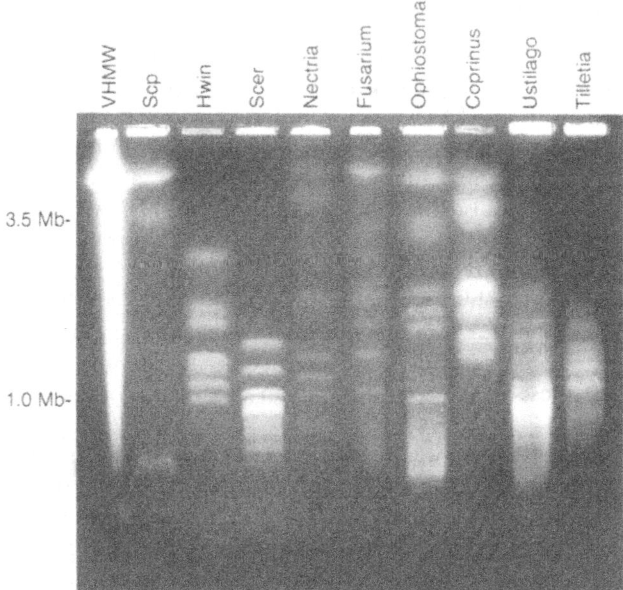

Figure 2.5 Optimizing electrophoretic karyotypes. Samples as in Fig. 3. **PFGE conditions:** a 1% SeaKem GTG agarose gel in 0.5× TBE chilled to 14°C was run at 2 V/cm with an included field angle of 106° for 96 hr using a linearly increasing switch time ramp of 15 to 30 min.

2.4C). A subsequent lengthening of the switch time to 30 min, and the use of 1X TAE buffer instead of 0.5X TBE buffer, resulted in an even higher size range ($\approx$3.0 to >4.6 Mb) of DNA molecules being resolved (Fig. 2.4D).

For our example of an optimized electrophoretic karyotype, we chose to improve the separations in the region of 1.0 to 3.5 Mb. While this would not be adequate for the resolution of all chromosomes of all strains, it would be helpful in further clarifying genome structure of the *Nectria, Fusarium, Ophiostoma,* and *Coprinus* samples. The protocol we derived was based on the following criteria:

(i) We wished to continue using 1% SeaKem GTG in 0.5X TBE at 14°C;

(ii) We wished to continue using a voltage gradient of 2 V/cm and an included field angle of 106°;

(iii) As we were interested in resolving chDNAs over a range of sizes, and the chromosomes of interest were distributed throughout this range, we chose to employ a linearly increasing switch time ramp;

(iv) We chose to extend the run time to 96 hr to allow an increased separation of the chDNAs we were capable of resolving.

The actual switch times we used were derived from information available in Fig. 2.4. Whereas a 10-min switch time (Fig. 2.4B) was insufficient, a 30-min switch time (with 1X TAE) was too much (Fig. 2.4D). Thus we settled on a ramp of 15 to 30 min. Our hope was that this would allow resolution of a size range slightly higher than that of the 20-min fixed switch time (Fig. 2.4C), and also that the separations of chDNAs at the upper and lower limits of the resolution zones would be improved. As shown in Fig. 2.5, the optimized conditions worked as predicted. ChDNAs from 1.0 to >3.5 Mb were resolved, but most importantly, these conditions led to improved separations for some of the unknown samples. For both the *Ophiostoma* and *Coprinus* samples, we were able to identify more chDNA bands under these PFGE conditions than any of the other conditions tested.

The purpose of this exercise was to demonstrate how the principles of PFGE can be put to use in the construction of fungal electrophoretic karyotypes. We used a variety of samples to show that this approach is not limited to a particular organism, but relies on an understanding of PFGE. Considering chromosomes of less than $\approx$6 Mb, a PFGE program to ameliorate the resolution for any of the samples can be derived from the results of Fig. 2.3 and 2.4. Although separations of >6 Mb were not attempted in this work, the information in Fig. 2.4 provides a starting point for further tries. Finally, we stress that it is the approach, not the reagents or PFGE system, that is important. A similar exercise could have been performed using different agaroses, buffers, temperatures, and PFGE systems, yet the outcome would have been similar.

V. Applications of Electrophoretic Karyotyping

While a full coverage of all the applications of electrophoretic karyotyping lies outside of the scope of this chapter, a brief overview is presented here (Table 2.2). We have divided the applications into four classes: (1) electrophoretic karyotyping, (2) comparative electrophoretic karyotyping, (3)

Table 2.2
Applications of Electrophoretic Karyotyping

Application	Species
1. Obtaining complete electrophoretic karyoyptes	
(i) By correspondence with linkage groups	
Carle and Olson (1985)	*Saccharomyces cerevisiae*
Magee *et al.* (1988)	*Candida albicans*
Orbach *et al.* (1988)	*Neurospora crassa*
Brody and Carbon (1989)	*Aspergillus nidulans*
Debets *et al.* (1990)	*Aspergillus niger*
Javerzat *et al.* (1993)	*Podospora anserina*
Kerrigan *et al.* (1993)	*Agaricus bisporus*
(ii) By correspondence with cytological observations	
Boehm and Bushnell (1992)	*Melampsora lini*
Borbye *et al.* (1992)	*Erysiphe graminis*
2. Comparative Electrophoretic Karyotyping	
(i) Chromosomal rearrangements during mitotic growth	
Smith *et al.* (1991)	*Acremonium chrysogenum*
Walz and Kück (1991)	*A. chrysogenum*
Rustchenko-Bulgac (1991)	*C. albicans*
Suzuki *et al.* (1991)	*Candida tropicalis*
(ii) rDNA expansion/contraction	
Butler and Metzenberg (1990)	*N. crassa*
Maleszka and Clark-Walker (1990)	*Kluyveromyces lactis*
Pasero and Marilley (1993)	*S. cerevisiae*
Pasero and Marilley (1993)	*Schizosaccharomyces pombe*
Pukkila and Skrzynia (1993)	*Coprinus cinereus*
(iii) Fingerprinting/taxonomy	
Carruba *et al.* (1991)	*Candida parapsilosis*
Steensma *et al.* (1988)	*Kluyveromyces* spp.
Taylor *et al.* (1991)	*Leptosphaeria maculans*
Boekhout *et al.* (1991)	*Itersonilia* spp.
Boekhout *et al.* (1992)	*Tilletiopsis* spp.
Cansado *et al.* (1992)	*S. cerevisiae*

continues

Table 2.2   *continued*

Application	Species
Merz *et al.* (1992)	*Candida lusitaniae*
Stimberg *et al.* (1992)	*Tolypocladium inflatum*
Doebbeling *et al.* (1993)	*C. tropicalis*
Boehm *et al.* (1994)	*Fusarium oxysporum*
Russel and Mills (1994)	*Tilletia* spp.
(iv) Measuring genome plasticity	
Magee and Magee (1987)	*C. albicans*
de Jonge *et al.* (1986)	*Kluyveromyces* spp.
Kinscherf and Leong (1988)	*Ustilago maydis*
Cooley and Caten (1991)	*Septoria nodorum*
Iwaguchi *et al.* (1990)	*C. albicans*
Masel *et al.* (1990)	*Colletotrichum gloeosporioides*
McDonald and Martinez (1991)	*Septoria tritici*
Talbot *et al.* (1991)	*Cladosporium fulvum*
Janderova and Sanca (1992)	*Schwanniomyces* spp.
Keller *et al.* (1992)	*Aspergillus* section Flavi
Lehmann *et al.* (1992)	*Kluyveromyces marxianus*
Naumov *et al.* (1992)	*Saccharomyces* spp.
Passoth *et al.* (1992)	*Candida shehatae*
Passoth *et al.* (1992)	*Pichia stipitis*
Shimuzu *et al.* (1992)	*Metarhizium anisopliae*
Stoltenburg *et al.* (1992)	*Candida utilis*
Dewar and Bernier (1993)	*Ophiostoma ulmi* s.l.
Morales *et al.* (1993)	*Leptosphaeria maculans*
Naumova *et al.* (1993)	*Yarrowia lipolytica*
Talbot *et al.* (1993)	*Magnaporthe grisea*
Nagy *et al.* (1994)	*Mucor circinelloides*
Martin (1995b)	*Pythium sylvaticum*
(v) Genome plasticity vs phenotype	
Fasullo and Davis (1988)	*S. cerevisiae*
Wickes *et al.* (1991a)	*Candida stellatoidea*
Kistler and Benny (1992)	*Nectria haematococca*
Thrash-Bingham and Gorman (1992)	*C. albicans*
Masel *et al.* (1993a)	*Colletotrichum gloeosporioides*
McCluskey *et al.* (1994)	*Ustilago hordei*
(vi) Genome plasticity vs ploidy level	
Kobori *et al.* (1991)	*Candida* spp.
Schillberg *et al.* (1991)	*Saccharomyces* spp.
Longo and Vezinhet (1993)	*S. cerevisiae*
(vii) Inheritance of chromosome length polymorphisms	
Ono and Ishino-Arao (1988)	*S. cerevisiae*
McCluskey and Mills (1990)	*Ustilago hordei*
Plummer and Howlett (1993)	*Leptosphaeria maculans*
Russell and Mills (1993)	*Tilletia* spp.
Zolan *et al.* (1994)	*Coprinus cinereus*
Dewar and Bernier (1995)	*Ophiostoma ulmi* s.l.
Martin (1995a)	*Pythium sylvaticum*

continues

Table 2.2 continued

Application	Species
(viii) Supernumerary/mimichromosomes	
Miao *et al.* (1991a)	*Nectria haematococca*
Tzeng *et al.* (1992)	*Cochliobolus heterostrophus*
Masel *et al.* (1993b)	*Colletotrichum gloeosporioides*
3. Assignment of markers to chromosomes	
Pretorious and Marmur (1988)	*S. cerevisiae*
Osiewacz *et al.* (1990)	*Podospora anserina*
Horton and Raper (1991)	*Schizophyllum commune*
Kayser and Schulz (1991)	*Penicillium janthinellum*
Kayser and Wöstemeyer (1991)	*Absidia glauca*
Miao *et al.* (1991b)	*Nectria haematococca*
Osiewacz and Ridder (1991)	*Curvularia lunata*
Royer *et al.* (1991)	*Ophiostoma ulmi s.l.*
Wickes *et al.* (1991b)	*C. albicans*
Carter *et al.* (1992)	*Trichoderma reesei*
Montenegro *et al.* (1992)	*A. nidulans*
Royer *et al.* (1992)	*Agaricus bisporus*
Herrera-Estrella *et al.* (1993)	*Trichoderma* spp.
Walz and Kück (1993)	*A. chrysogenum*
Ásgeirsdóttir *et al.* (1994)	*Schizophyllum commune*
Bowden *et al.* (1994)	*Ophiostoma ulmi s.l.*
4. Enriched libraries and physical mapping	
Zolan *et al.* (1992)	*Coprinus cinereus*
Vollrath *et al.* (1988)	*S. cerevisiae*
Strobel and Dervan (1991)	*S. cerevisiae*
Thierry *et al.* (1991)	*S. cerevisiae*
Thierry and Dujon (1992)	*S. cerevisiae*
Schwartz *et al.* (1993)	*S. cerevisiae*
Wang *et al.* (1995)	*Candida albicans*

the assignment of genes and markers to their respective chromosomes, and (4) enriched libraries and physical mapping. These categories are not mutually exclusive, but serve to point out an increasing level of complexity—what can be done using PFGE alone, using PFGE and Southern hybridizations, and using PFGE to obtain very-high-molecular-weight DNA for other types of physical genome analysis.

At its most basic, the purpose of an electrophoretic karyotype is to provide information on the number and sizes of chromosomes within a fungal genome. This is of intrinsic interest for any study of genome organization, but can also aid other projects [karyotype information allows an estimation of genome size, which permits a degree of confidence to be associated with calculations of the number of genomic clones required for genome coverage (Clarke and Carbon, 1976)]. A complete electrophoretic karyotype, however, is very difficult to achieve. Not only must the highest molecular weight chromosomes be resolved from the compression zone, there must also be confidence in the ability to detect chDNA doublets or multiplets. The identification of comigrating chromosomes, based on differences in relative EtBr staining, is not always reliable. Obtaining complete electrophoretic karyotypes requires a confirmation between the number of chDNA bands observed on the gel and another measure of genome structure. This confirmation has usually been done using Southern hybridizations to show correspondence between electrophoretic karyotypes and genetically defined linkage groups, although matching electrophoretic karyotypes to cytological observations can also be useful.

Comparative electrophoretic karyotyping uses PFGE to tudy differences in genome organization. It can be used to monitor chromosomal rearrangements in a single strain following selection, mutation, or rDNA expansion/contraction. It can also be used as a fingerprinting tool for aiding taxonomic classification or strain identification. Due to the high level of chromosome length polymorphisms in fungi, comparative electrophoretic karyotyping has been used widely in measuring genome plasticity: in measuring genome plasticity in wild-type populations, in investigating the effects of genome plasticity on phenotype, in determining how ploidy level and meiotic transmission affect the level of genome plasticity, and in studying the development, transmission, and roles of supernumerary and minichromosomes.

Once an electrophoretic karyotype has been established, hybridization to membranes blotted with pulsed-field gels permits the rapid assignment of genes, or markers linked to genes, to specific chromosomes. Southern hybridizations of pulsed-field gels also allow the monitoring of integration events following random or targeted insertional mutagenesis. Combined with comparative electrophoretic karyotyping, Southern hybridizations can further elucidate the extent of genome plasticity, and can help identify homologous chromosomes between strains comporting variable karyotypes.

Electrophoretic karyotyping procedures can also be used in a variety of ways to select or enrich libraries specific to particular chromosomes or groups of chromosomes. Probing Southern blots of PFGE-resolved chromosomes with clones from existing libraries can be used to build chromosome-specific sublibraries, or chDNAs can be excised and probed

---

[6]We distinguish between *specific* and *enriched* libraries since it is only the first case that allows the selection of clones verifiably specific to a chromosome. In enriched libraries, clones specific to a chromosome will be identified, as will (i) clones of repetitive DNA also present elsewhere in the genome, (ii) clones from other chromosomes containing similar sequence (i.e., clones from gene families), and (iii) clones misidentified due to the inherently difficult task of purifying chDNA bands without additional comigrating DNA fragments present as contaminants.

against existing libraries to create enriched sublibraries[6] (Brody *et al.*, 1991; Chapter 9). A third approach is to excise individual chDNA bands and use them as a source for DNA cloning (Zolan *et al.*, 1992). Chromosome "subtraction" libraries are a fourth option, where PFGE can be used to create compression zones containing all the chDNAs excepting the smallest chromosome(s).

While localizing genes, markers, or clones to particular chromosomes is a first step in genome mapping or chromosome walking, more detailed analyses can be performed by combining electrophoretic karyotyping with sequence-specific cleavage techniques. Chromosome fragmentation can be used not only to identify the chromosome carrying the marker, but also to provide preliminary evidence as its location on the chromosome (Fig. 2.6).

The preparation of chDNAs for electrophoretic karyotyping is also a procedure for the preparation of material for large DNA cloning. For fungal genomes (20–40 Mb), entire chromosomes or entire genomes may now be mapped by series of contiguous clones of large insert size. While

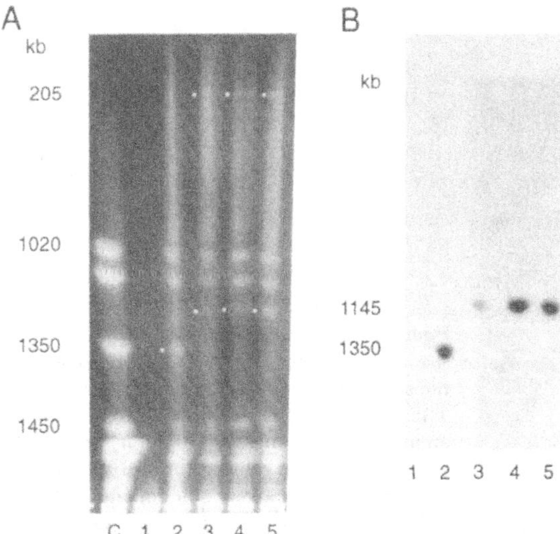

Figure 2.6 RecA-assisted restriction enzyme cleavage (RARE). **Panel A:** C, ChDNAs of *C. albicans* FC18 (ATCC 62376). 1, The same sample after digestion with *Eco*RI. 2, The same sample after *Eco*RI digestion following treatment with *Eco*RI methylase. 3–5, The same sample after RARE using an *ERG16* oligonucleotide. PFGE conditions: using an ED device, a 1% agarose gel in 0.5× TBE was run at 10 V/cm for 24 hr using a constant switch time of 180 sec, then run at 10 V/cm for 24 hr using a constant switch time of 120 sec. **B:** Autoradiogram of the same gel after Southern hybridization with a *C. albicans* chromosome V specific probe. (From Wang *et al.*, 1995. Reprinted with permission.)

cosmids have been used to map the *S. pombe* genome (Hoheisel *et al.*, 1993; Mizukami *et al.*, 1993), larger insert cloning systems including P1 bacteriophage (Sternberg, 1990; Pierce *et al.*, 1992), bacterial and P1 artificial chromosomes (Shizuya *et al.*, 1992; Ioannou *et al.*, 1994), and yeast artificial chromosomes (Burke *et al.*, 1987) will also be useful. The BAC and PAC systems have enormous potential in the mapping of fungal genomes: insert sizes can range from that of cosmids ($\approx$45 kb) to that of P1 ($\approx$95 kb) to that of medium sized YACs (300 kb); the cloning procedure is not as technically daunting as for YACs, and the clones are stable and easy to analyze.

VI. Conclusion

We believe that pulsed-field gels are magic, and fortunately, this magic can be harnessed. The power to resolve DNA molecules of millions of base pairs in length not only has made possible the separation and determination of the lengths of intact fungal chromosomes, but also has permit-

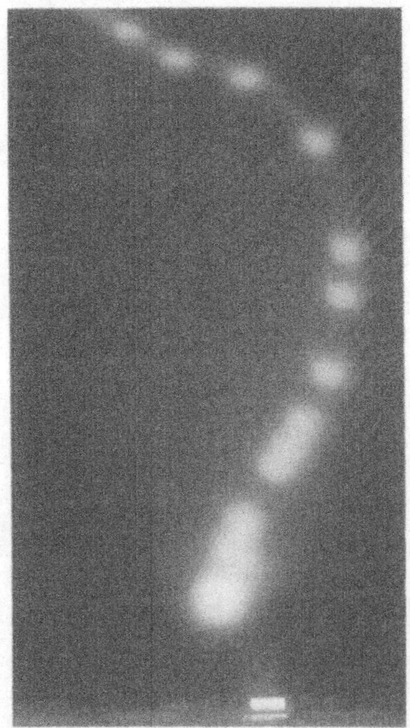

Figure 2.7 Nonlinear migration of *S. cerevisiae* chromosomes by PACE. **PFGE conditions:** a 1% SeaKem GTG agarose gel in 0.5× TBE chilled to 14°C was run at 6 V/cm with an included field angle of 120° for 50 hr. For the first 10 hr the gel was run with a 70-sec constant switch time and a net migration to 12°. For the second 10 hr the gel was run with a 60-sec constant switch time and a net migration to −81°. For the third 10 hr the gel was run with a 45-sec constant switch time and a net migration to −12°. For the fourth 10 hr the gel was run with a 30-sec constant switch time and a net migration to 57°. For the fifth 10 hr the gel was run with a 20-second constant switch time and a net migration to 128°.

ted detailed genomic analyses by permitting the purification, dissection, and cloning of large tracts of those chromosomes. As demonstrated by *S. cerevisiae* (Fig. 2.7), these techniques are now indispensable in allowing us to ask and answer new questions about fungal genomes and their structure, organization, and regulation.

Acknowledgments

We thank Corby Kistler and Ulla Benny (University of Florida), Dallice Mills and Brian Russell (Oregon State University), Mimi Zolan (Indiana University), Ed Smith (Tuskegee University), Yu-Ker Wang and David Schwartz (New York University), Hiroaki Shizuya (California Institute of Technology), and Alan Bakalinsky (Oregon State University) for their contributions. K.D. also thanks the participants of the 1993–95 CSH courses on the "Cloning and Analysis of Large DNA Molecules." Portions of this work were supported by the Natural Sciences and Engineering Research Council of Canada (NSERC), the Fonds FCAR of Québec, the Medical Research Council of Canada (MRC), and the Canadian Bacterial Diseases Network (CBDN).

References

Ásgeirsdóttir, S. A., Schuren, F. H. J., and Wessels, J. G. H. (1994). Assignment of genes to pulse-field separated chromosomes of *Schizophyllum commune. Mycol. Res.* **98,** 689–693.

Bakalinsky, A. T. (1990). "Electrophoretic Karyotypes of Wine Strains of *Saccharomyces cerevisiae,* " Bull. No. 1647. Bio-Rad Laboratories, Richmond, CA.

Bernier, L., and Hubbes, M. (1990). Mutations in *Ophiostoma ulmi* induced by *N*-methyl-*N*'-nitro-*N*-nitrosoguanidine. *Can. J. Bot.* **68,** 225–231.

Birren, B., and Lai, E. (1993). "Pulsed Field Gel Electrophoresis: A Practical Guide." Academic Press, San Diego, CA.

Boehm, E. W. A., and Bushnell, W. R. (1992). An ultrastructural pachytene karyotype for *Melampsora lini. Phytopathology* **82,** 1212–1218.

Boehm, E. W. A., Ploetz, R. C., and Kistler, H. C. (1994). Statistical analysis of electrophoretic karyotype variation among vegetative compatibility groups of *Fusarium oxysporum* f. sp. *cubense. Mol. Plant-Microbe Interact.,* **7,** 196–207.

Boekhout, T., Poot, G., Hackman, P., and Steensma, H. Y. (1991). Genomic characteristics of strains of *Itersonilia:* Taxonomic consequences and life cycle. *Can. J. Microbiol.* **37,** 188–194.

Boekhout, T., van Gool, J., van den Boogert, H., and Jille, T. (1992). Karyotyping and G+C composition as taxonomic criteria applied to the systematics of *Tilletiopsis* and related taxa. *Mycol. Res.* **95,** 331–342.

Borbye, L., Linde-Laursen, I., Christiansen, S. K., and Giese, H. (1992). The chromosome complement of *Erysiphe graminis* f. sp. *hordei* analysed by light microscopy and field inversion gel electrophoresis. *Mycol. Res.* **96,** 97–102.

Bowden, C. G., Hintz, W. E., Jeng, R. S., Hubbes, M., and Horgen, P. A. (1994). Isolation and characterization of the cerato-ulmin toxin gene of the Dutch elm disease pathogen, *Ophiostoma ulmi. Curr. Genet.* **25,** 323–329.

Brody, H., and Carbon, J. (1989). Electrophoretic karyotype of *Aspergillus nidulans. Proc. Natl. Acad. Sci. U.S.A.* **86,** 6260–6263.

Brody, H., Griffith, J., Cuticchia, A. J., Arnold, J., and Timberlake, W. E. (1991). Chromosome-specific recombinant DNA libraries from the fungus *Aspergillus nidulans. Nucleic Acids Res.* **19,** 3105–3109.

Burke, D. T., Carle, G. F., and Olson, M. V. (1987). Cloning of large segments of exogenous DNA into yeast using artificial-chromosome vectors. *Science* **236,** 806–812.

Burmester, A., and Wöstemeyer, J. (1994). Variability in genome organization of the zygomycete *Parasitella parasitica. Curr. Genet.* **26,** 456–460.

Butler, D. K., and Metzenberg, R. L. (1990). Expansion and contraction of the nucleolus organizer region of *Neurospora crassa:* Changes originate in both proximal and distal segments. *Genetics* **126,** 325–333.

Cansado, J., Velazquez, J. B., Calo, P., Sieiro, C., Longo, E., and Villa, T. G. (1992). Characterization of killer-resistant strains of *Saccharomyces cerevisiae* isolated from spontaneous fermentations. *FEMS Microbiol. Lett.* **97,** 13–18.

Carle, G. F., and Olson, M. V. (1984). Separation of chromosomal DNA molecules from yeast by orthogonal-field-alteration gel electrophoresis. *Nucleic Acids Res.* **12,** 5647–5664.

Carle, G. F., and Olson, M. V. (1985). An electrophoretic karyotype for yeast. *Proc. Natl. Acad. Sci. U.S.A.* **82,** 3756–3760.

Carruba, G., Pontieri, E., de Bernardis, F., Martino, P., and Cassone, A. (1991). DNA fingerprinting and electrophoretic karyotype of environmental and clinical isolates of *Candida parapsilosis. J. Clin. Microbiol.* **29,** 916–922.

Carter, G. L., Allison, D., Rey, M. W., and Dunn-Coleman, N. S. (1992). Chromosomal and genetic analysis of the electrophoretic karyotype of *Trichoderma reesei:* Mapping of the cellulase and xylanase genes. *Mol. Microbiol.* **6,** 2167–2174.

Clarke, L., and Carbon, J. (1976). A colony bank containing synthetic *Col El* hybrid plasmids representative of the entire *E. coli* genome. *Cell (Cambridge, Mass.)* **9,** 91–99.

Cooley, R. N., and Caten, C. E. (1991). Variation in electrophoretic karyotype between strains of *Septoria nodorum. Mol. Gen. Genet.* **228,** 17–23.

Cooley, R. N., Shaw, R. K., Franklin, F. C. H., and Caten, C. E. (1988). Transformation of the phytopathogenic fungus *Septoria nodorum* to hygromycin B resistance. *Curr. Genet.* **13,** 383–389.

Debets, A. J. M., Holub, E. F., Swart, K., van den Broek, H. W. J., and Bos, C. J. (1990). An electrophoretic karyotype of *Aspergillus niger. Mol. Gen. Genet.* **224,** 264–268.

de Jonge, P., de Jongh, F. C. M., Meijers, R., Steensma, H. Y., and Scheffers, W. A. (1986). Orthogonal-field-alternation gel electrophoresis banding patterns of DNA from yeasts. *Yeast* **2,** 193–204.

Dewar, K., and Bernier, L. (1993). Electrophoretic karyotypes of the elm tree pathogen *Ophiostoma ulmi (sensu lato). Mol. Gen. Genet.* **238,** 43–48.

Dewar, K., and Bernier, L. (1995). Inheritance of chromosome length polymorphisms in *Ophiostoma ulmi (sensu lato). Curr. Genet.* **27,** 541–549.

Doebbeling, B. N., Lehmann, P. F., Hollis, R. J., Wu, L.-C., Widmer, A. F., Voss, A., and Pfaller, M. A. (1993). Comparison of pulsed-field gel electrophoresis with isoenzyme profiles as a typing system for *Candida tropicalis. Clin. Infect. Dis.* **16,** 377–383.

Fan, J.-B., Chikashige, Y., Smith, C. L., Niwa, O., Yanagida, M., and Cantor, C. R. (1988). Construction of a *Not* I restriction map of the fission yeast *Schizosaccharomyces pombe* genome. *Nucleic Acids Res.* **17,** 2801–2818.

Fasullo, M. T., and Davis, R. W. (1988). Direction of chromosome rearrangements in *Saccharomyces cerevisiae* by use of *his*3 recombinational substrates. *Mol. Cell. Biol.* **8,** 4370–4380.

Ferrin, L. J., and Camerini-Otero, R. D. (1991). Selective cleavage of human DNA: RecA-assisted restriction endonuclease (RARE) cleavage. *Science* **254,** 1494–1497.

Ganal, M. W., Young, N. D., and Tanksley, S. D. (1989). Pulsed field gel electrophoresis and physical mapping of large DNA fragments in the *Tm-2a* region of chromosome 9 in tomato. *Mol. Gen. Genet.* **215,** 395–400.

Herrera-Estrella, A., Goldman, G. H., Van Montagu, M., and Geremia, R. A. (1993). Electrophoretic karyotype and gene assignment to resolved chromosomes of *Trichoderma* spp. *Mol. Microbiol.* **7,** 515–521.

Higashiyama, T., and Yamada, T. (1991). Electrophoretic karytotyping and chromosomal gene mapping of *Chlorella. Nucleic Acids Res.* **19,** 6191–6195.

Hightower, R. C., Callahan, T. M., and Upchurch, R. G. (1995). Electrophoretic karyotype of *Cercospora kikuchii. Curr. Genet.* **27,** 290–292.

Hoheisel, J. D., Maier, E., Mott, R., McCarthy, L., Grigoriev, A. V., Schalkwyk, L. C., Nizetic, D., Francis, F., and Lehrach, H. (1993). High resolution cosmid and P1 maps spanning the 14 Mb genome of the fission yeast *S. pombe. Cell (Cambridge, Mass.)* **73,** 109–120.

Horton, J. S., and Raper, C. A. (1991). Pulsed-field gel electrophoretic analysis of *Schizophyllum commune* chromosomal DNA. *Curr. Genet.* **19,** 77–80.

Howlett, B. J. (1989). An electrophoretic karyotype for *Phytophthora megasperma. Exp. Mycol.* **13,** 199–202.

Iaonnou, P. A., Amemiya, C. T., Garnes, J., Kroisel, P. M., Shizuya, H., Chen, C., Batzer, M. A., and de Jong, P. J. (1994). A new bacteriophage P1-derived vector for the propagation of large human DNA fragments. *Nat. Genet.***6,** 84–89.

Iwaguchi, S.-I., Homma, M., and Tanaka, K. (1990). Variation in the electrophoretic karyotype analysed by the assignment of DNA probes in *Candida albicans. J. Gen. Microbiol.* **136,** 2433–2442.

Janderova, B., and Sanca, A. (1992). Electrophoretic karyotyping within the genus *Schwanniomyces. Syst. Appl. Microbiol.* **15,** 590–592.

Javerzat, J. P., Jacquier, C., and Barreau, C. (1993). Assignment of linkage groups to the electrophoretically-separated chromosomes of the fungus *Podospora anserina. Curr. Genet.* **24,** 219–222.

Kayser, T., and Schulz, G. (1991). Electrophoretic karyotype of cellulolytic *Penicillium janthinellum* strains. *Curr. Genet.* **20,** 289–291.

Kayser, T., and Wöstemeyer, J. (1991). Electrophoretic karyotype of the zygomycete *Absidia glauca:* Evidence for differences between mating types. *Curr. Genet.* **19,** 279–284.

Keller, N. P., Cleveland, T. E., and Bhatnagar, D. (1992). Variable electrophoretic karyotypes of members of *Aspergillus* section *Flavi. Curr. Genet.* **21,** 371–375.

Kemp, D. J., Corcoran, L. M., Coppel, R. L., Stahl, H. D., Bianco, A. E., Brown, G. V., and Anders, R. F. (1985). Size variation in chromosomes from independent cultured isolates of *Plasmodium falciparum. Nature (London)* **315,** 347–350.

Kenwrick, S., Patterson, M., Speer, A., Fischbeck, K., and Davies, K. (1987). Molecular analysis of the Duchenne muscular dystrophy region using pulsed field gel electrophoresis. *Cell (Cambridge, Mass.)* **48,** 351–357.

Kerrigan, R. W., Royer, J. C., Baller, L. M., Kohli, Y., Horgen, P. A., and Anderson, J. B. (1993). Meiotic behaviour and linkage relationships in the secondarily homothallic fungus *Agaricus bisporus. Genetics* **133,** 225–236.

Kinscherf, T. G., and Leong, S. A. (1988). Molecular analysis of the karyotype of *Ustilago maydis. Chromosoma* **96,** 427–433.

Kistler, H. C., and Benny, U. (1992). Autonomously replicating plasmids and chromosome rearrangement during transformation of *Nectria haematococca. Gene* **117**, 81–89.

Kistler, H. C., and Miao, V. P. W. (1992). New modes of genetic change in filamentous fungi. *Annu. Rev. Phytopathol.* **30**, 131–152.

Kobori, H., Takata, Y., and Osumi, M. (1991). Interspecific protoplast fusion between *Candida tropicalis* and *Candida boidini:* Characterization of the fusants. *J. Ferment. Bioeng.* **72**, 439–444.

Kwan, H. S., Li, C. C., Chiu, S. W., and Cheng, S. C. (1991). A simple method to prepare intact yeast chromosomal DNA for pulsed field gel electrophoresis. *Nucleic Acids Res.* **19**, 1347.

Lehmann, P. F., Khazan, U., Wu, L.-C., Wickes, B. L., and Kwon-Chung, K. J. (1992). Karyotype and isozyme profiles do not correlate in *Kluyveromyces marxianus* var. *marxianus. Mycol. Res.* **96**, 637–642.

Longo, E., and Vezinhet, F. (1993). Chromosomal rearrangements during vegetative growth of a wild strain of *Saccharomyces cerevisiae. Appl. Environ. Microbiol.* **59**, 322–326.

Magee, B. B., and Magee, P. T. (1987). Electrophoretic karyotypes and chromosome numbers in *Candida species. J. Gen. Microbiol.* **133**, 425–430.

Magee, B. B., Koltin, Y., Gorman, J. A., and Magee, P. T. (1988). Assignment of cloned genes to the seven electrophoretically separated *Candida albicans* chromosomes. *Mol. Cell. Biol.* **8**, 4721–4726.

Maleszka, R., and Clark-Walker, G. D. (1990). Magnification of the rDNA cluster in *Kluyveromyces lactis. Mol. Gen. Genet.* **223**, 342–344.

Martin, F. (1995a). Meiotic instability of *Pythium sylvaticum* as demonstrated by inheritance of nuclear markers and karyotype analysis. *Genetics* **139**, 1233–1246.

Martin, F. (1995b). Electrophoretic karyotype polymorphisms in the genus *Pythium. Mycologia* **87**, 333–353.

Masel, A. M., Braithwaite, K., Irwin, J., and Manners, J. (1990). Highly variable molecular karyotypes in the plant pathogen *Colletotrichum gloeosporioides. Curr. Genet.* **18**, 81–86.

Masel, A. M., Irwin, J. A. G., and Manners, J. M. (1993a). DNA addition or deletion is associated with a major karyotype polymorphism in the fungal phytopathogen *Colletotrichum gloeosporioides. Mol. Gen. Genet.* **237**, 73–80.

Masel, A. M., Irwin, J. A. G., and Manners, J. M. (1993b). Mini-chromosomes of *Colletotrichum* spp. infecting several host species in various countries. *Mycol. Res.* **97**, 852–856.

McCluskey, K., and Mills, D. (1990). Identification and characterization of chromosome length polymorphisms among strains representing fourteen races of *Ustilago hordei. Mol. Plant-Microbe Interact.* **3**, 366–373.

McCluskey, K., Russell, B. W., and Mills, D. (1990). Electrophoretic karyotyping without the need for generating protoplasts. *Curr. Genet.* **18**, 385–386.

McCluskey, K., Agnan, J., and Mills, D. (1994). Characterization of genome plasticity in *Ustilago hordei. Curr. Genet.* **26**, 486–493.

McDonald, B. A., and Martinez, J. P. (1991). Chromosome length polymorphisms in a *Septoria tritici* population. *Curr. Genet.* **19**, 265–271.

Merz, W. G., Khazan, U., Jabra-Rizk, M. A., Wu, L.-C., Osterhout, G. J., and Lehmann, P. F. (1992). Strain delineation and epidemiology of *Candida (Clavispora) lusitaniae. J. Clin. Microbiol.* **30**, 449–454.

Miao, V. P. W., Covert, S. F., and VanEtten, H. D. (1991a). A fungal gene for antibiotic resistance on a dispensable ("B") chromosome. *Science* **254**, 1773–1776.

Miao, V. P. W., Matthews, D. E., and VanEtten, H. D. (1991b). Identification and chromosomal locations of a family of cytochrome P-450 genes for pisatin detoxification in the fungus *Nectria haematococca. Mol. Gen. Genet.* **226**, 214–223.

Mills, D., McCluskey, K., Russell B. W., and Agnan, J. (1995). Electrophoretic karyotyping: Method and applications. *In* "Molecular Methods in Plant Pathology" (R. P. Singh and U. S. Singh, eds.). CRC Press, Boca Raton.

Mizukami, T., Chang, W. I., Garkavtsev, I., Kaplan, N., Lombardi, D., Matsumoto, T., Niwa, O., Kounosu, A., Yanagida, M., Marr, T., and Beach, D. (1993). A 13 kb resolution cosmid map of the 14 Mb fission yeast genome by nonrandom sequence-tagged site mapping. *Cell (Cambridge, Mass.)* **73,** 121–132.

Montenegro, E., Fierro, F., Fernandez, F. J., Gutiérrez, S., and Martin, J. F. (1992). Resolution of chromosomes III and VI of *Aspergillus nidulans* by pulsed-field gel ectrophoresis shows that the penicillin biosynthetic pathway genes *pcbAB, pcbC,* and *penDE* are clustered on chromosome VI (3.0 Megabases). *J. Bacteriol.* **174,** 7063–7067.

Morales, V. M., Séguin-Swatz, G., and Taylor, J. (1993). Chromosome size polymorphism in *Leptosphaeria maculans. Phytopathology* **83,** 503–509.

Nagy, A., Vagvölgi, C., Balla, É., and Ferenczy, L. (1994). Electrophoretic karyotype of *Mucor circinelloides. Curr. Genet.* **26,** 45–48.

Naumov, G. I., Naumova, E. S., Lantto, R. A., Louis, E. J., and Korhola, M. (1992). Genetic homology between *Saccharomyces cerevisiae* and its sibling species *S. paradoxus* and *S. bayanus:* Electrophoretic karyotypes. *Yeast* **8,** 599–612.

Naumova, E., Naumov, G., Fournier, P., Nguyen, H.-V., and Gaillardin, C. (1993). Chromosomal polymorphism of the yeast *Yarrowia lipolytica* and related species: Electrophoretic karyotyping and hybridization with cloned genes. *Curr. Genet.* **23,** 450–454.

Ono, B.-I., and Ishino-Arao, Y. (1988). Inheritance of chromosome length polymorphisms in *Saccharomyces cerevisiae. Curr. Genet.* **14,** 413–418.

Orbach, M. J., Vollrath, D., Davis, R. W., and Yanofsky, C. (1988). An electrophoretic karyotype of *Neurospora crassa. Mol. Cell. Biol.* **8,** 1469–1473.

Osiewacz, H. D., and Ridder, R. (1991). Genome analysis of imperfect fungi: Electrophoretic karyotyping and characterization of the nuclear gene coding for glyceraldehyde-3-phosphate dehydrogenase (gpd) of *Curvularia lunata. Curr. Genet.* **20,** 151–155.

Osiewacz, H. D., and Weber, A. (1989). DNA mediated transformation of the filamentous fungus *Curvularia lunata* using a dominant selectable marker. *Appl. Microbiol. Biotechnol.* **30,** 375–380.

Osiewacz, H. D., Clairmont, A., and Huth, M. (1990). Electrophoretic karyotype of the ascomycete *Podospora anserina. Curr. Genet.* **18,** 481–483.

Overhauser, J., and Radic, M. Z. (1987). Encapsulation of cells in agarose beads for use with pulsed-field gel electrophoresis. *Focus* **9,** 8–9.

Pasero, P., and Marilley, M. (1993). Size variation of rDNA clusters in the yeasts *Saccharomyces cerevisiae* and *Schizosaccharomyces pombe. Mol. Gen. Genet.* **236,** 448–452.

Passoth, V., Hansen, U., and Emeis, C. C. (1992). The electrophoretic banding pattern of the chromosomes of *Pichia stipitis* and *Candida shehatae. Curr. Genet.* **22,** 429–431.

Pfiefer, T. A., and Khachatourians, G. G. (1993). Electrophoretic karyotype of the entomopathogenic deuteromycete *Beauveria bassiana. J. Invertebr. Pathol.* **61,** 231–235.

Pierce, J. C., Sauer, B., and Sternberg, N. (1992). A positive selection vector for cloning high molecular weight DNA by the bacteriophage P1 system: Improved cloning efficacy. *Proc. Natl. Acad. Sci. U.S.A.* **89,** 2056–2060.

Plummer, K., and Howlett, B. J. (1993). Major chromosome length polymorphisms are evident after meiosis in the phytopathogenic fungus *Leptosphaeria maculans. Curr. Genet.* **24,** 107–113.

Powell, W. A., and Kistler, H. C. (1990). In vivo rearrangement of foreign DNA by *Fusarium oxysporum* produces linear self-replicating plasmids. *J. Bacteriol.* **172,** 3163–3171.

Pretorious, I. S., and Marmur, J. (1988). Localization of yeast glucoamylase genes by PFGE and OFAGE. *Curr. Genet.* **14**, 9–13.

Pukkila, P. J., and Skrzynia, C. (1993). Frequent changes in the number of reiterated ribosomal RNA genes throughout the life cycle of the basidiomycete *Coprinus cinereus*. *Genetics* **133**, 203–211.

Rollo, F., Ferracuti, T., and Pacilli, A. (1989). Separation of chromosomal DNA molecules from *Phoma tracheiphila* by orthogonal-field-alteration gel electrophoresis. *Curr. Genet.* **16**, 477–479.

Rommens, J. M., Iannuzzi, M. C., Kerem, B.-S., Drumm, M. L., Melmer, G., Dean, M., Rozmahel, R., Cole, J. L., Kennedy, D., Hidaka, N., Zsiga, M., Buchwald, M., Riordan, J. R., Tsui, L.-C., and Collins, F. S. (1989). Identification of the cystic fibrosis gene: Chromosome walking and jumping. *Science* **245**, 1059–1065.

Royer, J. C., Dewar, K., Hubbes, M., and Horgen, P. A. (1991). Analysis of a high frequency transformation system for *Ophiostoma ulmi*, the causal agent of Dutch elm disease. *Mol. Gen. Genet.* **225**, 168–176.

Royer, J. C., Hintz, W. E., Kerrigan, R. W., and Horgen, P. A. (1992). Electrophoretic analysis of the button mushroom, *Agaricus bisporus*. *Genome* **35**, 694–698.

Russell, B. W., and Mills, D. (1993). Electrophoretic karyotypes of *Tilletia caries, T. controversa,* and their F_1 progeny: Further evidence for conspecific status. *Mol. Plant-Microbe Interact.* **6**, 66–74.

Russell, B. W., and Mills, D. (1994). Morphological, physiological and genetic evidence in support of a conspecific status for *Tilletia caries, T. controversa,* and *T. foetida*. *Phytopathology* **84**, 576–582.

Rustchenko-Bulgac, E. P. (1991). Variations of *Candida albicans* electrophoretic karyotypes. *J. Bacteriol.* **173**, 6586–6596.

Schillberg, S., Zimmerman, M., and Emeis, C.-C. (1991). Analysis of hybrids obtained by rare-mating of *Saccharomyces* strains. *Appl. Microbiol. Biotechnol.* **35**, 242–246.

Schwartz, D. C., and Cantor, C. R. (1984). Separation of yeast chromosome-sized DNAs by pulsed field gradient gel electrophoresis. *Cell (Cambridge, Mass.)* **37**, 67–75.

Schwartz, D. C., Saffran, W., Welsh, J., Haas, R., Goldenberg, M., and Cantor, C. R. (1982). New techniques for purifying large DNAs and studying their properties and packaging. *Cold Spring Harbor Symp. Quant. Biol.* **47**, 189–195.

Schwartz, D. C., Li, X., Hernandez, L. I., Ramnarain, S. P., Huff, E. J., and Wang, Y.-K. (1993). Ordered restriction maps of *Saccharomyces cerevisiaie* chromosomes constructed by optical mapping. *Science* **262**, 110–114.

Shimuzu, S., Arai, Y., and Matsumoto, T. (1992). Electrophoretic karyotype of *Metarhizium anisopliae*. *J. Invertebr. Pathol.* **60**, 185–187.

Shizuya, H., Birren, B., Kim, U.-J., Mancino V., Slepak, T., Tachiiri, Y., and Simon, M. (1992). Cloning and stable maintenance of 300-kilobase-pair fragments of human DNA in *Escherichia coli* using an F-factor-based vector. *Proc. Natl. Acad. Sci. U.S.A.* **89**, 8794–8797.

Skatrud, P. L., and Queener, S. W. (1989). An electrophoretic molecular karyotype for an industrial strain of *Cephalosporium acremonium*. *Gene* **78**, 331–338.

Smith, A. W., Collis, K., Ramsden, M., Fox, H. M., and Peberdy, J. F. (1991). Chromosome rearrangements in improved cephalosporin C-producing strains of *Acremonium chrysogenum*. *Curr. Genet.* **19**, 235–237.

Smith, C. L., Econome, J. G., Schutt, A., Klco, S., and Cantor, C. R. (1987). A physical map of the *Escherichia coli* K12 genome. *Science* **236**, 1448–1453.

Snell, R. G., and Wilkins, R. J. (1986). Separation of chromosomal DNA molecules from *C. albicans* by pulsed field gel electrophoresis. *Nucleic Acids Res.* **14**, 4401–4406.

Steensma, H. Y., de Jongh, F. C. M., and Linnekamp, M. (1988). The use of electrophoretic karyotypes in the classification of yeasts: *Kluyveromyces marxianus* and *K. lactis*. *Curr. Genet.* **14**, 311–317.

Sternberg, N. (1990). A bacteriophage P1 cloning system for the isolation, amplification and recovery of DNA fragments as large as 100 Kilobase pairs. *Proc. Natl. Acad. Sci. U.S.A.* **87**, 103–107.

Stimberg, N., Walz, M., Schörgendorfer, K., and Kück, U. (1992). Electrophoretic karyotyping from *Tolypocladium inflatum* and six related strains allows differentiation of morphologically similar species. *Appl. Microbiol. Biotechnol.* **37**, 485–489.

Stoltenburg, R., Klinner, U., Ritzerfeld, P., Zimmermann, M., and Emies, C. C. (1992). Genetic diversity of the yeast *Candida utilis. Curr. Genet.* **22**, 441–446.

Strobel, S. A., and Dervan, P. B. (1991). Single-site enzymatic cleavage of yeast genomic DNA mediated by triple helix formation. *Nature (London)* **350**, 172–174.

Suzuki, T., Miyamae, Y., and Ishida, I. (1991). Variation of colony morphology and chromosomal rearrangement in *Candida tropicalis* pK233. *J. Gen. Microbiol.* **137**, 161–167.

Talbot, N. J., Oliver, R. P., and Coddington, A. (1991). Pulsed field gel electrophoresis reveals chromosome length differences between strains of *Cladosporium fulvum* (syn. *Fulvia fulva*). *Mol. Gen. Genet.* **229**, 267–272.

Talbot, N. J., Salch, Y. P., Ma, M., and Hamer, J. E. (1993). Karyotypic variation within clonal lineages of the rice blast fungus, *Magnaporthe grisea. Appl. Environ. Microbiol.* **59**, 585–593.

Taylor, J. L., Borgmann, I., and Séguin-Swartz, G. (1991). Electrophoretic karyotyping of *Leptosphaeria maculans* differentiates highly virulent from weakly virulent isolates. *Curr. Genet.* **19**, 273–277.

Thierry, A., and Dujon, B. (1992). Nested chromosomal fragmentation in yeast using the meganuclease I-*Sce* I: A new method for physical mapping of eukaryotic genomes. *Nucleic Acids Res.* **20**, 5625–5631.

Thierry, A., Perrin, A., Boyer, J., Fairhead, C., Dujon, B., Frey, B., and Schmitz, G. (1991). Cleavage of yeast and bacteriophage T7 genomes at a single site using the rare cutter endonuclease I-*Sce* I. *Nucleic Acids Res.* **19**, 189–190.

Thrash-Bingham, C., and Gorman, J. A. (1992). DNA translocations contribute to chromosome length polymorphisms in *Candida albicans. Curr. Genet.* **22**, 93–100.

Tzeng, T.-H., Lyngholm, L. K., Ford, C. F., and Bronson, C. R. (1992). A restriction length polymorphism map and electrophoretic karyotype of the fungal maize pathogen *Cochliobolus heterostrophus. Genetics* **130**, 81–96.

Upchurch, R. G., Ehrenshaft, M., Walker, D. C., and Sanders, L. A. (1991). Genetic transformation system for the fungal soybean pathogen *Cercospora kikuchii. Appl. Environ. Microbiol.* **57**, 2935–2939.

Van der Ploeg, L. H. T., Schwartz, D. C., Cantor, C. R., and Borst, P. (1984). Antigenic variation in *Trypanosoma brucei* analyzed by electrophoretic separation of chromosome-sized DNA molecules. *Cell (Cambridge, Mass.)* **37**, 77–84.

Vollrath, D., and Davis, R. W. (1987). Resolution of DNA molecules greater than 5 megabases by contour-clamped homogeneous electric fields. *Nucleic Acids Res.* **15**, 7865–7876.

Vollrath, D., Davis, R. W., Connelly, C., and Hieter, P. (1988). Physical mapping of large DNA by chromosome fragmentation. *Proc. Natl. Acad. Sci. U.S.A.* **85**, 6027–6031.

Walz, M., and Kück, U. (1991). Polymorphic karyotypes in related *Acremonium* strains. *Curr. Genet.* **19**, 73–76.

Walz, M., and Kück, U. (1993). Targeted integration into the *Acremonium chrysogenum* genome: Disruption of the *pcb* C gene. *Curr. Genet.* **24**, 421–427.

Wang, Y.-K., and Schwartz, D. C. (1993). Chopped inserts: A convenient alternative to aga-rose/DNA inserts or beads. *Nucleic Acids Res.* **21,** 2528.

Wang, Y.-K., Huff, E. J., and Schwartz, D. C. (1995). Optical mapping of site-directed cleavages on single DNA molecules by the RecA-assisted restriction endonuclease technique. *Proc. Natl. Acad. Sci. U.S.A.* **92,** 165–169.

Wickes, B. L., Golin, J. E., and Kwon-Chung, K. J. (1991a). Chromosomal rearrangement in *Candida stellatoidea* results in a positive effect on phenotype. *Infect. Immun.* **59,** 1762–1771.

Wickes, B., Staudinger, J., Magee, B. B., Kwon-Chung, K.-J., Magee, P. T., and Scherer, S. (1991b). Physical and genetic mapping of *Candida albicans:* Several genes previously assigned to chromosome 1 map to chromosome R, the rDNA-containing linkage group. *Infect. Immun.* **59,** 2480–2484.

Yoder, O. C. (1988). *Cochliobolus heterostrophus,* cause of southern corn leaf blight. *Adv. Plant Pathol.* **6,** 93–112.

Zolan, M. E., Crittenden, J. R., Heyler, N. K., and Seitz, L. C. (1992). Efficient isolation and mapping of rad genes of the fungus *Coprinus cinereus* using chromosome-specific librar-ies. *Nucleic Acids Res.* **20,** 3993–3999.

Zolan, M. E., Heyler, N. K., and Stassen, N. Y. (1994). Inheritance of chromosome-length polymorphisms in *Coprinus cinereus. Genetics* **137,** 87–94.

Isolation and Analysis of High-Molecular-Weight DNA from Plants

Martin Ganal

I. Introduction

In the last several years detailed genetic maps based on restriction fragment length polymorphisms (RFLPs) have been constructed for a large number of plants (O'Brien, 1993). For *Arabidopsis* (Hauge *et al.*, 1993), tomato/potato (Tanksley *et al.*, 1992; Gebhardt *et al.*, 1991), and rice (McCouch *et al.*, 1988; Nagamura *et al.*, 1993), these maps have a density of markers in excess of one marker per megabase and thus make physical mapping of defined regions of the genome possible. Techniques are now available that allow the saturation of defined regions of the genome with markers using PCR techniques, such as random amplified polymorphic sequences (RAPDs) (see Chapter 4). Furthermore, map-based cloning of interesting genes is feasible not only for *Arabidopsis* (Arondel *et al.*, 1992), but also for genomes in the size range of tomato (approx. 1 pg or 1000 Mbp per haploid genome) (Martin *et al.*, 1993a). An important prerequisite for map-based cloning is the correlation of genetic and physical distances by means of pulsed-field gel electrophoresis (Ganal *et al.*, 1989; Martin *et al.*, 1993b). In addition, the isolation of good high-molecular-weight DNA from plants is the most important step in the construction of DNA libraries in yeast artificial chromosomes suitable for chromosome walking (Guzman and Ecker, 1988; Ward and Jen, 1990; Grill and Sommerville, 1991; Martin *et al.*, 1992; Kleine *et al.*, 1993).

Construction of long-range restriction maps, spanning hundreds or thousands of kilobases, requires the availability of very-high-molecular-weight DNA as starting material. The isolation of large quantities of very-high-molecular-weight DNA from plants is difficult due to several factors: Plant cells have a very stable cell wall that is not easily removed by physical means without damaging the contents of the cell. Also, the largest organelle of a plant cell is the vacuole, which is full of degradative enzymes, including DNases and large amounts of secondary metabolites, such as phenolics, which can damage DNA by oxidation.

The isolation and analysis of high-molecular-weight DNA from plants can be separated into several steps: The first step is the initial preparation of protoplasts. This step is the most crucial because it must be adopted to a given species. The subsequent steps, namely embedding the DNA into agarose, purification of the DNA by digestion with proteinase K, and digestion of the DNA with restriction enzymes, are the same for all organisms.

The procedures presented in this chapter have been optimized during the course of the last several years and successfully used for the physical mapping of regions of the tomato and potato genome (Ganal *et al.*, 1989, 1991; Martin *et al.*, 1993b). The same techniques for DNA isolation were applied for the construction of a yeast artificial chromosome (YAC) library (Martin *et al.*, 1992). Additionally, this protocol has been adapted by other groups for physical mapping of plants such as *Brassica*, soybean, barley, and wheat. In this chapter, recently published modifications of this preparation for special purposes such as the preparation of microbeads (Wing *et al.*, 1993) and DNA for YAC cloning will also be mentioned.

II. Materials

(1) Protoplast isolation buffer: 0.5 M mannitol, 20 mM MES (2[N-morpholino]ethanesulfonic acid, pH 5.6–5.8, adjusted with concentrated KOH. For storage, this solution should be autoclaved.

(2) Digestion buffer: Protoplast isolation buffer with 1% cellulase Onozuka RS (Yakult Honsha Co., Ltd., Tokyo, Japan) and 0.05% pectolyase Y-23 (Seishin Pharmaceutical Co., Tokyo, Japan). It is possible to replace the 1% cellulase Onozuka RS with 1.5% Cellulysin (Calbiochem, La Jolla, CA). Dissolve the enzymes by stirring vigorously. The solution can be stored frozen in aliquots at −20°C for at least 6 months. Thaw the enzyme in a water bath at 30°C and mix the solution well immediately before use to dissolve the precipitated enzyme. Excess digestion buffer can be frozen again.

(3) Embedding solution: 1% low-melting agarose (BRL ultrapure) in protoplast isolation buffer (in SCE for microbeads). Boil and keep at 50°C until used.

(4) SCE: 1 M sorbitol, 0.1 M sodium citrate, 10 mM EDTA, pH 7.0.

(5) ESP: 0.5 M EDTA, pH 9–9.5, 1% sarkosyl, 1 mg/ml proteinase K (Boehringer Mannheim). Mix equal volumes of a 1 M EDTA, pH 9–9.5, stock solution and 2% sarkosyl, then add proteinase K. It is difficult to prepare a 1 M EDTA solution. Dissolve EDTA in some water, while slowly adding solid NaOH. When the EDTA is completely dissolved the pH of the solution is close to pH 9. Fill to the desired volume.

(6) TE 10/10: 10 mM Tris, 10 mM EDTA and adjust to pH 8 with 1 N NaOH.

(7) PMSF: Prepare fresh 100 mM PMSF (phenylmethylsulfonyl fluoride) in isopropanol. PMSF is very toxic and should be prepared using gloves and a fume hood.

(8) TE 10/1: 10 mM Tris, 1 mM EDTA adjust to pH 8.0 with 1 N HCl.

(9) Light mineral oil (Sigma).

(10) Spermidine: 40 mM Spermidinetrihydrochloride (Sigma) in sterile water. Store at -20°C.

(11) ES: 0.5 M EDTA, pH 9–9.5, 1% sarkosyl. Prepare as described for ESP.

(12) Equilibration buffer: 10 mM Tris, 100 mM NaCl, 10 mM EDTA, adjust to pH 7.5 with 1 N HCl.

III. Procedures

A. Isolation of High-Molecular-Weight DNA from Protoplasts of Tomato Leaves

(1) Harvest young, expanding leaves into a 150-mm petri dish with wet filter paper. If they cannot be processed directly, they can be stored at room temperature overnight.

(2) Transfer leaves in a petri dish with 20–40 ml protoplast isolation buffer and cut the leaves in 1- to 2-mm strips parallel to the veins with razor blades starting near the midvein while leaving the morphology of the leaf intact.

(3) Transfer the cut leaves into a new 150-mm petri dish with 75 ml of digestion buffer. Approximately 2–3 g of trimmed and cut leaves is necessary to cover such a petri dish.

(4) Shake gently (approx. 50 rpm) at room temperature. Monitor digestion occasionally under an inverted microscope. For tomato leaves from greenhouse-grown plants, the digestion requires approximately 6 hrs.

(5) After 6 hrs, most of the leaf should be digested away. Collect the residual leaf pieces with forceps, shake them extensively in the digestion buffer to release additional protoplasts, and discard them.

(6) Pass the protoplasts in the digestion buffer sequentially through a 80-μm sieve to remove large residual debris and then through a 30- to 40-μm sieve. Rinse the petri plate and all other vessels with a few milliliters of protoplast isolation buffer since the protoplasts settle quickly at the bottom of the vessels.

(7) Transfer the protoplast solution into conical 50-ml centrifuge tubes and spin at 200g for 5–10 min. Pour off the supernatant carefully and resuspend the green pellet in 20 ml of protoplast isolation buffer with gentle shaking. Resuspension of the protoplasts should be done very carefully and requires several minutes. After combination of the tubes, repeat the centrifugation and resuspension step once to further purify and concentrate the isolated protoplasts.

(8) Count an aliquot of the resuspended protoplasts in a yeast hemocytometer (0.2 mm depth) and calculate the total number of protoplasts isolated. The expected yield is approximately 5 to 10 $\times$ 10^6 protoplasts/gram of leaf tissue.

1. Preparation of Agarose Blocks

(1) Blocks are prepared at different concentrations of protoplasts depending on their intended use. Spin the protoplasts again at 200g for 5–10 min and remove the supernatant completely by pipetting. Resuspend the protoplasts with protoplast isolation buffer to the following densities: restriction enzyme digestions, approx. 3 $\times$ 10^7/ml; large-scale DNA isolation, approx. 6 $\times$ 10^7/ml.

(2) Add a volume of embedding solution equal to the total volume of protoplasts and mix carefully by inverting the tube a few times. Then transfer the solution to a plug mold. Keep the mold at 4°C for 10–15 min to solidify the agarose.

(3) Transfer the agarose blocks into 10 times their volume of ESP and then incubate the solution at 50°C with gentle shaking.

(4) After approximately 36 hr discard the dark-green ESP solution. Add fresh ESP and incubate for another 24 hrs. Agarose blocks can be stored in this solution for several months.

2. Preparation of Microbeads (Adapted from Koob and Szybalski, 1992; Wing *et al.*, 1993)

(1) Spin protoplasts at 200*g* for 5–10 min and carefully remove the protoplast isolation buffer completely by pipetting. Add 25 ml of SCE and shake gently to resuspend the protoplasts. Spin again at 200*g* for 5–10 min and resuspend the protoplasts in SCE to the same concentrations as described above.

(2) Warm the resuspended protoplasts to 40–45°C and mix with an equal volume of embedding solution in SCE. Immediately, add 2 vol of light mineral oil (at 40–45°C) and shake vigorously for a few seconds. Pour this entire mixture into 200 ml of ice-cold SCE that is rapidly stirring in a beaker sitting in an ice bath. Continue stirring for 5 min. Microbeads are collected from the aqueous phase by centrifugation (500*g*, 5 min) in 50-ml conical centrifuge tubes.

All further steps are the same as described for the agarose blocks, except that the beads have to be collected by centrifugation for each washing step.

B. Procedure for Restriction Enzyme Digestions

(1) All handling of agarose blocks should be done with sterile equipment. We use freshly bent pasteur pipettes for this purpose. Microbeads can be pipetted with cut-off pipet tips (3–5 mm tip diameter). Transfer blocks for restriction enzyme digestion into a 50-ml conical centrifuge tube with 20 times the block volume of TE 10/10, and add PMSF to a concentration of 1 m*M*. Incubate at 50°C for 60–90 min. Discard the solution and repeat this washing step one more time with PMSF and an additional time without PMSF. After this, the plugs can be stored overnight in this solution if necessary.

(2) Wash plugs two more times with TE 10/1 at 50°C for 60–90 min.

(3) Depending on the block size of your mold, for a 50- to 100-µl block (or the same volume of beads) a total volume of 150–250 µl is used for digestion with restriction enzymes. If it is necessary to cut your blocks to smaller sizes prior digestion, use a sterile coverslip. Set up the reaction as indicated by the manufacturer of the restriction enzyme with 5–10 units of enzyme per microgram of DNA. Spermidine at a final concentration of 4 m*M* (1/10 vol) significantly improves the digestibility of high-molecular-weight DNA in agarose blocks or microbeads. Digest for 4–15 hrs.

(4) Add 1 ml ES to stop digestions with blocks. If necessary, the blocks can be stored at 4°C. Digestions with microbeads can be stopped by the addition of 1/5 to 1/10 vol of a conventional gel loading buffer or 1/10 vol of ES.

C. Preparation of High-Molecular-Weight DNA for Cloning

(1) Transfer the agarose blocks or microbeads into a 50-ml conical centrifuge tube and add 20 times their volume TE 10/10 with 1 mM PMSF. Incubate at 50°C for 60–90 min. Discard the solution and repeat this washing step one more time with PMSF and an additional time without. After this step the plugs can be stored overnight in this solution.

(2) Incubate blocks twice in 20 times their volume of equilibration buffer at 50°C for 1 hr.

(3) Transfer the blocks into a new centrifuge tube and barely cover them with equilibration buffer. Melt blocks at 68°C for 10–15 min in a water bath.

(4) Cool the solution to 37°C and then add 25–50 units of agarase/ml (New England Biolabs). Mix the solution by slowly inverting the tube a few times. Incubate at 37°C for 12–16 hrs.

(5) Add another 15 units agarase/ml, mix, and continue incubation for 6 hr at 37°C.

(6) Carefully pour the extremely viscous solution into a sterilized dialysis bag and dialyze the DNA at 4°C for 16 hr against 1–2 liters of TE 10/1 with one change of the buffer.

(7) After dialysis the DNA can be stored at 4°C. DNA isolated in that manner is suitable for use in standard cloning procedures with yeast artificial chromosomes, bacteriophage P1, or bacterial artificial chromosomes.

IV. Pitfalls

A. Protoplast Quality and Isolation

The isolation of good quality protoplasts in a sufficiently high number is the most critical step in the whole procedure. If this is successful, then the protocols presented here should work without any problems. If the protoplast isolation is of low yield and/or low quality, the best thing is to start all over because there is no way to remedy this problem. Specifically, the following points require attention:

1. Growth Conditions

The quality of the leaf material for protoplast isolation is essential. There are basically two ways to grow plants for protoplast isolation. Each of these has its own advantages and disadvantages:

(1) *In vitro*-grown plants. Such plants are good for protoplast isolation because they are grown under uniform conditions and high humidity. The major disadvantage is the large amount of work required to maintain such plants in sufficient numbers under sterile conditions. In our experience, two points are critical with *in vitro*-grown plants. One is that such plants usually contain large amounts of starch (see below). The other point is that leaves from such plants are very tender (and small) and require only short times (1–3 hr) of incubation with protoplasting enzymes. Thus, the digestion should be monitored very carefully to avoid overdigestion.

(2) Greenhouse and growth chamber plants. The main advantage of such plants is that large amounts of tissue are readily available at any given time without major efforts. This is essential for large-scale isolation of protoplasts. The disadvantage is that the material is not entirely uniform in its digestibility due to varying growth conditions. Such plants require longer times of incubation with protoplasting enzymes. For tomato, the optimal stage for harvesting leaves is 4–6 weeks after germination. Older plants have leaves that are difficult to digest and contain more starch. High humidity keeps the leaves generally more tender and uniform. Expanding leaves should be preferred over mature leaves because the vacuole is smaller, the leaf softer, and the amount of starch lower.

2. Protoplast Isolation

This procedure has been used extensively for the isolation of high-molecular-weight DNA from tomato. However, the same procedure and buffers can be used for many other dicotyledonous plants (e.g., potato, cucumber, *Brassica*). The following pilot experiments should be performed if the procedure described for tomato does not work the first time: Digest some leaves in a small petri dish with the buffers described above and watch them under an inverted microscope in 30-min intervals for several hours. If no protoplasts are released during that time, the concentration of the protoplasting enzymes should be increased (up to 2% for the cellulase; up to 0.25% for the pectolyase). If you do not observe free protoplasts but increasing amounts of debris, the fate of individual cells should be observed during protoplasting. If the cells increase in size during digestion and subsequently burst, the concentration of osmoticum (mannitol) is too low. Conversely, if the size of the cells decreases dramatically during digestion, the concentration of the osmoticum is too high. Mannitol concentrations can be varied from 0.2 to 0.7 M.

It is important to note that the isolation of reasonable amounts of high-molecular-weight DNA from protoplasts compromises the protoplast quality. Do not try to get high-quality protoplasts, such as those used for regeneration experiments, because you will never achieve the necessary amounts. A preparation with large amounts of protoplasts usually has some debris from lysed protoplasts. A yield of approximately $5–10 \times 10^6$ protoplasts/gram of leaf tissue is typical for tomato and many other plants. If your yield is 1×10^6/gram or lower, you should continue optimizing your procedures to get higher yield rather than using more tissue.

3. Starch

Starch is the most problematic plant component in these procedures. If the cells contain large starch grains, they can destroy the protoplasts during centrifugation. Furthermore, large amounts of starch in agarose blocks prevent the restriction enzymes from digesting the high-molecular-weight DNA. Large amounts of starch are accumulated if the plants are kept under high-intensity lights or if tissue is used that is too old. In tomato, a good indicator of the starch amount and age is whether the lower side of the leaves contains a high accumulation of anthocyanin. Starch accumulation can be reduced by growing the plants under lower intensity lights, by keeping the plants in the dark for 1–2 days before protoplast isolation, or by harvesting leaves from very young plants whose leaves are still expanding. Excess starch in intact protoplasts results in agarose blocks that are deeply white rather than translucent. Do not use such blocks because the starch often interferes with the action of restriction enzymes and thus gives incomplete digestions. Furthermore, we always observed a lower than expected amount of DNA in such blocks.

B. Other Useful Hints for Protoplast Isolation and Embedding in Agarose

(1) Cutting the leaves as indicated reduces dramatically the amount of debris from the vascular system of the leaves and makes filtration easier.

(2) Agarose blocks should have a volume less than 200 μl. Larger blocks will not get completely penetrated by the solutions.

(3) Protoplasts should be pipetted only with cut-off pipet tips to avoid destruction by shearing.

(4) Gentle shaking during digestion with ESP is essential to remove the digestion products.

(5) For the preparation of microbeads, the protoplasts have to be resuspended in SCE in the final step because the protoplast isolation buffer prevents the proper formation of microbeads.

C. Restriction Enzyme Digestions

1. DNA Quality in General

The size of high-molecular-weight DNA isolated from plant protoplasts is usually in the range of a few million base pairs. However, it will never be the large size and high quality of DNA isolated from animal cells or anything close to chromosome size. Usually, high-molecular-weight DNA isolated from plant protoplasts is of sufficient quality for physical mapping by pulsed-field gel electrophoresis. However, the size of the high-molecular-weight DNA can be a problem if one works with restriction enzymes that create very large DNA fragments ($>$1.5 million base pairs). Some degraded DNA ($<$100 kb) is usual.

2. DNA Amount

The amount of DNA per agarose block is dependent on the number of protoplasts and the genome size. For hybridization, only the number of genomes, i.e., the number of protoplasts per block, is of importance. Thus, it is necessary to isolate the same number of protoplasts independent of genome size. For tomato with a haploid genome size of approximately 1 pg, a 100-μl block with 3×10^7/ml protoplasts contains $3 \times 10^6 \times 1$ pg $\times 2 = 6$ μg of DNA.

3. Digestion of DNA in Agarose

If the DNA isolation from protoplasts was successful and the agarose blocks are fairly clear after all washing steps, there are usually no problems digesting the DNA. If problems with digestions occur they are mostly due to the following:

• The digestion with ESP was not complete. This happens when not enough ESP is used or the digestion was not completed due to problems with accessibility for the ESP (e.g., at high concentrations of protoplasts in the blocks or large amounts of starch).

• One of the solutions was contaminated. This is the most common problem observed. A few molecules of DNase are sufficient to degrade high-molecular-weight DNA to sizes that are no longer useful for pulsed-field gel electrophoresis. This means that solutions (restriction enzymes, buffers, spermidine, etc.) that are successfully used for conventional DNA digests might not work for high-molecular-weight DNA.

A simple test for the quality of your solutions is to process your blocks in parallel with commercially available blocks from yeast chromosomes through the steps outlined in the methods section. Then run the processed sample parallel with unprocessed samples (directly from the ESP solution) on a pulsed-field gel. If the unprocessed samples (yeast and plant) are fine and the processed samples are degraded, most likely, one of your solutions was contaminated. If only the treated plant sample is degraded, the digestion in ESP was not complete. If the untreated and treated plant samples are degraded, the problem is with your protoplast preparation.

A good indication of whether your restriction digestions are complete is if you observe one or a few (depending on the restriction enzyme) prominent bands in the low-molecular-weight range (between 25 and 100 kb) of a pulsed-field gel. These bands are derived from the chloroplast genome and provide a nice internal control (see Fig. 3.1).

Often some high-molecular-weight DNA remains undigested and trapped in the blocks. Usually, this is no reason for concern since this DNA does not interfere with the digested DNA. There seems to be a correlation between the amount of trapped DNA and the amount of starch.

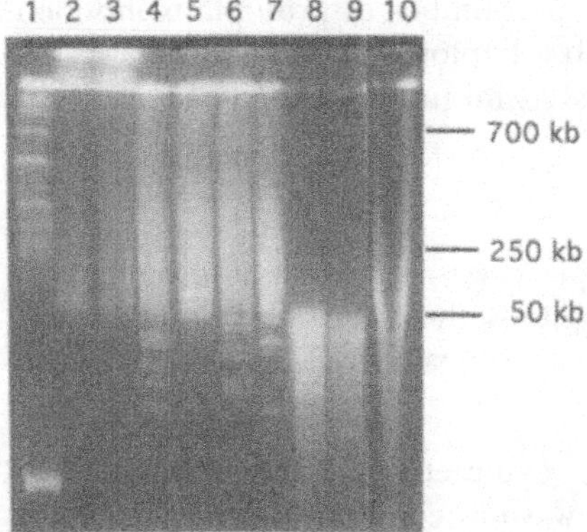

Figure 3.1 Separation of intact and digested high-molecular-weight plant DNA on a pulsed-field gel. High-molecular-weight DNA was separated on a CHEF-gel (90 sec pulse time, 125 V). Lane 1, chromosomes of yeast strain AB 972 as marker; lanes 2 and 3, two different preparations of undigested high-molecular-weight DNA. Some degraded DNA in the range above 50 kb is normal. Lanes 4–7 show high-molecular-weight DNA digested with some rare-cutting restriction enzymes. Lane 4, *Nru*I; Lane 5, *Mlu*I; Lane 6, *Sma*I; Lane 7, *Sal*I. Lane 8 shows high-molecular-weight DNA digested to completion with *Eco*RI and lane 9 with *Bam*HI. Degraded high-molecular-weight DNA usually looks very similar to the samples shown in lanes 8 and 9. Lane 10 shows partially degraded high-molecular-weight DNA which displays a smear over the entire separation range.

The size range of restriction enzyme digestion products is very variable in plants. In tomato, the usual infrequent cutting restriction enzymes give large DNA fragments. This is not the case, for example, in rice. Here physical mapping by pulsed-field gel electrophoresis is almost impossible because all restriction enzymes tested result in extremely small fragments (<300 kb on average). Thus, if one starts with a new species, the suitability of the available rare-cutting restriction enzymes has to be evaluated first. The average fragment size is dependent on the methylation status of the genome and the general genome organization (Wu and Tanksley, 1993).

D. Preparation of High-Molecular-Weight DNA for Cloning

1. The most important point is that the protoplasts should be embedded at the highest possible concentration. However, this requires special care since a concentration that is too high prevents the formation of good agarose blocks. Do not try to resuspend protoplasts directly in the embedding solution. For tomato, the preparation of high-molecular-weight DNA in liquid does not work because of the high content of nucleases in the protoplasts.

(2) For a quick test to see whether the agarase treatment is complete, take an aliquot of the solution and place it on ice for a few minutes. The solution should stay liquid.

(3) When the blocks are melted you will observe two different phases in your tube. At the bottom is a whitish aggregate that represents the high-molecular-weight DNA; the clear supernatant does not contain significant amounts of DNA. If you do not observe these two phases, your DNA concentration is probably too low. A low concentration of DNA is invariably associated with lower quality because high concentrations seem to stabilize DNA against shearing. Always handle the high-molecular-weight DNA carefully and with cut-off pipet tips. DNA isolated in this manner is at least 1–2 million base pairs in size and readily digestible with restriction enzymes.

E. Isolation of High-Molecular-Weight DNA from Other Plants

The following table gives a summary of some procedures published for a number of major crop plants for further reference:

Plant species	Reference
Arabidopsis	Guzman and Ecker (1988)
	Grill and Sommerville (1991)
	Ward and Jen (1990)
Maize	Edwards *et al.* (1992)
Pea	Ellis *et al.* (1988)

Plant species	Reference
Potato	Ganal *et al.* (1991)
Rice	Sobral *et al.* (1990)
	Wu and Tanksley (1993)
Soybean	Honeycutt *et al.* (1992)
Sugar beet	Jung *et al.* (1990)
Tomato	Ganal *et al.* (1989)
	Van Daelen *et al.* (1989)
	Wing *et al.* (1993)
Wheat, barley	Cheung and Gale (1990)
	Röder *et al.* (1992)
	Siedler and Graner (1991)
	Kleine *et al.* (1993)

References

Arondel, V., Lemieux, B., Hwang, I., Gibson, S., Goodman, H., and Somerville, C. R. (1992). Map-based cloning of a gene controlling omega-3 fatty acid desaturation in *Arabidopsis*. *Science* **258**, 1353–1355.

Cheung, W. Y., and Gale, M. D. (1990). The isolation of high molecular weight DNA from wheat, barley, and rye for analysis by pulsed-field gel electrophoresis. *Plant Mol. Biol.* **14**, 881–888.

Edwards, K. J., Thompson, H., Edwards, D., Saizieu, A. D., Sparks, C., Thompson, J. A., Greenland, A. J., Eyers, M., and Schuch, W. (1992). Construction and characterization of a yeast artificial chromosome library containing three haploid maize genome equivalents. *Plant Mol. Biol.* **19**, 299–308.

Ellis, T. H. N., Lee, D., Thomas, C. M., Simpson, P. R., Cleary, W. G., Newman, M.-A., and Burcham, K. W. G. (1988). 5S rRNA genes in *Pisum*: sequence, long range and chromosomal organization. *Mol. Gen. Genet.* **214**, 333–342.

Ganal, M. W., Young, N. D., and Tanksley, S. D. (1989). Pulsed field gel electrophoresis and physical mapping in the *Tm2a* region of chromosome 9 in tomato. *Mol. Gen. Genet.* **215**, 395–400.

Ganal, M. W., Bonierbale, M. W., Röder, M. S., Park, W. D., and Tanksley, S. D. (1991). Genetic and physical mapping of the patatin genes in potato and tomato. *Mol. Gen. Genet.* **225**, 501–509.

Gebhardt, C., Ritter, E., Barone, A., Debener, T., Walkemeier, B., Schachtschnabel, U., Kaufmann, H., Thompson, R. D., Bonierbale, M. W., Ganal, M. W., Tanksley, S. D., and Salamini, F. (1991). RFLP maps of potato and their alignment with the homeologous tomato genome. *Theor. Appl. Genet.* **83**, 49–57.

Grill, E., and Sommerville, C. R. (1991). Construction and characterization of a yeast artificial chromosome library of *Arabidopsis* which is suitable for chromosome walking. *Mol. Gen. Genet.* **226**, 484–490.

Guzman, P., and Ecker, J. R. (1988). Development of large DNA methods for plants: Molecular cloning of large segments of *Arabidopsis* and carrot DNA into yeast. *Nucleic Acids Res.* **16**, 11091–11105.

Honeycutt, R. J., Sobral, B. W. S., McCelland, M., and Atherly, A. G. (1992). Analysis of large DNA from soybean (*Glycine max* L. Merr.) by pulsed-field gel electrophoresis. *Plant J.* **2**, 133–135.

Hauge B. M., Hanley, S. M., Cartinhour, S., Cherry, J. M., Goodman, H. M., Koornneef, M., Stam, P., Chang, C., Kempin, S., Medrano, L., and Meyerowitz, E. M. (1993). An integrated genetic/RFLP map of the *Arabidopsis thaliana* genome. *Plant J.* **3**, 745–754.

Jung, C., Kleine, M., Fischer, F., and Herrmann, R. G. (1990). Analysis of DNA from a *Beta procumbens* chromosome fragment in sugar beet carrying a gene for nematode resistance. *Theor. Appl. Genet.* **79**, 663–672.

Kleine, M., Michalek, W., Graner, A., Herrmann, R. G., and Jung, C. (1993). Construction of a barley (*Hordeum vulgare* L.) YAC library and isolation of a *Hor1*-specific clone. *Mol. Gen. Genet.* **240**, 265–272.

Koob, M., and Szybalski, W. (1992). Preparing and using agarose microbeads. *In* "Methods in Enzymology" (R. Wu, ed.), Vol. 216, pp. 13–20. Academic Press, San Diego, CA.

Martin, G. B., Ganal, M. W., and Tanksley, S. D. (1992). Construction of a yeast artificial chromosome library of tomato and identification of cloned segments linked to two disease resistance loci. *Mol. Gen. Genet.* **223**, 25–32.

Martin, G. B., Brommonschenkel, S. H., Chungwonse, J., Frary, A., Ganal, M. W., Spivey, R., Wu, T., Earle, E. D., and Tanksley, S. D. (1993a). Map-based cloning of a protein kinase gene conferring disease resistance in tomato. *Science* **262**, 1432–1436.

Martin, G. B., Vicente, M. C., and Tanksley, S. D. (1993b). High-resolution linkage analysis and physical characterization of the *Pto* bacterial resistance locus in tomato. *Mol. Plant-Microbe Interact.* **6**, 26–34.

McCouch, S. R., Kochert, G., Yu, Z. H., Wang, Z. Y., Kush, G. S., Coffman, W. R., and Tanksley, S. D. (1988). Molecular mapping of rice chromosomes. *Theor. Appl. Genet.* **76**, 815–829.

Nagamura, Y., Yamamoto, K., Harushima, Y., Wu, J., Antonio, B. A., Sue, N., Shomura, A., Lin, S.-Y., Miyamoto, Y., Toyama, T., Kirihara, T., Shimizu, T., Wang, Z.-X., Tamura, Y., Ashikawa, I., Yano, M., and Kurata, N. (1993). A high density STS and EST linkage map of rice. *Rice Genome* **2**, 3.

O'Brien, S. J., ed. (1993). "Genetic Maps," 6th ed. Cold Spring Harbor Lab. Press, Cold Spring Harbor, NY.

Röder, M. S., Sorrells, M. E., and Tanksley, S. D. (1992). 5S ribosomal gene clusters in wheat: Pulsed field gel electrophoresis reveals a high degree of polymorphism. *Mol. Gen. Genet.* **232** 215–220.

Siedler, H., and Graner, A. (1991). Construction of physical maps of the *Hor1* locus of two barley cultivars by pulsed field gel electrophoresis. *Mol. Gen. Genet.* **226**, 177–181.

Sobral, B. W. S., Honeycutt, R. J., Atherly, A. G., and McClelland, M. (1990). Analysis of rice (*Oryza sativa* L.) genome using pulsed-field electrophoresis and rare cutting restriction endonucleases. *Plant Mol. Biol. Rep.* **8**, 253–275.

Tanksley, S. D., Ganal., M. W., Prince, J. P., deVincente, M. C., Bonierbale, M. W., Broun, P., Fulton, T. M., Giovannoni, J. J., Grandillo, S., Martin, G. B., Messeguer, R., Miller, J. C., Miller, L., Paterson, A. H., Pineda, O., Röder, M. S., Wing, R. A., Wu, W., and Young, N. D. (1992). High density molecular linkage maps of the tomato and potato genomes. *Genetics* **132**, 1141–1160.

Van Daelen, R. A. J., Jonkers, J. J., and Zabel, P. (1989). Preparation of megabase-sized tomato DNA and separation of large restriction fragments by field inversion gel electrophoresis (FIGE). *Plant Mol. Biol.* **12**, 341–352.

Ward, E. R., and Jen, G. C. (1990). Isolation of single-copy-sequence clones from a yeast artificial chromosome library of randomly-sheared *Arabidopsis thaliana* DNA. *Plant Mol. Biol.* **14**, 561–568.

Wing, R. A., Rastogi, V. K., Zhang, H.-B., Paterson, A. H., and Tanksley, S. D. (1993). An improved method for plant megabase DNA isolation in agarose microbeads suitable for physical mapping and YAC cloning. *Plant J.* **4**, 893–898.

Wu, K.-S., and Tanksley, S. D. (1993). PFGE analysis of the rice genome: Estimation of fragment sizes, organization of repetitive sequences and relationships between genetic and physical distances. *Plant Mol. Biol.* **23**, 243–254.

4

Generating and Using DNA Markers
in Plants

J. Antoni Rafalski, Julie M. Vogel, Michele Morgante,[1] Wayne Powell,[2]

Chaz Andre,[3] and Scott V. Tingey

I. Introduction

Scarcely a decade ago, the repertoire of plant genetic markers and hence the ability to perform detailed genetic mapping were extremely limited. Classical phenotypic markers were plentiful in only a few well-characterized species such as maize, and their utility was restricted by the difficulties involved in constructing multiply marked lines, by the low resolution of the maps produced, and by the large amount of labor required to generate and use these markers. Although isozyme markers, which allow identification of allelic differences, were well-established as a potentially useful class of genetic marker, they were not numerous enough to permit high-resolution mapping. Applications involving the first DNA marker system, Restriction Fragment Length Polymorphism (RFLP) analysis (Botstein *et al.*, 1980), began to supplement phenotypic markers in some instances, but had distinct limitations, as discussed below, and the need for additional marker systems remained.

[1]Permanent address: Dipartimento di Produzione Vegetale e Tecnologie Agrarie, Università di Udine, Via della Scienze 208, I-33100 Udine, Italy.

[2]On leave from Scottish Crop Research Institute, Invergowrie, Dundee, Scotland DD2 5DA.

[3]Present address: Perkin Elmer Corp. Applied Biosystems Div. 350, Lincoln Centre Drive, Foster City, CA 94404.

NONMAMMALIAN GENOMIC ANALYSIS: A PRACTICAL GUIDE

The advent of the polymerase chain reaction (PCR) technique (Mullis *et al.*, 1986) accelerated development of new DNA marker systems. As a result, a potentially bewildering array of marker systems, including Random Amplified Polymorphic DNA (RAPD), Simple Sequence Repeat Polymorphism (SSR), Cleavable Amplified Polymorphic Sequences (CAPS), Amplified Fragment Length Polymorphism (AFLP), and Inter-SSR Amplification (ISA), now is available for genetic mapping in plants.

In this chapter we describe and compare these and other DNA marker systems. This discussion is intended to help guide the choice of a suitable marker system for a given genome analysis application. We provide a general discussion of all marker systems, and give relatively brief protocols for the more commonly known marker types and detailed experimental protocols for those that are lesser known. For all marker systems, we provide numerous references.

II. Comparison of DNA Marker Systems

DNA markers are being applied to a wide variety of problems central to plant genome analysis (Table 4.1). Although each marker system is characterized by a unique combination of advantages and disadvantages, the choice of a marker system is dictated to a significant extent by the application. Factors to consider in choosing a marker system include the amount of available plant material (and consequently the amount of available DNA), the quality of the DNA, the availability of public collections

Table 4.1
Applications of DNA Markers

Creation of genetic maps
Mapping of simple traits
Mapping of quantitative trait loci (QTLs)
Mapping of mutations
Characterization of transformants
Genetic diagnostics for plant breeding
Population genetics
Molecular taxonomy and evolution
Identification of individuals
Forensic analysis
Germplasm characterization
Identification of proprietary germplasm
Analysis of herbarium samples
Estimation of genome size

of DNA markers for the species being examined, the resources and skills available to develop new markers or to use existing markers, the cost on a per marker or per locus basis, and any restrictions on the use of radioactive materials.

Marker systems can be classified conveniently according to two criteria: information content and multiplex ratio.

A. Information Content

Simply speaking, the more informative a given class of markers, the easier it becomes to detect a polymorphism between two individuals. The relative information content of any two classes of markers may be evaluated by comparing the mean heterozygosity (or the effective number of alleles) in a given set of individuals for a number of randomly chosen markers in each of the marker classes.[4] The higher the heterozygosity value, the greater the chance that a polymorphism can be detected.

The information content of a given class of markers also depends on the type of nucleotide changes to be distinguished, the length of DNA to be examined, and the technical limitations of the method (gel resolution, for example). For RFLP markers, each DNA band is produced by two restriction endonuclease cleavages; for a hexanucleotide recognition sequence this corresponds to 12 nucleotides. Any mutation within these 12 nucleotides will prevent enzyme recognition and can be viewed as a polymorphism. Similarly, any gel-resolvable insertion or deletion between these sites also will be detected, if the gel resolution is adequate. In contrast, production of a RAPD band requires oligonucleotide primer binding to two 8–10 nucleotide sites, and any mutation within these 16–18 nucleotides should prevent primer binding and lead to the absence of a band. Therefore the RAPD assay is 1.3–1.5× more efficient than RFLP (on a per band basis) in identifying sequence polymorphisms at a locus. SSR markers are associated with the highest levels of polymorphism, a result of the high evolution rate of SSR loci (Morgante and Olivieri, 1993; Tautz, 1989), making them potentially the most informative class of markers.

[4]Heterozygosity is the probability, P, that two gametes chosen at random from a population will carry different alleles at a locus

$$P = 1 - \Sigma p_n^2,$$

where p_n is allele frequency of allele n in the population. The same formula also describes the probability that two inbred individuals, chosen at random from a germplasm collection for which allele frequencies are known, will differ in the alleles present at a locus. Effective number of alleles, n, is the inverse of homozygosity:

$$n = 1/\Sigma p_n^2.$$

Table 4.2
Comparison of marker systems

	RFLP	RAPD	SSR	AFLP	ISA	CAPS
Principle	Restriction endonuclease digestion, Southern blotting, hybridization	DNA amplification with random primers	PCR of simple sequence repeat regions	PCR of a subset of restriction fragments from extended adapter primers	PCR amplification between closely spaced simple sequence repeats	Restriction endonuclease digestion of PCR products
Nature of polymorphism	Single base changes, insertions, deletions	Single base changes, insertions, deletions	Repeat length changes	Single base changes, insertions, deletions	Insertions, deletions	Single base changes, insertions, deletions
Abundance in the genome	High	Very high	Medium	High	Genome-dependent, medium	High
Level of polymorphism	Medium	Medium	High	Medium	High	Medium
Dominance	Codominant	Dominant	Codominant	Mixed	Dominant and codominant	Codominant
Multiplex ratio	1–2	5–20, adjustable	1	30–100, adjustable	20–100, organism-dependent and adjustable	Low
DNA amt. required	2–10 μg	10–25 ng	50–100 ng	1–2 μg	10–100 ng	50–100 ng
DNA sequence information requirement	No	No	Yes	No	No	Yes
Radioactive detection	Yes/no	No	No/yes	Yes/no	Yes	No
Development cost	Medium	Low	High	Medium	Low	Medium–high
Start-up costs	Medium/high	Low	High	Medium	Low	High

B. Multiplex Ratio

The multiplex ratio is the number of different genetic loci that may be simultaneously analyzed per experiment (that is, per gel lane). The multiplex ratio of each RFLP marker is 1, or at most 2 or 3, since only one or two loci typically are detected within a diploid genome with each RFLP probe (this value may be higher for polyploid species). In contrast, AFLP or ISA assays allow the simultaneous examination of many (up to 50 or more) genetic loci. The RAPD assay carries an intermediate multiplex ratio; 5 to 20 bands may be scored per primer, depending upon the precise experimental conditions and detection technology used. For both the RAPD and AFLP assays, some adjustment of the multiplex ratio is possible by changing experimental conditions.

A comparison of the features of the commonly used marker systems, RFLP, RAPD, SSR, AFLP, ISA, and CAPS, is shown in Table 4.2.

III. Materials

Most of the elementary molecular biology procedures and materials are described in Sambrook *et al.* (1989) or Ausubel *et al.* (1987–1993) and will not be repeated here. PCR methodology is discussed in Innis *et al.* (1990). Commercially available kits, which come with detailed instructions, simplify many of the commonly used procedures. Most of the reagents are available from multiple suppliers, only some of which are listed. More detailed lists of materials are provided in each section.

DNA was prepared according to Murray and Thompson (1980) or Dellaporta *et al.* (1985).

Custom oligonucleotides were ordered from Operon Technologies (Alameda, CA). Stock solutions prepared at 500 μM by dissolving the oligonucleotide in deionized water were kept frozen at $-20°C$. Working solutions at 10 μM were also stored frozen.

RadPrime random-primer labeling kit (Life Technologies, Gaithersburg, MD) was used according to the manufacturer's directions.

Agarose (NuSieve, Metaphor) was from FMC Bioproducts (Rockland, ME) or (molecular biology grade) from Life Technologies.

DNA size standards: bacteriophage lambda DNA digested with *Hind*III and phage ΦX174 DNA digested with *Hae*III (Life Technologies) was used at concentrations recommended by the supplier.

Nylon membranes, GeneScreen Plus, Colony/Plaque Screen (DuPont NEN Research Products, Boston, MA), Hybond N (Amersham, Arlington Heights, IL).

20× SSPE was prepared according to Sambrook *et al.* (1989) and diluted as required.

dGTP, dATP, dTTP, dCTP solutions, 10 μM each, diluted from 100 mM stock (Pharmacia Biotech, Piscataway, NJ).

Amplitaq DNA polymerase and Amplitaq Stoffel Fragment (Perkin–Elmer, Norwalk, CT).

Glogos II fluorescent autorad markers (Stratagene, La Jolla CA) may be used to facilitate alignment of autoradiograms with hybridization membranes or gels.

Autoradiographic films: XAR 5 (Eastman Kodak, Rochester, NY) or Hyperfilm MP (Amersham).

Thermocyclers, Original and Model 9600 (Perkin–Elmer).

Biomek 1000 and 2000 Automated Laboratory Workstation (Beckman Instruments, Palo Alto, CA).

Shaking incubator, LabLine Orbit Environ-Shaker (Lab-Line Instruments Inc., Melrose Park, IL).

IV. Restriction Fragment Length Polymorphism Markers

Genetic mapping was revolutionized by the realization that length polymorphisms in restriction fragments between individuals could be detected on DNA blots using radioactively labeled DNA probes that hybridize to a single target sequence in the genome (Botstein *et al.*, 1980). The utility of RFLP markers for plant genome analysis was soon recognized (Beckmann and Soller, 1983) and has been applied successfully to a wide range of plant species.

RFLP markers are generally codominant, allowing detection and characterization of multiple alleles at a given RFLP locus among individuals in a population. Several types of polymorphism can be detected, including single base substitutions, insertions, and deletions. One clear disadvantage to using RFLP markers is the large amount of high-quality genomic DNA required from each individual (typically 1–10 μg per gel lane, depending on the genome size). Nevertheless, this method remains useful for a number of reasons. RFLP marker analysis involves only the simple techniques of agarose gel electrophoresis and DNA blotting, requires little special instrumentation, and involves relatively low startup costs. A large selection of high-quality and inexpensive restriction enzymes is now available, increasing the probability of identifying a useful polymorphism. RFLP markers have traditionally been used as radiolabeled probes; however, safer, nonradioactive probing methods are becoming increasingly sensitive and widely available (Martin *et al.*, 1990; Anonymous, 1992). Whatever the detection method, large numbers of individuals may be analyzed simultaneously, DNA blots can be reprobed multiple times with different RFLP probes, and RFLP results are quickly scored and easily interpretable. Although developing a set of RFLP probes is labor-intensive (see below), mapped RFLP clones are frequently and increasingly available from other investigators. Since RFLP probes from one species are likely to work in a wide range of related species (with some adjustment of hybridization stringency), there may be no need to generate markers for a new plant species. Such use of heterologous RFLP probes across species boundaries also permits analysis of genome synteny.

RFLP mapping involves the following general steps:

(1) Generation of probes
(2) Generation of a segregating population
(3) Identification of polymorphisms between mapping parents or individuals of interest
(4) Genotype determination of the individuals in the segregating population, using probes that detect polymorphisms
(5) Analysis of segregation data to produce a genetic map

Some of these steps (2,4,5) may be eliminated if the objective is not mapping, but rather assessment of the genetic diversity in a population.

A. Probe Considerations

Both short, single-, or low-copy number genomic fragments and cDNA clones are useful as RFLP probes (Landry *et al.*, 1987). In a number of plant species such as maize, soybean, and *Arabidopsis*, public collections of RFLP probes are available from one or multiple sources. These generally can be identified in various public genome databases, and appropriate curators can be contracted to request these clones.[5]

New RFLP markers can be generated in the following way: 50–100 µg of high-quality, high-molecular-weight genomic DNA obtained from the organism of interest is digested with a restriction endonuclease such as *Pst*I. Since *Pst*I is sensitive to methylation, only a small fraction of DNA is digested, preferentially producing fragments from single-copy sequences (Burr *et al.*, 1988). The digest is fractionated on a preparative agarose gel and fragments between 500 and 2000 bp are excised and eluted. These

---

[5]Index of plant genome data bases is available at the following Internet address: "gopher://probe.nalusda.gov/11/.plant/.species." For example, Arabidopsis genome data base is at "gopher://weeds.mgh.harvard.edu/11/arabidopsis," soybean data base is at "gopher://mendel.agron.iastate.edu/1/," and maize database is at "gopher://theosinte.agron.missouri.edu."

are cloned into a *Pst*I-digested and alkaline phosphatase-treated plasmid vector, such as pUC18, allowing blue–white selection of clones containing inserts. *Pst*I digests of plasmids prepared from white colonies are examined on agarose gels to verify the presence and estimate the size of the genomic DNA insert. These gels are Southern-blotted, and the membranes probed with a sheared genomic DNA probe from the same organism from which the *Pst*I fragments were generated. This hybridization serves to exclude plasmids with inserts containing high and medium copy repeated sequences which would not be useful as probes. Probes from sheared genomic DNA are best generated by nick translation or random priming (Feinberg and Vogelstein, 1983), and can be either radiolabeled (e.g., with ^{32}P) or nonradioactively labeled (e.g., with digoxygenin). Plasmids showing no hybridization to the whole-genome probe then are labeled either radioactively or nonradioactively and used as candidate low-copy RFLP probes. The insert can be released with *Pst*I, and purified by agarose gel electrophoresis, if desired. In general, it is not necessary to purify the insert DNA away from the vector, but this process may increase the signal strength and produce a cleaner hybridization. After removing the enzyme by phenol extraction or microcolumn purification, the DNA concentration is adjusted to 25 ng/μl and then labeled by random priming. One labeling reaction requires 25–50 ng of DNA. Commercial kits may be conveniently used for this purpose. Alternatively, the RFLP probe can be generated by PCR amplification from 1 μl of 1:100–1:1000 dilution of the clone stock using flanking vector primers. After PCR, unincorporated dNTPs should be removed by spin column filtration or ultrafiltration. More detailed preparation of labeled RFLP probes is described under *Experimental Considerations*.

B. Mapping Populations

The choice of a mapping population will vary with the biological system under investigation. In plants, particularly if codominant markers are used, it is convenient to use an F2 population. This is obtained by crossing the two parental individuals, selfing a single F1 plant, collecting all the F2 seeds, and growing out a suitable number (100 or more) of F2 individuals to collect tissue for DNA extraction. If a larger amount of tissue is needed, individuals within F3 families, each of which represents a single F2 individual, may be pooled. Alternatively, recombinant inbred (RI) populations (Burr *et al.*, 1988) can be used. Recombinant inbreds are convenient because they may be propagated in perpetuity and shared with collaborators wishing to contribute to a common map. If a dominant marker system is employed, it may be most convenient to use a backcross or RI population rather than an F2 population.

The choice of parents for the mapping population is important. The parents should be as different genetically as possible, to maximize the probability of finding a polymorphism with each RFLP probe. Interspecific crosses are sometimes used, but this also increases the possibility for chromosomal translocations or other genetic incompatibilities between the parents, and leads to the loss of accurate map information. Therefore, the use of maximally different parents within a species is recommended. If mapping specific traits is one of the objectives of the project, parental lines that differ for one or a number of these traits may be used for the initial cross, thus maximizing the number of traits of interest that will be segregating in the mapping population. The nature and size of the mapping population will affect map resolution (Allard, 1956). In general, the larger the population the higher the map resolution. The exact order of closely linked markers frequently will be uncertain, and great caution should be taken when using recombinational mapping information either to guide physical mapping or to direct chromosome walking. However, a detailed discussion of this subject is beyond the scope of this review

C. Polymorphism Screen

Once the mapping parents have been chosen and low-copy RFLP probes obtained, those probes that detect useful polymorphisms in the segregating population must be identified. "Parental" DNA blots, containing size-fractionated genomic DNA from the two parents digested with a range of restriction endonucleases (usually with hexanucleotide recognition sites), are taken through repeated rounds of hybridization with individual RFLP probes (Fig. 4.1A). Each probe is scored for those enzyme(s) that reveal useful polymorphisms between the parents. Twenty or more duplicate parental blots may be probed simultaneously by one worker, and each membrane may be stripped and reprobed numerous (8–10) times. For record-keeping, it is useful to enter all data into a simple microcomputer database or a spreadsheet. The fraction of probes that reveal useful polymorphisms varies widely between species and among individuals. For species whose genomes show low levels of polymorphism, it may be necessary to use 5–15 different restriction endonucleases to increase the probability of finding useful polymorphisms. In species whose genomes are highly polymorphic (e.g., maize), 2 or 3 restriction enzymes may be adequate for revealing a polymorphism for almost any RFLP marker.

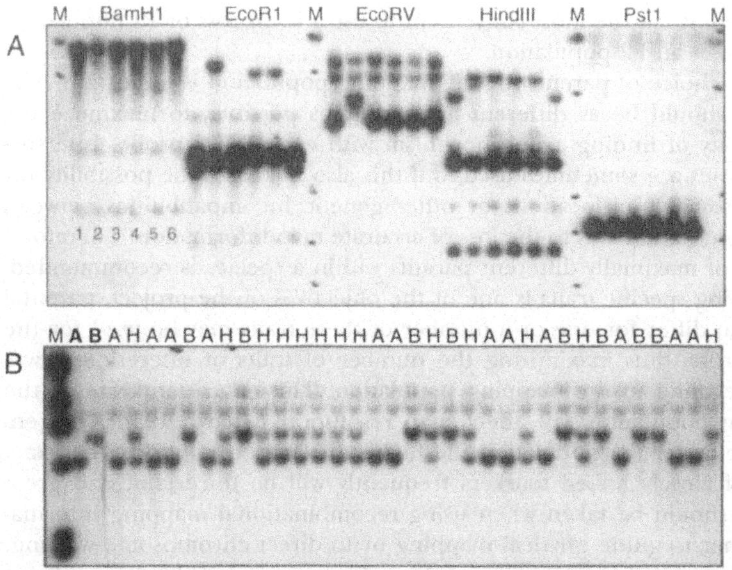

Figure 4.1 Restriction fragment length polymorphism. (A) Example of a parent blot used to identify polymorphisms. DNA (5 μg/lane) from six lines of soybean: cv. Bonus, *G. soja* PI81.762, PI416.937, N85-2176, PI153.293, PI230.970 was digested with restriction endonucleases identified on the figure and separated on 0.8% agarose gel. The DNA was transferred to a nylon membrane, and the blot was probed with anonymous *Pst*1 clone 40-21 as described in the text and autoradiographed. Several polymorphisms are visible, including, in the *Eco*RV lanes, the one also shown in (B). M, DNA size markers. (B) Progeny blot with scores shown. DNA from 35 F2 individuals derived from a *G. max* cv. Bonus X *G. soja* PI81.762 cross was digested with endonuclease *Eco*RV, separated on gel, blotted, and probed as described above. The bold **(A,B)** indicates lanes containing parental DNA. Progeny is scored as A (homozygous for the Bonus allele), B (homozygous for the PI81.762 allele), and H (heterozygous). M, DNA size markers, phage lambda DNA digested with restriction endonuclease *Hind*III.

D. Progeny Genotyping

Once polymorphic probe/enzyme combinations have been identified, progeny blots should be prepared. Several sets of progeny blots should be generated using each of the restriction enzymes that reveals a polymorphism in the parental screen. Each progeny blot must contain DNA from both parents, from the F1 individual (if available), and from all of the F2 progeny. In addition, these blots should carry suitable DNA size markers with a range spanning 0.5–20 kb. Each progeny blot should contain individual DNA samples arranged in precisely the same order, and the orientation of the blot should be clearly marked, for example by providing two size marker lanes on the right side. RFLP probes that identify polymorphisms in the parental screen are hybridized to the appropriate membranes, and the resulting autoradiograms are scored for segregation of the polymorphic hybridizing bands (Fig. 4.1B), using a code required by the mapping software (for example, Lander *et al.*, 1987).[6] The results are entered into a spreadsheet or database.

E. Data Analysis

Analysis of the segregation data to obtain recombination frequencies between markers may be performed best using mapping software such as Mapmaker (Lander *et al.*, 1987). Software manuals should be consulted for detailed instructions. In general, first two-point distances for all possible pairs of markers are calculated, and used to infer possible linkage group assignments. For a locus to be assigned to a linkage group, maximum recombination fraction to another locus in the group should be ≤0.4, and the LOD score (logarithm of the ratio of probability of linkage to probability of no linkage) should be at least 3.0. Next, a combination of three-point and multipoint approaches may be used to infer the most probable marker order (Lander *et al.*, 1987). Attention should be paid to the quality of the data and the resulting maps. It is a prudent practice to place on the "core" map only markers whose order can be established with a high confidence level (LOD≥3, that is, the logarithm of the likelihood ratio of the most likely map order and the next most likely order is at least 3.0). Some of the software packages produce computer-editable, publication-quality map images.

It should be emphasized that map orders are not absolute, and that there is always a finite probability that the true physical order of loci is different from the most likely map. This is especially true for maps determined using small populations. Therefore extreme caution should be exercised when using genetic map information to infer physical loci orders and distances (for example, for applications to positional cloning).

F. Experimental Considerations: Materials

Hybridization solution: 1 *M* NaCl, 50 m*M* Tris-HCl, pH 7.5, 1% SDS, 5% dextran sulfate. Dissolve 58.5 g NaCl in 700 ml of deionized water, add 50 ml of 1 *M* Tris-HCl, pH 7.5, then add 50 g dextran sulfate (Phar-

[6]Mapmaker software is available from Dr. E. Lander, Whitehead Institute, Nine Cambridge Center, Cambridge MA 02142 (anonymous ftp address "genome.wi.mit.edu"). The Macintosh version is available from Dr. Scott Tingey, DuPont Agricultural Biotechnology, PO Box 80402, Wilmington, DE 19880-0402, e-mail "tingeysv@esvax.dnet.dupont.com."

macia Biotech), while stirring with a magnetic stirrer and heating to 60–70°C. After all of the dextran sulfate has dissolved add water to 950 ml; add 50 ml 20% sodium dodecyl sulfate (SDS) last and mix well. The solution may solidify partly if kept at low temperatures; preheat to 65°C and mix before use.

Probe denaturing solution: 75% formamide, 1% SDS, 10 mg/ml boiled, sheared salmon testes DNA.

Salmon testes DNA, Cat. No. D-1626 (Sigma Chemical Co., St. Louis, MO) was prepared as described in Sambrook *et al.* (1989).

Stackable Rubbermaid (Wooster, OH) Keeper drawer organizers (No. 2916, 9 × 6 × 2 in.), available from supermarkets, are useful for hybridization.

Plastic folders (clear polypropylene project folders, No. 62127, C-Line Products Inc., Des Plaines, IL) are available from stationery suppliers.

Most of the procedures used in RFLP mapping are discussed in standard laboratory manuals (Sambrook *et al.*, 1989; Ausubel *et al.*, 1987–1993). The following hybridization conditions work well with most neutral and positively charged nylon membranes.

The membranes are prehybridized in an air shaker or water bath at 65°C in 1 M NaCl, 50 mM Tris-HCl, pH 7.5, 1% SDS, 5% dextran sulfate (Pharmacia), for 1–4 hrs. From one to five filters, up to 13 × 20 cm each, can be conveniently prehybridized and hybridized in an air shaker in stackable Rubbermaid drawer organizers using about 40–60 ml of the hybridization solution per box. Multiple boxes may be stacked. The probe is labeled with ^{32}P or digoxygenin using random hexamer priming (Feinberg and Vogelstein, 1983); precise methods vary according to the labeling kit manufacturer's instructions (we have had success with both the RadPrime (Life Technologies, Gaithersburg, MD) and the Genius (Boehringer Mannheim, Indianapolis, IN) labeling systems). The labeled probe (50 µl) is mixed with 250 µl of denaturing solution (75% formamide, 1% SDS, 10 mg/ml boiled, sheared salmon testes DNA as a carrier), denatured by incubation at 95°C for 10 min, and added directly to the tray containing the filters. Small amounts of radioactive probe prepared from size marker DNA should also be added to visualize the size markers on the blots. Typically, bacteriophage lambda DNA digested with *Hind*III may be used as size markers in the gels, and 0.25 ng lambda DNA may be included with 25–50 ng probe DNA in the labeling mixture, to provide enough probe for the visualization of size markers on the autoradiogram. After hybridization at 65°C for 8–16 hrs, the probe solution is removed, and the membranes are rinsed for 5 min in 2× SSPE 0.1% SDS at room temperature and then washed to the desired stringency. Typically the first two stringency washes (15–30 min) are in a solution of 2× SSPE, 1% SDS, followed by 15–30 min wash in 0.5× SSPE, 0.1% SDS (all stringency washes are at 65°C, with shaking, using at least 50 ml wash solution per 10 × 20-cm filter). The damp membranes are wrapped in plastic wrap or placed inside plastic folders (clear polypropylene project folders, No. 62127, C-Line Products Inc.) for autoradio- or chemiluminographic exposure to X-ray film. For ^{32}P-detection, detection is best carried out at −70°C with an intensifying screen. Fluorescent stickers, Glogos II (Stratagene) may be affixed to the folder prior to exposure to facilitate orientation of the autoradiogram with respect to the blot.

Blots may be reused up to 10 times after stripping the probe. We have found that an alkaline stripping works well: 30 min at 42°C in 50 ml per blot of 0.4 N NaOH, followed by 30 min at 42°C in 50 ml per blot of 0.2 M Tris-HCl, pH 7.5, 0.1% SDS, 0.1× SSPE.

Nonradioactive probes, for example those labeled with digoxygenin (Anonymous, 1992), may be prepared and used following the manufacturers' instructions. Hybridization and washing conditions may require some adjustment, and hybridization signals are best visualized by chemiluminescence detection.

V. Cleavable Amplified Polymorphic Sequences

CAPS analysis (Konieczny and Ausubel, 1993), also known as PCR-RFLP and MAPREC (Lu *et al.*, 1993), is a technique related to RFLP analysis in which restriction site differences define polymorphisms between individuals, but these differences are visualized within locus-specific PCR amplification products. Briefly, a segment of DNA is PCR-amplified from two or more individuals, the products are cleaved with one or more restriction enzymes, and the cleavage polymorphisms are detected on ethidium bromide-stained gels. Variability with respect to the restriction cleavage site itself, or to amplicon size differences resulting from length polymorphisms within the amplified fragment, can produce either dominant or codominant polymorphisms. The information content of this method is somewhat lower than that for standard RFLPs, since the PCR amplification allows visualization of only relatively small (<2kb) amplicons. A large number of restriction endonucleases may have to be screened before a polymorphism is found. Locus-specific primers for the generation of CAPS amplicons can be derived from DNA sequences found in public databases or from DNA sequencing of random low-copy number genomic clones. The caveat to the latter approach is that most of the DNA sequences found in databases are derived from protein-encoding regions of genes, and are likely to be less polymorphic then noncoding sequences. It may be advantageous to design primers that amplify introns or 3′-nontranslated regions, which are less likely to be under strict evolutionary constraint. CAPS analysis also may be performed on RAPD fragments (see next section) (Paran and Michelmore, 1993).

VI. Random Amplified Polymorphic DNA

Williams and co-workers (1990) and Welsh and McClelland (1990) have described the use of single short oligonucleotide primers of arbitrary sequence for the amplification of randomly distributed segments of genomic DNA. The RAPD (Williams *et al.*, 1990) and AP-PCR (Welsh and McClelland, 1990) techniques are based on the amplification of the DNA segments between pairs of small, inverted DNA sequences scattered throughout a genome, and provide an innovative technology for DNA mapping, fingerprinting, and related research (Rafalski *et al.*, 1991; Waugh and Powell, 1992; Hadrys *et al.*, 1992; Tingey and del Tufo, 1993; Williams *et al.*, 1993a). This advance has resulted in a DNA marker technology that can be readily employed because of its simplicity and because of the wide availability of synthetic oligonucleotides. RAPD polymorphisms result from mutations or rearrangements at or between oligonucleotide primer binding sites in a genome. These polymorphisms can be analyzed on either agarose or acrylamide gels, and manifest themselves in the presence or absence of an amplification product(s). RAPDs are visualized as dominant markers. Although this feature carries some disadvantages, such as the inability to identify heterozygotes in F2 populations, these drawbacks are balanced by numerous advantages. The ease of use, low cost, accessibility by nonspecialists, and potential for automation are features that have attracted widespread application for both plant and animal genome mapping. In contrast to RFLP markers, RAPD markers not only require extremely small amounts of genomic DNA which can be of low quality, but also eliminate the need for DNA blotting and the use of radioactivity. To assure reproducibility of the assay, however, special attention must be paid to the optimization of the RAPD experimental conditions.

A. Experimental Considerations

RAPD experiments are usually performed in several stages. The choice of a population to study will be influenced by the dominant nature of RAPD markers (Williams *et al.*, 1993a). Initially, the RAPD amplification conditions should be optimized using DNA from the organism and strain of interest, with high-quality primer stocks (there is, however, no need to HPLC-purify primer oligonucleotides). Precisely the same experimental conditions identified from the optimization should then be used consistently throughout the study (see Protocol). Next, it is necessary to identify RAPD primers that identify polymorphisms between the parents of the mapping populations being analyzed. Large numbers of primers (several hundred to a few thousand) are usually screened, until a satisfactory number of polymorphisms is found. This number will vary with the intended application. Typically 10-mers are used for RAPD mapping experiments, and collections of suitable short oligonucleotides are available commercially (for example, from Operon Technologies). Finally, primers identified as producing polymorphisms are used for amplifications from all individuals in the population of interest. Scoring and analysis of the resulting data have been described above in the RFLP discussion. Detailed reviews are available (Tingey and del Tufo, 1993; Williams *et al.*, 1993a) and should be consulted before planning RAPD experiments.

B. Reproducibility of RAPDs

The reproducibility of RAPD markers is affected by several factors, including the genomic DNA concentration, the temperature profile of the thermocycler, the magnesium ion concentration, and the choice of thermostable polymerase. It is extremely important to optimize the amount of genomic DNA used in the PCR amplifications. The addition of too much genomic DNA results in smeary patterns, while the use of too little leads to irreproducibility. Usually, 5–25 ng of genomic DNA per 20-µl reaction is satisfactory. Larger reaction volumes, for example 50 µl, may also be used. There appears to be no clear linear relationship between the amount of DNA required and genome size. An estimation of the genomic DNA concentration can be made by measuring UV absorption at 260 nm, but may not be satisfactory, since genomic DNA preparations often contain RNA. Therefore, it is essential that the amount of template DNA to be used be determined empirically for every DNA preparation by performing RAPD amplification reactions with several primers on a dilution series of the DNA. For example, a series of twofold dilutions from 256 ng DNA/reaction to 0.25 ng DNA per reaction is convenient for this purpose.

Individual thermocyclers, especially those from different manufacturers, often differ in their actual temperature profiles even if programmed to nominally identical settings. Therefore, to assure reproducibility of RAPD profiles from one thermocycler to another the actual temperature profiles on each machine should be assessed using a thermocouple inserted into the test tube and the thermal cycle recorded. Modifications to the thermal cycling profile should be tested empirically.

It is worthwhile to optimize the magnesium ion concentration in the amplification reaction. Different DNA preparations may contain different amounts of EDTA, and therefore adjustments from the standard reaction conditions (below) may be necessary to produce a desired product(s).

RAPD and AP-PCR amplifications using Taq DNA polymerase were originally described. It was later discovered that the Stoffel fragment of Taq polymerase (Lawler *et al.*, 1993) provides better reproducibility in some laboratories and also shifts the distribution of the amplification products toward the lower molecular weight range. Compared to Taq polymerase, Stoffel is at least twofold more thermostable, exhibits optimal activity over a broader range of magnesium ion concentrations, and lacks an intrinsic 5′-3′ exonuclease activity (Lawler *et al.*, 1993). Careful optimization for the minimal amounts of these expensive enzymes can result in substantial cost savings.

Finally, it is important to note that errors in pipetting small volumes of liquids may result in apparent RAPD irreproducibility. It is recommended that reaction cocktails be used when performing multiple reactions, and that reaction components be formulated to avoid pipetting small volumes (less then 3 μl), especially when the investigator pipets "blindly," that is, without visually confirming the delivery of each pipetted aliquot.

C. Targeted Polymorphism

Several approaches to identify polymorphisms in specific regions of the genome have been used extensively with RAPD markers. In general, these strategies have been based on the exploitation of the genetics of the biological starting material or germplasm. We describe two of the most common strategies here.

D. Near Isogenic Lines (NILs)

Near isogenic lines are frequently the product of plant breeding programs where the objective has been the introgression of a desirable character or trait (gene) from a donor parent into an otherwise agronomically acceptable cultivar (recurrent parent). Near isogenic lines are generated by a process of repeated backcrossing with selection for the desired character at each round of crossing. After seven or eight backcrosses, individual selections are selfed to identify homozygotes at the target locus. The result is a pair of genotypes (the recurrent parent and the NIL) that are essentially identical at all genetic loci with the exception of the region surrounding the gene under selection. It is likely that any polymorphisms detected between the NIL and recurrent parent arise from genetic differences between the donor parent and the recurrent parent at or around the selected locus. Screening for RAPD polymorphisms between NILs therefore represents a quick and effective way of identifying markers potentially linked to the trait of interest. The linkage relationships should always be verified by segregation analysis, since polymorphic regions, unlinked to the trait, also could be present in the introgressed NIL line. The approach has proved particularly successful for isolating markers linked to plant disease resistance loci. By screening a pair of tomato NILs with 144 random primers, Martin *et al.* (1991) were able to identify three markers tightly linked to a *Pseudomonas syringae* resistance locus (PTO) on tomato chromosome 5. Similarly, Carland and Staskawicz (1993) used a different pair of NILs to isolate seven RAPD markers linked to the PTO locus. One of the primers used by both groups was identical and thus allowed a comparative ordering of the markers flanking the PTO locus.

The efficiency with which NILs can be used to identify closely linked markers depends largely on the level of variation between the donor and recurrent parents surrounding the locus under study. While the success rate in the above examples was relatively high, Ronald *et al.* (1992) had to examine 985 RAPD primers in order to identify two RAPD loci linked to the rice bacterial blight resistance locus Xa21. Identification of a single marker linked to the oat stem rot gene Pg 3 required screening with 204 primers (Penner *et al.*, 1993). Paran *et al.* (1991) screened 212 primers in two sets of lettuce NILs to identify four markers linked closely to the Dm1 and Dm3 region, and six markers linked to Dm11. Finally, Barua *et al.* (1993) screened a pair of NILs with 300 primers to identify a single marker loosely linked to a *Rhyncosporium secalis* resistance locus in barley. In all these examples, linkage was confirmed by segregation analysis.

E. Pooling Strategies

One of the drawbacks of using NILs is the large number of backcrosses required to generate these lines. An alternative strategy was initially described by Arnheim *et al.* (1985), in which pooled DNA samples from a segregating population are used to identify polymorphic markers at a specific chromosomal location. The approach is based on the linkage disequilibrium that exists for markers surrounding a target locus in the individuals used to construct the pools (Fig. 4.2). Only markers that are closely linked to the trait will be shared by all individuals in a pool. The pools can be constructed based upon phenotypic categorizations of individuals from an appropriate segregating population. Equal quantities of DNA from each member of each category are pooled. The limitation of the approach described by Arnheim *et al.* (1985) is that it relies on RFLP technology. Michelmore *et al.* (1991) adapted the technique by using RAPDs in an approach termed bulked segregant analysis. An F_2 population segregating for resistance to downy mildew was divided into two groups based on the resistance phenotype, and DNA samples from each group were pooled. Screening the two bulked DNAs and the two parents with approximately 100 RAPD primers identified three markers linked to the resistance locus.

This approach was extended by Giovanonni *et al.* (1991), Reiter *et al.* (1992), and Williams *et al.* (1993b), who used genotypic information (mapped molecular markers) to identify individuals to be pooled. In all cases, a "genetic interval" encompassing the trait of interest could be defined based upon the existing DNA marker genotypes of all the individuals in a population. The defined interval could be as small as the distance between two closely linked DNA markers or as large as the entire

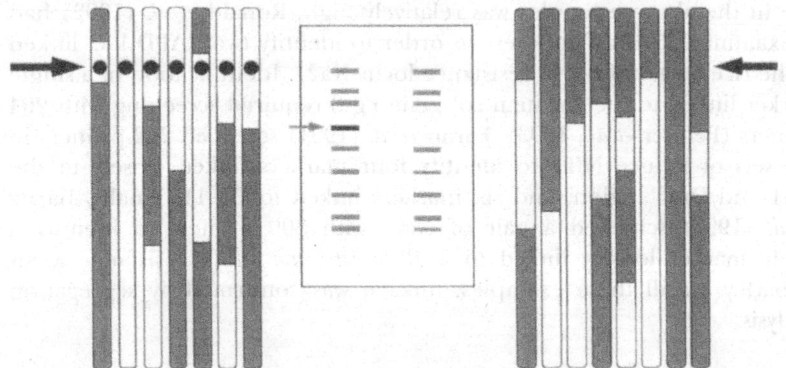

Figure 4.2 Schematic representation of Bulk Segregant Analysis (BSA). Chromosomes from several individuals selected for the phenotypic expression of a gene of interest (left arrow, black dots), and from several individuals selected for the lack of expression of the gene (right arrow). Genomic segments derived from one parent are shown in gray, and from the other parent in white. An agarose gel showing a RAPD polymorphism between the DNA bulk derived from the expressing individuals (left gel lane) and the DNA bulk derived from the nonexpressing bulk (right gel lane) is represented in the center of the figure. The only polymorphism detectable between the two gel lanes (small arrow) is the amplified fragment derived from the genomic region linked to the gene of interest.

chromosome (Reiter *et al.* 1992). While the majority of the published applications of pooling strategies have focused on qualitative characters (e.g., Barua *et al.*, 1993; Haley *et al.*, 1993; Martin *et al.*, 1993; Miklas *et al.*, 1993), it may be possible to apply bulk segregant analysis toward mapping quantitative characters using members of a population that exhibit the extremes of the distribution. Chalmers *et al.* (1993), for example, were able to identify markers linked to genetic factors controlling the milling energy requirement of barley, a polygenic character.

Pooling strategies are attractive for marker identification because these approaches are less likely than other strategies (i.e., NIL analysis) to identify markers that are unlinked to the target locus, since many segregating individuals are used to create each pool. Michelmore *et al.* (1991) provided a simple formula, $P = 2(1-(1/4)^n)(1/4)^n$ for calculating the probability that polymorphic bands are unlinked to the target locus in pools of n individuals created from an F_2 population. A simple calculation illustrates the value of using bulk populations; if two pools are created, each originating from just six individuals, there is an extremely low probability (5×10^{-4}) that a polymorphic band will come from an unlinked locus.

F. Conversion of RAPD Markers to Sequence Tagged Sites

The general applicability of RAPD marker systems is restricted somewhat by the dominant nature of RAPDs and the sensitivity of the amplifications to precise reaction conditions. Nevertheless, some of these problems can be overcome by cloning and sequencing polymorphic RAPD products, to create longer, locus-specific oligonucleotide primers. Paran and Michelmore (1993), converted eight polymorphic RAPD bands into more reliable PCR-based markers which they termed SCARs (sequence-characterized amplified regions). Five of the eight SCARs amplified both parental alleles (i.e., converted the dominant RAPD reaction into a codominant assay) and three continued to amplify the dominant allele. If SCAR primers reveal no amplification polymorphisms between the alleles of interest, then sequencing different SCAR-amplified alleles may provide the information needed to construct allele-specific amplification primers.

G. Experimental Protocol: Materials

Genomic DNA may be isolated as described by Dellaporta *et al.* (1985). Amount to be used per amplification reaction should be optimized (see *Experimental Considerations*).

10-mer primers may be purchased as kits from Operon Technologies and diluted with sterile distilled water to 4 µ*M*. Alternatively, custom-synthesized primers may be obtained.

AmpliTaq DNA polymerase and AmpliTaq DNA polymerase, Stoffel Fragment, Perkin–Elmer.

10X PCR Buffer II (Perkin–Elmer) is supplied with AmpliTaq DNA polymerase, or may be prepared by mixing 500 µl 1 *M* KCl with 400 µl sterile distilled water and 100 µl 1 *M* Tris-HCl, pH 8.3.

10x Stoffel buffer is supplied with AmpliTaq Stoffel fragment DNA polymerase (Perkin–Elmer).

dNTP mix (2 m*M* each dGTP, dATP, dTTP, dCTP) is prepared by diluting concentrated solutions of the four deoxynucleotide triphosphates (dNTP set, 100 m*M*, Pharmacia Biotech). Twenty microliters of each dNTP is added to 920 µl sterile distilled water, mixed well, and stored frozen in 50-µl aliquots.

25 m*M* MgCl$_2$ is supplied with AmpliTaq polymerase, or may be obtained by mixing 25 µl of 1 *M* MgCl$_2$ with 975 µl sterile distilled water.

Specific reaction conditions are provided in Tables 4.3 and 4.4. It is to be stressed that precise amplification conditions, including cycling parameters, template and primer concentrations, magnesium concentration, and polymerase amount, should be optimized.

The RAPD assay is performed essentially as described previously (Williams *et al.*, 1990; 1993a). Diluted genomic DNA is placed in a test tube compatible with the thermocycler. A RAPD primer is added to the DNA, followed by the remaining reaction components. The reaction is mixed, centrifuged briefly, overlayed with a drop of mineral oil (if recommended

Table 4.3
RAPD Conditions for Use with Taq Polymerase

Reagent	Stock	Final conc.	Amount used
DNA	1–25 ng/µl	5–25 ng	5 µl
PCR Buffer II	10× Cetus Buffer II, 500 m*M* KCl, 100 m*M* Tris-HCl, pH 8.3	1× 50 m*M* KCl, 10 m*M* Tris-HCl, pH 8.3	2.5 µl
MgCl$_2$	25 m*M*	1.7–1.9 m*M*	1.7–1.9 µl
Primer	4 µ*M*	0.2–0.4 µ*M*	1.25–2.5 µl
dNTP mix	2 m*M* each	0.1 m*M* each	1.25 µl
AmpliTaq	5 U/µl	20–40 U/ml	0.1–0.2 µl
Water			Make up to 25 µl

Table 4.4
RAPD Conditions for Use with Stoffel Fragment of Taq Polymerase

Reagent	Stock	Final conc.	Amount used
DNA	1–25 ng/μl	5–25 ng	5 μl
Stoffel Buffer	10× 100 mM KCl, 100 mM Tris-HCl, pH 8.3	1× 10 mM KCl, 10 mM Tris-HCl, pH 8.3	2.5 μl
MgCl$_2$	25 mM	2.5 mM	2.5 μl
Primer	4 μM	0.4 μM	2.5 μl
dNTP mix	2 mM each	0.1 mM each	1.25 μl
Stoffel	10 U/μl	2 U	0.2 μl
Water			Make up to 25 μl (11.05 μl)

by the thermocycler manufacturer), and placed in the thermocycler. Depending upon the nature of the experiment, all but one of the reaction components may be added to all sample tubes first (leaving out the template DNA or the primer), and the final component then added just prior to cycling. The concentration of the genomic DNA as well as the magnesium concentration should be optimized for reproducibility and band intensity. For further discussion of reaction conditions see (Williams *et al.*, 1993a).

Typically, random 10-mers (50–70 % GC) are used as RAPD primers. For standard thermocyclers, 40–45 amplification cycles (94°C 1 min, 35°C 1 min, 72°C 2 min, with 1-sec transition times) are performed, followed by a 6-min extension at 72°C. If a fast-cycling machine is available (for example, the Perkin–Elmer 9600), shortened cycling conditions may be used (40 cycles of 15 sec at 94°C, 30 sec at 35°C, 60 sec at 72°C, followed by 7 min extension at 72°C).

An aliquot of the sample is analyzed next to suitable DNA size standards (a mixture of phage lambda DNA digested with *Hind*III and phage ΦX174 DNA digested with *Hae*III works well) by electrophoresis through 1.2 or 1.4% agarose in 1X TBE gels using standard conditions (Sambrook *et al.*, 1989), and the gels are photographed or scanned on a UV transilluminator (Sambrook *et al.*, 1989). Instead of standard grade agarose, NuSieve or Metaphor agaroses (FMCBioproducts) may be used for increased resolution. The presence or absence of bands of interest is noted and the segregation of these bands is scored. If the data are used for genetic mapping, segregation of the RAPD bands in the expected ratio should be confirmed.

H. Modifications of the Standard Protocol

The level of resolution of agarose gels is generally sufficient to detect up to five or so individual products from a single RAPD reaction. However, more than this number of bands can be visualized if the RAPD reaction products are resolved on silver-stained polyacrylamide gels (Caetano-Anolles *et al.*, 1991; Caetano-Anolles, 1993). This method may be particularly useful for fingerprinting, providing reproducibility is maintained. Higher resolution may also be obtained by using radioactively labeled (with ^{33}P or ^{32}P) primers, following by fractionation of the products on polyacrylamide gels (Welsh *et al.*, 1991). An automated DNA sequencer may be used to resolve the products of RAPD reactions performed with fluorescently labeled primers; such sequencers are particularly useful for accurately determining fragment sizes, for generating fingerprints, and for multiplexing several individual RAPD reactions in a single "lane." A detailed discussion of other modifications to the standard RAPD method is described by Williams and co-workers (1993a).

A computer bulletin board for RAPD users has been established. To subscribe, send an e-mail message with your computer address to "biosci@net.bio.net" (within the U.S.A.) or "biosci@daresbury.ac.uk" (countries outside the U.S.A.).

VII. Microsatellite Markers (Simple Sequence Repeats, SSR)

Simple sequence repeats (SSRs), also called microsatellite repeats, consist of stretches of tandemly repeated mono-, di-, tri-, tetra-, penta-, or hexanucleotide motifs. SSR loci harbor considerable length variation, are extremely abundant, and are randomly distributed throughout eukaryotic genomes. Consequently, SSRs have found wide application as markers in human and mammalian genetics, for the construction of highly informative and saturated genetic maps. SSR loci are individually amplified by the polymerase chain reaction (Mullis *et al.*, 1986), using pairs of oligonucleotide primers specific to the unique DNA flanking the SSR sequence (Fig. 4.3). The amplified products usually exhibit high levels of length polymorphism, which result from variation between alleles in the number of tandemly repeating units at the locus (Litt and Luty, 1989; Tautz, 1989; Weber and May, 1989). Generally, SSR markers are codominant, revealing polymorphic amplification products from all individuals in a population. SSR-based maps have been obtained for humans (Weissenbach, 1992), mouse (Dietrich *et al.*, 1992), and rat (Serikawa *et al.*, 1992), and are under construction for other animal species. Their high information content,

```
351   GTCACATGCT  TACCTTTCTT  GTGAGTCGTA  GTCTAGCTAC  ATGACACAAT

401   TCTTAGGGAC  CATATTCTTG  ATATCACTTC  TCTGACTCAT  AAACAAAGAT

451   AATATATATA  TATATATATA  TATATATATG  ATCAGTTTTT  TTATTGATCT

501   ACAATATCCA  TCTCAATATT  TTAAATATAT  CATTGCTTCA  CATATTCCAC

551   TGATTTCCAA  AGAAGTCAAT  GTAGAATGAC  AATGTGAAAA  TCATTTTTTA
```

Figure 4.3 Simple sequence repeat present in the GenBank sequence GB_pl:SoySC514 (Accession No. X56139, Lipoxygenase; Shibata *et al.*, 1991) (AT)$_{14}$ SSR is shown in bold. Primers designed to amplify the SSR-containing segment of SC514 locus are underlined.

which is directly related to the effective number of alleles at each locus, and the ease of automating the PCR assays for identifying simple sequence repeat polymorphisms (SSRPs), make SSRs ideal genetic markers. These features compensate for concerns that derive from the relative difficulty of generating these markers compared to other types of DNA markers (Rafalski and Tingey, 1993b). Because primer sequences are easy to share, it is expected that public SSRP maps for many plant species will soon become available.

The most frequently occurring and thus most widely utilized simple repetitive dinucleotide motif in mammalian genomes is poly(dA-dC)$_n$/(dG-dT)$_n$ (hereafter referred to as AC repeats) (Beckmann and Weber, 1992). Other simple repeats, composed of both di- and trinucleotide motifs, also have been used occasionally for mapping in animal species. SSRs have just started to be used as markers for plant genetics. Initial reports showing the presence and hypervariability of such markers in soybean (Akkaya *et al.*, 1992; Morgante and Olivieri, 1993) were soon followed by others describing SSR polymorphisms in different *Brassica* species (Lagercrantz *et al.*, 1993), grapevine (Thomas and Scott, 1993), *Arabidopsis* (Bell and Ecker, 1994), maize (Lynn Senior and Heun, 1993) and rice (Wu and Tanksley, 1993a; Zhao and Kochert, 1993). In all cases, the amplified SSR loci were confirmed to be highly variable (in length), inherited in a codominant Mendelian fashion, and somatically stable. When a direct comparison between variability data for SSRPs and RFLPs was made, SSRPs in both rice and soybean were shown to be significantly more informative than RFLPs (Wu and Tanksley, 1993a; Morgante *et al.*, 1994). Screening the public DNA sequence databases for SSRs revealed that the most frequent dinucleotide repeat in plants is AT, with AG and then AC (Lagercrantz *et al.*, 1993; Morgante and Olivieri, 1993) as the second and third most frequent SSRs. The relative scarcity of AC, the most common mammalian dinucleotide SSR, seems to be a general feature of plant genomes, even when frequency estimations from genomic blot hybridizations are considered. On the other hand, the unexpectedly high abundance of AT repeats in plant genomes is technically difficult to confirm by hybridization analysis, because of the difficulties in using these self-complementary AT repeated sequences as probes.

Many of the studies on plants mentioned above have used SSR loci that were identified through gene sequences cataloged in public databases. The amount of DNA sequence information available even for the most intensely studied plant species is so low that only a limited number of SSR markers can be derived. A considerable amount of effort and expense is therefore required to identify additional SSR loci to generate a sufficient number of SSR markers to support genetic and breeding applications. In this section we will describe the procedures that we use for the discovery of soybean and maize SSRP markers, compare these to other published methods, and discuss possible improvements in methodology.

The most commonly followed method for the identification of SSRPs includes the following steps:

    (A) Genomic DNA library construction
    (B) Library screening by hybridization
    (C) DNA sequence determination of positive clones
    (D) Locus-specific primer design
    (E) PCR analysis and identification of polymorphisms
    (F) Polymorphic marker mapping

A. Materials

Lambda ZAPII vector (Stratagene)
NA-45 DEAE membrane (Schleicher and Schuell, Keene, NH)
Poly(dA-dC)·Poly(dG-dT) (Pharmacia Biotech)
Poly(dA-dG)·Poly(dC-dT) (Pharmacia Biotech)
Microcon-100 concentrators (Amicon Inc., Beverly, MA)
T3 and T7 promoter-specific primers (Stratagene)
Synthetic linkers and adaptors (New England Biolabs, Beverly, MA)
Custom synthesized primers, Operon Technologies, or other supplier. Dilute in water to 10 μM and store frozen. HPLC purification is not necessary.
Ziplock bags, available from supermarkets in several sizes
PCR reagents—see RAPDs Materials
Primer selection software, Oligo 4.0 for Macintosh (National Biosciences, Plymouth, MN) or Primer (available from Dr. Eric Lander, Whitehead Institute, Nine Cambridge Center, Cambridge MA 02142).

B. Genomic DNA Library Construction

Identification of SSR loci requires characterization of unique sequences flanking each side of the repeat region to design locus-specific primers for PCR amplification of a given SSR. Two approaches have generally been used. The first is to screen a large-insert genomic library with the repeat probe (usually, random-primed probes from synthetic polymers representing the repeat). Once positive clones are identified, they are purified and sequenced to determine the DNA sequence flanking the SSR. Often, multiple rounds of subcloning may be required before the DNA sequence can be determined confidently. This multistep approach is not suited for large-scale projects where hundreds of new markers are desired, and so streamlining the process is crucial. A second, more common approach is to construct and screen a genomic library harboring extremely small inserts (200–500 bp). This allows direct sequencing of positive clones, but requires screening large numbers of plates in order to obtain a large set of positive clones, especially when the frequency of the repeat in the genome is low. Nevertheless, screening a small insert library is the method of choice for obtaining a large number of markers.

Standard procedures are used for library construction (Sambrook *et al.*, 1989). Genomic DNA fragments of the desired size range may be obtained by treatment with DNase I or by sonication (Sambrook *et al.*, 1989), with an aim toward cleaving the majority of the genome into fragments within the desired size range. These random shearing methods are preferable to using a single or a combination of restriction enzymes, which allow only a subset of the repeats to be present in the correct size fraction. The DNA is then rendered blunt-ended through the use of the Klenow fragment of *Escherichia coli* polymerase I or T4 DNA polymerase (Sambrook *et al.*, 1989), and then size fractionated. In constructing a soybean small-insert library, we digested DNA with a range of restriction enzymes (with both 4- and 6-bp recognition sites) that generate blunt ends (*Alu*I, *Rsa*I, *Dpn*I, *Dra*I, *Hae*III, *Hpa*I, *Ssp*I).

To isolate size-selected DNA fragments, we prefer fractionation on agarose gels followed by recovery of the DNA on a NA-45 DEAE membrane (Schleicher and Schuell). Two requirements determine the size of the DNA fraction to be recovered. DNA fragments should be small enough to allow flanking regions of the repeat to be sequenced directly using common, vector-specific primers, but should not be so small that huge numbers of them must be screened in order to obtain the desired number of positive clones. In addition, inserts that are too small will increase the probability that the SSR sequence is too close to one of the cloning sites, thus eliminating the possibility that sequences flanking the

SSR can be determined. Once the desired genomic DNA size fraction is obtained, typically in the size range of 350–500 bp, the fragments are ligated to synthetic linkers and cloned into a suitable vector (Sambrook *et al.*, 1989). Bacteriophage lambda insertion vectors, bacteriophage M13, or plasmid vectors may be used. The advantages of lambda vectors are that a representative library can be generated easily and that screening of phage plaques by hybridization is technically more straightforward than screening bacterial colonies. This is important because AC and AG repeats are relatively rare in plants, as compared with mammals, so that the screening of a large number of clones is required. Some phage lambda vectors provide for convenient excision of a plasmid subclone (for example Lambda ZAPII, Stratagene), thus allowing the clone, once identified as a positive, to be maintained as a plasmid.

In order to reduce the number of clones to be screened, genomic libraries may be enriched for simple sequence repeats. Several enrichment schemes have been proposed, but most of these approaches have not been tested rigorously, and should be used with caution. The selection steps for the enrichment may be performed either before or after library construction. Enrichment following fragment cloning, as proposed by Ostrander *et al.* (1992), is based on selective second-strand synthesis, initiated on a closed circular single-stranded phagemid DNA template, from an oligonucleotide primer containing the target repeated motif. Only clones that contain repeats should produce double-stranded molecules. Selection against single-stranded DNA molecules is then applied following transformation into a bacterial host. Ostrander and co-workers achieved a 50-fold enrichment for $(CA)_n$ sequences using this approach.

Some protocols involving selection after cloning take advantage of the ability of DNA molecules to form stable triple-stranded complexes under the appropriate conditions. Triple-stranded DNA complexes between a synthetic biotinylated SSR oligonucleotide and plasmids containing the SSR can be obtained by two different methods. In one method (Ito *et al.*, 1992), triple helix DNA formation results from the stable binding of single-stranded homopyrimidine tracts to double-stranded homopurine–homopyrimidine helices under low pH conditions. This method limits what can be enriched for, however, since the homopyrimidine dinucleotide $(TC)_n$ repeat, used as a capture probe, is able to enrich only for double-stranded DNAs containing AG repeats. In a second enrichment method (Rigas *et al.*, 1986), which has yet to be applied to SSR isolation, stable triple stranded complex formation is mediated by the *E. coli* RecA protein. This method places no restrictions on the DNA sequence of the capture oligonucleotide, and therefore no restriction on the type of SSR selected. Once the complexes are formed by either method, they can be selectively isolated by binding a biotin moiety present on the capture oligonucleotide, to streptavidin-coated paramagnetic beads. SSR-enriched DNA fragments are then recovered and transformed into bacterial hosts.

The main concern underlying any enrichment scheme involving selection after cloning is that the complexity of the resulting library might be too low to generate a sufficient number of independent clones that contain SSR repeats. This is especially true if the clones are small and the frequency of the selected repeat in the genome is not very high. This problem may be overcome partially, however, using methods where selection is applied before the DNA fragments are cloned. Three similar methods based on this principle have been published recently (Karagyozov *et al.*, 1993; Lyall *et al.*, 1993; Kandpal *et al.*, 1994). All involve the production of DNA fragments which have a known sequence at both ends, achieved by ligating adaptors to the fragments (Fig. 4.4), or by obtaining the fragments through PCR amplification using degenerate primers. After modifying the ends by either method, fragments containing SSRs are then selected by hybridization to a biotinylated SSR oligonucleotide. The DNA hybrids are purified by selection on streptavidin-coated paramagnetic beads. DNA fragments containing the desired SSR repeats are eluted as single strands from the beads, and the second strand is generated by a PCR reaction using adaptor-specific primers. Several cycles of amplification with the adaptor primers will generate multiple copies of these enriched fragments, which are then cloned following their digestion with a restriction enzyme whose recognition site was part of the adaptor sequence (Fig. 4.4). Since the selected fragments are PCR-amplified following selection, concerns may arise for possible amplification artifacts and for possible competition between fragments during amplification.

C. Library Screening by Hybridization

Clones containing SSR sequences are usually identified by plaque or colony hybridization with an SSR probe (Sambrook *et al.*, 1989). If the desired repeats are frequent, such as for AC repeats in humans (AC-positive clones compose 1% of all 400- to 500-bp inserts, or one repeat per 50 kb (Lyall *et al.*, 1993), then low-density plating (100–200 plaques per 100-mm plate) at the primary screen stage can produce single positive plaques or colonies. If the repeats are relatively rare, as for AC and AG repeats in soybean (one repeat occurs every 500–700 kb), then initial plating at high density (10,000—20,000 plaques per 150-mm plate) followed by one or more rounds of rescreening is more efficient. The total number of plaques to be screened will depend on the desired number of SSRs and

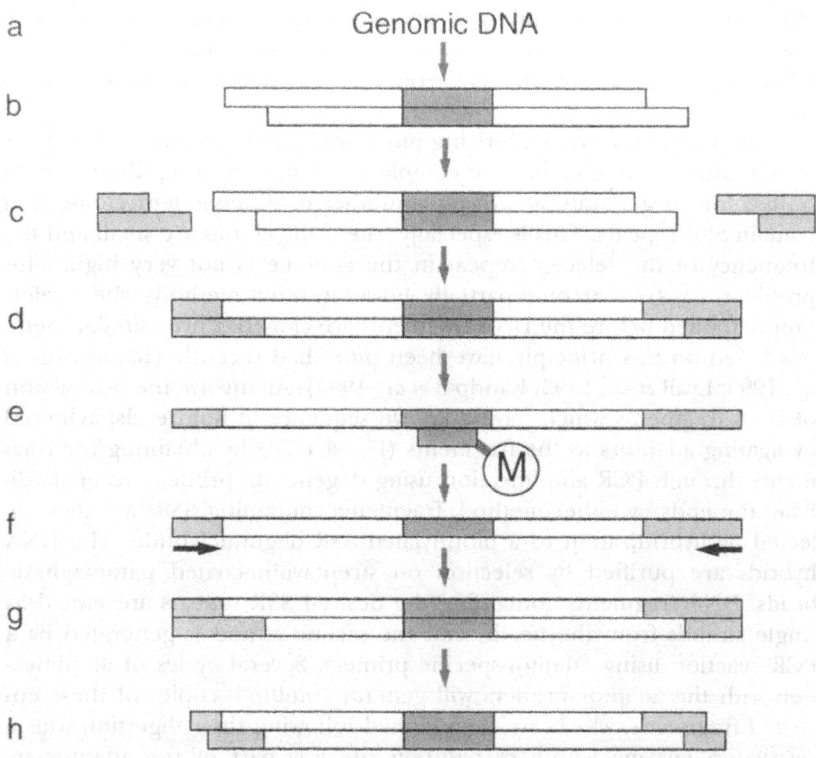

Figure 4.4 Schematic representation of the DNA capture SSR enrichment method (Kandpal *et al.*, 1994). Genomic DNA (a) is digested with a restriction endonuclease (b), adaptors are attached (c, d), the DNA is denatured, hybridized with a biotin-tagged probe, and bound to streptavidin-coated beads, M (e), single stranded DNA (f) is eluted and PCR-amplified using adaptor-directed primers (f, g), and the resulting double-stranded DNA is digested with restriction endonuclease, to produce DNA fragments suitable for cloning (h).

their abundance in the genome of interest. For most applications, a small subset of all repeats present in the genome is needed, and therefore there is no need to screen several genome equivalents of the library.

Suitable probes for these screens can be made from commercially available double-stranded polynucleotides, such as $poly(dA\text{-}dC)\cdot(dG\text{-}dT)$, which are labeled with ^{32}P or digoxygenin through random priming (Feinberg and Vogelstein, 1983). Alternatively, oligonucleotide probes such as $(AC)_n$, with $n = 10$ to 15, may be end-labeled with ^{32}P using T4 polynucleotide kinase. Both methods result in a probe with specific activity up to 1×10^9 dpm/μg; however, the specific activity calculated on a per mole basis is much higher for random primer probes, because multiple radioactive phosphates are incorporated per nucleotide chain.

AT repeats represent the most abundant type of repeats within the plant sequences in the Genbank database (Lagercrantz *et al.*, 1993; Morgante and Olivieri, 1993), but these are difficult to screen for because of the self-complementarity of the probe sequence. Nonetheless, successful hybridization schemes using AT probes have been described (Ermak *et al.*, 1990; Greaves and Patient, 1985).

It has been demonstrated in humans (Weber, 1990), and confirmed in other species, that the level of polymorphism of the SSR markers is proportional to the length of the repeat region, and that SSR markers with fewer than 10 repeats exhibit little or no length polymorphism. Therefore, the stringency for the screen should be optimized to reduce the number of very short, noninformative SSR markers that are identified. When screening for AC and AG repeats with random-primed poly(AG/TC) or poly (AC/GT) probes, we hybridize 50 μCi probe with 10 137-mm filters in Ziplock-bag enclosed 150-mm culture dishes in ca. 50 ml of 1 M NaCl, 50 mM Tris-HCl, pH 7.5, 1% SDS, 5% dextran sulfate at 56°C overnight, briefly rinse the filters twice in 2× SSPE 0.1% SDS and follow with two higher stringency washes (15 min each) at 65°C in 2× SSPE, 1% SDS (at least 50 ml per filter). DNA sequence analysis confirmed that only 2 of 100 clones that we isolated by this method contained fewer than 10 repeats.

D. DNA Sequence Determination of Positive Clones

Selected and purified clones containing SSRs must be sequenced in order to design PCR primers specific to the unique sequences flanking each repeat. Not every sequenced hybridization-positive clone will result in a useful SSR polymorphism. Typically, the yield may be 20–50%. Anchor PCR procedure (see below) improves the yield considerably by eliminating false positives and positives with repeats positioned too close to the vector. Fluorescence-based automated DNA sequencing machines have the requisite high-throughput capabilities, and facilitate sequence data acquisition and transfer to the primer-design software. For each clone, a single sequencing reaction, generating sequence from only one strand, is usually sufficient, as long as high-quality sequence can be obtained for both regions flanking the SSR. This is not always the case because the polymerase often has difficulty generating a clean sequence ladder across the SSR itself; either the polymerase falls off the template, causing false terminations (this is especially true for the palindromic AT repeats which we often find adjacent to AC and AG repeats in soybean), or it slips along the repeat during the extension (replication slippage; Levinson and Gutman, 1987), causing an overlap of peaks after the repeated region. We have identified several critical conditions that are required to obtain good DNA sequence from a single run. First, Taq polymerase cycle sequencing with dye terminator chemistry (Applied Biosystems) works better than dye primer chemistry on double-stranded SSR-containing templates, especially in the region downstream of the repeat. Also, it is extremely important to use a high-quality template DNA at a consistent concentration (1 µg/reaction) in the sequencing reaction. We have found it useful to concentrate our plasmid miniprep DNAs by ultrafiltration (Amicon Microcon-100 or similar) prior to sequencing. Finally, the location of the repeat within the insert affects how well it can be sequenced. It is very difficult to obtain good sequence information from the downstream flanking region if the repeat is more than 300 bp away from the priming site.

In order to rationalize large-scale sequencing efforts, we have found it useful to prescreen the clones for the presence and location of the repeats (Taylor *et al.*, 1992; M. Morgante and C. Andre, unpublished data). This can be done with purified lambda phage plaques after the secondary screen. For the prescreening, five PCR amplification reactions are performed on each clone, one with two vector primers (e.g., T3 and T7 promoter-specific primers) and four reactions each using one of the vector primers in combination with a primer specific to the repeat (AC and GT or AG and CT, depending on the repeat type present). A product representing the entire insert is expected from the two vector primers, and a shorter product is expected from two of the four vector-repeat combinations (Fig. 4.5). The location and orientation of the repeat within each insert can be determined by cataloging the presence and sizes of the resulting PCR products (M. Morgante and C. Andre, unpublished data). Clones containing no repeat can be recognized and immediately discarded, as they produce an amplification product only from the vector primer pair. Clones in which the repeat is either too far from both cloning sites, or much too close to one end of the insert (too little flanking DNA for good primer design), also can be discarded. Those clones that pass the PCR prescreen then are sequenced from the vector end that is closer to the repeat. The PCR screening process, including the preparation of the reactions and gel loading, may be automated using a robotic pipettor (Beckman Biomek or similar). We have confirmed that PCR prescreening on over 100 AC and 70 AG clones generates results that are in complete agreement with the DNA sequence data.

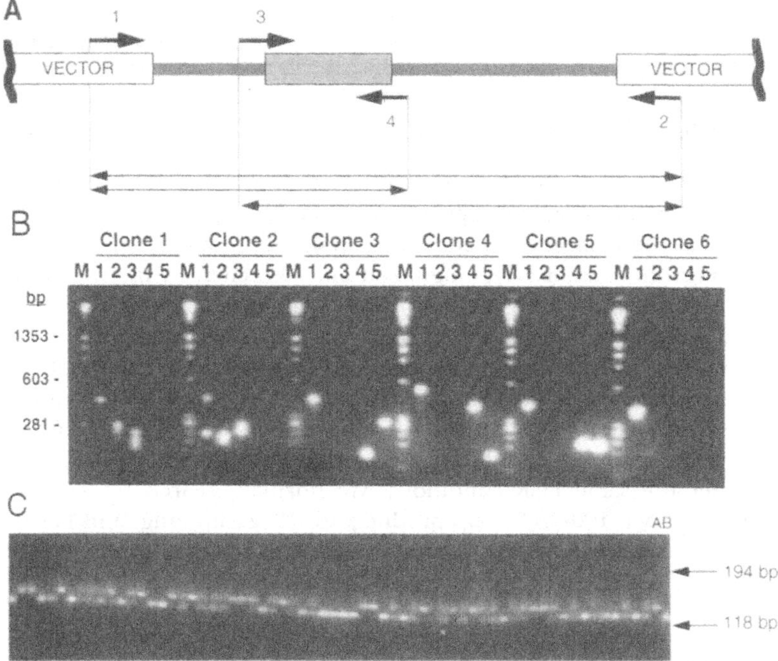

Figure 4.5 Simple sequence repeat analysis by PCR. (A) Primer arrangement for anchor PCR. Grey rectangle represent the SSR sequence. Primers 1 + 2 amplify the whole DNA insert. Primers 1 + 4 and 3 + 2 or, if the orientation of the SSR sequence is opposite, primers 1 + 3 and 4 + 2, amplify DNA segments corresponding in length to vector–SSR distances (see text). (B) Anchor PCR prescreening of six SSR-positive clones. Five microliters of a 1:10 dilution of the phage supernantant prepared from a SSR hybridization-positive plaque was used as a template for PCR amplification with the following primers: lane 1, vector forward and vector reverse primer; lane 2, vector forward and AG anchored primer mix; lane 3, vector reverse and CT anchored primer mix; lane 4, vector forward and CT anchored primer mix, lane 5, vector reverse and AG anchored primer mix, in the following conditions: 1× PCR buffer (10 mM Tris-HCl, pH 8.3, 50 mM KCl), 0.2 mM each dGTP, dATP, dTTP, and dCTP, 2.5 mM MgCl$_2$, 0.2 μM each primer as listed above, 0.5 U Taq polymerase, total volume 25 μl. The temperature cycling conditions were: denaturation 5 min at 95°C, 30 cycles of 1 min at 94°C, 1 min at 54°C, 1 min at 72°C, followed by 8 min at 72°C. The anchored primer mixtures were: anchored AG primer, equimolar mixture of HBH(AG)$_8$A and BHB(GA)$_8$G; anchored CT primer, equimolar mixture of VDV(CT)$_8$C and DVD(TC)$_8$T, where B is every base except A, D is every base except C, H is every base except G, and V is every base except T. M, DNA size marker lane. Amplification products appear in lane 1 (insert amplified from vector primers), and in either lanes 2, 3, or 4, 5, depending on the orientation of the AG simple sequence repeat in a given clone, with respect to the vector primers. (C) Separation of SSR alleles on agarose. DNA samples (100 ng) from 64 F2 progeny individuals from a cross between *G. max* cv. Bonus and *G. soja* PI81.762 were amplified under standard conditions (see text) using primers flanking an anonymous SSR locus (AG43). Lanes A (Bonus), B (PI81.762) correspond to the amplification products of parental DNA. Amplification products were resolved on 3.5% Metaphor (FMC) agarose and the image was recorded with a CCD camera.

E. Locus-Specific Primer Design

Once the DNA sequences for the SSR-containing clones have been obtained, primers flanking the repeats can be designed, most efficiently by using one of the computer-based primer design programs available [e.g., Primer (Lincoln *et al.*, 1991); Oligo, National Biosciences, Inc.]. The sequence files obtained from the automated sequencer are first corrected for obvious mistakes. If both strands have been sequenced, the differences are reconciled and the sequences merged. The flanking regions of the resulting sequences should then be compared to the data base of previously sequenced SSR, to identify and remove duplicates. Consistent criteria for primer design should be used, so that all primer pairs can be used under the same PCR conditions. We normally search for primers that will amplify a 100- to 200-bp product, are 19–24 bp long, and have a T_m between 56 and 61°C (58°C is optimal). Thus, a single set of PCR amplification conditions can be applied to every primer set.

F. PCR Analysis and Identification of Polymorphisms

To identify and characterize SSR alleles among genomes, the flanking primer pairs are used for PCR amplification of individual SSR loci from genomic DNA. Standard reaction and cycling parameters are used (for example, 100 ng of genomic DNA is amplified in 25 μl of 10 mM Tris-HCl, pH 8.3, 50 mM KCl, 2 mM MgCl2, 0.2 mM each dGTP, dATP, dTTP, dCTP, 0.2 μM each primer. After 5 min denaturation at 95°C, 30 cycles of 45 sec at 94°C, 45 sec at 58°C, and 45 sec at 72°C are performed, followed by 8 min for a final extension at 72°C). It is best to use a single set of amplification and reaction conditions for all primer pairs, without any additional optimization of these conditions. This is a necessity when large numbers of primers must be screened and then used in mapping and/or breeding applications. Several factors may lead to failure of these PCR reactions. First, nonspecific annealing might occur because of poor primer design, inaccurate sequence data, or homology of the primers to repeated regions in the genome. In addition, Bell and Ecker (1994) reported that AC repeats in the *Arabidopsis* genome were difficult to amplify and almost invariably produced two or more bands, even when extensive optimization of PCR conditions was attempted; AG repeats, in contrast, amplified well. We have also found that AG repeats are more efficiently amplified than AC repeats in soybean, but the difference is not as dramatic as it seems to be for *Arabidopsis* (M. Morgante, unpublished observations). Also, polyploid genomes may result in the amplification of more than one locus. Soybean is an ancient polyploid and single-copy RFLP probes usually reveal two bands, corresponding to the two homologous loci (Rafalski and Tingey, 1993a). Nevertheless, we seldom have observed the amplification of two loci when using primers flanking an SSR sequence. We attribute this to the considerable divergence of the two constituent genomes, especially in noncoding regions where many SSR loci are likely to reside.

High-resolution gel electrophoretic separation of SSR amplification products is required in order to identify alleles that may differ in size by 2 bp or more. Denaturing polyacrylamide sequencing gels (Sambrook *et al.*, 1989) offer the best resolution, but radioactive labeling or silver staining of the fragments is required, adding considerable labor and limiting throughput. For many applications, we use high-resolution agarose gels stained with ethidium bromide, with minimal sacrifice of resolving power. Although 3% Nusieve (FMC Bioproducts), 1% agarose gels (Morgante and Olivieri, 1993) are satisfactory, we have found that 3.5% Metaphor (FMC Bioproducts) gels offer higher resolution and are easier to prepare and handle. Other separation media (Synergel, Diversified Biotech, Boston, MA) have been used successfully. We screen our initial amplification products for polymorphisms using Metaphor gels; if no polymorphisms are visible, then we reanalyze them on sequencing gels. We have been able to resolve and map a 2-bp polymorphism on Metaphor gels, and 4-bp differences in 100- to 200-bp fragments routinely can be separated. Another advantage of agarose gels is that SSR fragments separated on them do not show the ladder-like "stuttering" pattern visible on sequencing gels. The causes and possible cures for the appearance of the shadow bands on sequencing gels have been discussed (Litt *et al.*, 1993; Mellersh and Sampson, 1993; Odelberg and White, 1993). As an alternative to standard sequencing gels, fluorescently labeled SSR amplification products can be analyzed in a fluorescence detection automated sequencer, which allows extremely accurate sizing of the fragments, especially if an internal size standard is used. Use of different fluorophors in different PCR amplifications allows for multiple reactions to be resolved in a single lane. Fluorescently labeled primers typically are used, but we have obtained good results on the ABI373 Sequencer by incorporating fluorescein-12-dUTP into the fragments during the PCR amplification (Mansfield and Kronick, 1993). Another alternative for the separation of the SSR fragments is capillary electrophoresis, which allows for very fast and accurate sizing (Marino *et al.*, 1994).

G. Polymorphic Marker Mapping

Mapping SSR loci follows the general principles discussed for RFLP markers. SSR markers are almost always codominant, although null alleles (i.e., lack of an amplified fragment) resulting from deletion or alteration of one of the priming sites are sometimes observed (Morgante *et al.*, 1994). While most well-designed primer pairs produce single amplification products, occasionally two or more amplification products are observed, presumably from duplicated loci. Both may be polymorphic. Given the higher information content of SSRP markers compared to RFLP markers, a larger proportion of polymorphic markers will segregate in each single cross, making it possible to obtain maps from crosses where few RFLP markers would be detectable. Once enough flanking sequence primers have been accumulated, automation of the SSRP PCR assay should make map construction fast and efficient. We have automated the reaction assembly and gel loading for mapping SSRPs using a Beckman Biomek 1000 Automated Laboratory Workstation and 3.5% Metaphor gels stained with ethidium bromide. Digital imaging of gels with a CCD camera (Fig. 4.5) allows semiautomated genotype scoring from the gels. The scores are then used as input data for mapping software (e.g., Mapmaker). This procedure makes it possible for an individual to map several loci per day. This throughput will be reduced, however, if the polymorphism cannot be detected on agarose and sequencing gels are required. Apparatus and procedures for the multiplexing of SSR markers on sequencing gels have been described (Vignal *et al.*, 1993).

VIII. Sequence-Based Polymorphism Assays

DNA sequencing is the ultimate polymorphism assay. If the complete DNA sequences of the genomes to be compared were known, then no further analysis (marker discovery, polymorphism characterization, segregation analysis, etc.) would be required. At present, of course, this ideal goal is not attainable, and instead limited sequence comparisons can be utilized to design relatively simple and automatable PCR-based assays. The degree of nucleotide sequence divergence between two randomly chosen individuals differs greatly between taxa. The maize genome exhibits about 1% nucleotide divergence (Shattuck-Eidens *et al.*, 1990); however, melons (Shattuck-Eidens *et al.*, 1990) show only 0.1% divergence. Therefore, sequencing of only a few hundred base pairs in maize is likely to reveal differences between individuals, whereas 10-fold more sequence information must be acquired to reveal differences in melon. Templates suitable for sequencing may be prepared by PCR amplification of the genomic regions of interest from two individuals. RAPD markers and SCAR markers provide an easy source of genomic fragments to be sequenced and compared among individuals (Paran and Michelmore, 1993).

The single-strand conformation polymorphism (SSCP) (Orita *et al.*, 1989; Beier *et al.*, 1992) approach may be used to screen amplified fragments for the existence of sequence polymorphisms. The method has recently been adapted for use in a commercial automated electrophoresis system (PhastSystem, Pharmacia; Mohabeer *et al.*, 1991) and may be used for mapping or genetic diagnostics. The SSCP method detects differences in the secondary structure of polymorphic DNA fragments. The DNA fragments are denatured and quickly cooled, thereby "freezing" their secondary structure. The samples are resolved on cool nondenaturing polyacrylamide gels, revealing differences in mobility due to differences in DNA sequence-based secondary structure.

Once sequence differences have been identified by direct sequencing, one of several methods may be applied to design a diagnostic assay. Here, we will briefly discuss allele-specific amplification, allele-specific ligation, and the nucleotide incorporation assay. The reader is encouraged to consult the primary literature for more details.

A. Allele-Specific Amplification

Allele-specific amplification (ASA) relies on the design of PCR primers that amplify one allele (sequence variant) and not the alternate allele (Wu *et al.*, 1989). Typically, PCR primers are designed that match one allele perfectly, but differ from the other allele by at least one nucleotide mismatch usually at the 3'-end of one or both of the primers. Some care should be taken in primer design, however, since not all mismatches are equally refractive to amplification (Kwok *et al.*, 1990; Huang *et al.*, 1992). A mismatched T at the 3'-most position in the primer, for example, has been reported not to reduce significantly amplification efficiency (Kwok *et al.*, 1990). It is critical to optimize the amplification conditions to promote the greatest primer specificity. The presence or absence of the reaction product is usually viewed on agarose gels. It is usually advantageous to produce several pairs of ASA primers for a locus, each primer specific for a given allele, thereby converting a dominant assay into a codominant one. It is also possible to automate the allele-specific amplification assay. One of the primers can be labeled with a fluorescent or antigenic tag, and the other biotinylated, to facilitate isolation of the amplification product away from the unincorporated reaction components using a streptavidin-coated solid support (multiwell assay plate or magnetic beads). The amount of amplification product is then assayed via specific detection of the fluorescent or antigenic tag on the bound PCR product.

B. Allele-Specific Ligation

Allele-specific ligation (ASL, Landegren *et al.*, 1988) and ligase chain reaction (Bárány, 1991) are similar to allele-specific amplification in that a pair of allele-specific oligonucleotides is used that discriminates between alleles that can differ by a single nucleotide. The 3'-nucleotide of the first oligonucleotide primer overlaps the mutation to be analyzed. The second oligonucleotide, which is 5'-phosphorylated and which carries a radioactive, fluorescent, or biotinylated tag, is positioned immediately adjacent to the nucleotide-specific 3'-end of the first primer. If both of the oligonucleotides create a perfect match with the DNA template, they can be ligated together with T4 DNA ligase. Any mismatch to the template at the 3'-most end of the first primer will severely reduce the ligation efficiency. Successful ligation is detected by analyzing labeled reaction products on a gel, or by using the biotin-capture and fluorescence-detection method described above for ASA (Nickerson *et al.*, 1990). To ensure adequate sensitivity, ASL is usually performed on a PCR-amplified template.

C. Nucleotide Incorporation Assay

The nucleotide incorporation assay may be used to identify the precise nucleotide present in the template DNA adjacent to the 3'-end of a locus-specific synthetic oligonucleotide, by template-guided incorporation of a single, labeled dideoxynucleotide (Sokolov, 1990). All four dideoxynucleotides may be used simultaneously in this incorporation assay if they are all differentially labeled, or four separate reactions may be performed, each with one labeled and three unlabeled dideoxynucleotides (Syvanen *et al.*, 1990; Kuppuswamy *et al.*, 1991). The single nucleotide extension products are collected (via capture of biotin tags at the 5' end of the primer), unincorporated reactants are removed, and the nature of the incorporated dideoxynucleotide is determined. The assay is best performed in a multiwell plate using PCR-amplified template. It is particularly suitable for the determination of a single base mutation at a defined and well-characterized genetic locus (Livak and Hainer, 1993).

IX. Higher Multiplex Ratio Assays: Amplified Fragment Length Polymorphism and Interrepeat Amplification

Both AFLP and ISA assays use PCR-based technology. Both are largely dominant, and each assay can produce a high multiplex ratio. In addition, both rely on the random arrangement of specific sequences in the genome. For AFLP, these sequences are restriction sites combined with the 1–10 nucleotides immediately adjacent to the restriction sites. For interrepeat amplifications, these are highly repetitive sequences that, ideally, are randomly dispersed throughout the genome. Examples include Alu repeats in mammals, and the more ubiquitous SSRs found in both animal and plant genomes.

A. Amplified Fragment Length Polymorphism

The AFLP approach to genome analysis (Zabeau and Vos, 1993) combines features from several of the previously discussed methods. The relatively high polymorphism inherent in the random placement of restriction sites between different genomes, combined with nucleotide sequence variability within a short stretch of DNA directly flanking these restriction sites, makes AFLP a highly informative assay for both plant and animal genomes. The AFLP reaction can be adjusted continually to produce a desired multiplex ratio. With appropriate conditions, for example, this assay can be used to simultaneously display more than 100 loci per gel lane, a high proportion of which may be polymorphic between genomes.

The method involves the selective amplification of an arbitrary subset of restriction fragments, generated by total digestion of the genome with a single- or double-enzyme combination. Prior to the amplification, however, the ends of every fragment in the digestion are modified by the addition of enzyme site-specific, double-stranded adaptors. In the selective amplification, pairs of end-labeled oligonucleotide primers are used, whose target sites span both the synthetic adaptors and the restriction sites, as well as the first few (1–3) nucleotides of the restriction fragment itself (Fig. 4.6). This amplification is selective because these few 3′-most nucleotides on each primer are chosen to be a random, nondegenerate sequence, and under the appropriate annealing conditions will discriminate target loci that differ by even a single-nucleotide within this short stretch of sequence. Thus, this assay is reminiscent of the RAPD assay (see above), in that sequences contained on an arbitrary amplification primer are used to select an arbitrary subset of amplification targets. Out of the entire population of restriction fragments derived from a given genome, only those with ends homologous to the primers' arbitrary 3′-sequences will be amplified.

The number and quality of nucleotides in the variable 3′-ends of the primers influence the multiplex ratio for the assay. Specifically, the degree of selectivity in the amplification reaction can be estimated using the formula 4^{2n}, where n = the number of selective bases. Pairs of primers having no selective nucleotides (i.e., primers corresponding to the adaptor and restriction site only) will synchronously coamplify every restriction fragment in the population. For the soybean genome (2C ~2 × 10^9 bp; Arumuganathan and Earle, 1991), for example, primers with $n = 0$ will coam-

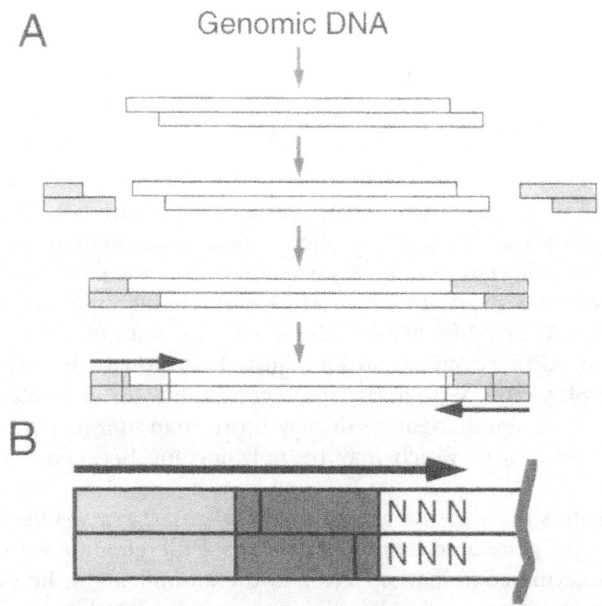

Figure 4.6 Schematic representation of the AFLP process. (A) Genomic DNA is digested with two restriction endonucleases, double-stranded adaptors are attached to the ends of every restriction fragment by ligation, and the resulting DNA fragments are PCR-amplified using primers corresponding to the synthetic adaptors but whose 3′-ends also extend into genomic fragment itself. (B) Each AFLP primer has a tripartite structure. The primer's 5′-end corresponds to the synthetic adaptor (light gray), its middle portion carries the restriction endonuclease recognition site (dark gray), and the 3′-end carries a short stretch of aribitrary, nondegenerate sequence that anneals to the first few nucleotides (NNN) of the genomic restriction fragment.

plify each and every one of the estimated $(1/4096)(2 \times 10^9) = 4.9 \times 10^5$ restriction fragments produced by a single enzyme with a 6-bp recognition site. One selective nucleotide on each primer will reduce the complexity by 1/16 (3.05×10^4 coamplified fragments), two selective bases will decrease the number by 1/256 (1.9×10^3 fragments), and three selective bases on both primers will reduce the complexity by 1/4096, to an estimated 120 fragments.[7]

[7]This simplified calculation does not take into account the base composition of the genome and of the recognition sites of the restriction endonucleases used to produce genomic fragments. These effects can be quite large. For example, if the genome is 60% GC and the restriction endonuclease recognizes TTTAAA (*Dra* I), the sites will occur on average every $1/(0.2)^6 = 15625$ bp, as compared to $1/(0.25)^6 = 4096$ bp for a 50% GC genome.

1. Experimental Method: Materials

Genomic DNA samples, purified away from most contaminants; requirement for 0.2–2.5 μg per template DNA preparation

10× Restriction-ligation (RL) buffer (compatible with
most restriction endonucleases and ligases)

10× concentration	To prepare 10 ml
100 mM Tris-acetate, pH 7.5	1 ml 1 M Tris-acetate, pH 7.5
100 mM Mg-acetate	1 ml 1 M Mg-acetate
100 mM potassium-acetate	1 ml 1 M potassium-acetate
50 mM dithiothreitol	0.5 ml 1 M dithiothreitol
	Sterile, ultrapure H_2O to 10 ml

Use at a 1× final concentration

Restriction endonucleases
    Available from several commercial suppliers
    Concentrations range from 1 to 50 Units/μl
    Should not be methylation-sensitive
    Use at 5 units/μg genomic DNA
T4 DNA ligase
    Available from several commercial suppliers
    Concentrations range from 1 Weiss unit/μl to 20 units/μl
    Use at 1 unit/reaction
ATP, 10 mM, pH 7 (made in sterile, ultrapure H_2O)
Dynabead M-280 Streptavidin paramagnetic beads (Dynal, Inc., Great Neck, NY)
STEX buffer

Final concentration	To prepare 1 liter
100 mM NaCl	25 ml 4 M stock
10 mM Tris-Cl, pH 8.0	10 ml 1 M stock
1 mM EDTA	2 ml 0.5 M stock
0.1% Triton ×-100	10 ml 10% stock
	Sterile, ultrapure H_2O to 1 liter

$TE_{0.1}$ buffer: For 1 liter, combine 10 ml of 1 M Tris-HCl stock (pH 8.0) with 0.05 ml 500 mM EDTA (pH 8.0) stock. Adjust volume to 1 liter with sterile, ultrapure H_2O. Autoclave.
    T4 polynucleotide kinase
        Available from several commercial suppliers
        Concentrations range from 5 units/μl to 20 units/μl
        Use 0.5 unit/reaction
    10× kinase buffer

10× concentration	To prepare 10 ml
600 mM Tris-HCl, pH 7.9	6.0 ml 1 M stock
100 mM MgCl$_2$	1.0 ml 1 M stock
150 mM dithiothreitol	1.5 ml 1 M stock
	Sterile, ultrapure H$_2$O to 10 ml

Use at a 1× final concentration.

[γ^{33}P]ATP or [γ^{32}P]ATP

Supplier	Catalog No. Concentration
Amersham	AH 9968 10mCi/ml
Du Pont NEN	NEG-302H 10mCi/ml

PCR amplification primers (see below): 50 ng/μl each, diluted in sterile, ultrapure H$_2$O

Deoxyribonucleotide (dNTP) mixture: 5 mM each of dATP, dCTP, dGTP, dTTP, pH 7.0, in sterile, ultrapure H$_2$O

AmpliTaq DNA polymerase
From Perkin–Elmer (Emeryville, CA)
Concentration: 5 units/μl
Use at 0.1 unit/μl final reaction volume
10× PCR buffer

10× concentration	To prepare 10 ml
100 mM Tris–HCl, pH 8.5	1.0 ml 1 M stock
15 mM MgCl$_2$	0.15 ml 1 M stock
500 mM KCl	5.0 ml 1 M stock
	Sterile, ultrapure H$_2$O to 10 ml

Use at a 1× final concentration

10× Tris-borate-EDTA (TBE) buffer

10× concentration	To prepare 1 liter
0.89 M Tris–HCl pH 8.3	108 g Tris base
0.89 M boric acid	55 g boric acid
0.02 M EDTA	40 ml 0.5 M EDTA, pH 8.0

Use at 1× final concentration

40% Acrylamide: N-, N-methylene bisacrylamide (19:1) stock solution (deionized, filtered). Available from several commercial suppliers (e.g., Boehringer Mannheim)

6% polyacrylamide gel (enough for one 30 × 40 × 0.04-cm gel): Combine 9 ml 40% acrylamide: (19:1) stock solution, 6.0 ml 10× TBE,

25.2 g urea (ultrapure), and 10 ml H_2O. Heat at 65°C until urea crystals are just dissolved, then bring volume to 60 ml with H_2O. Filter through Whatman 1 then degas for 5–10 min. Add 0.4 ml 10% freshly made ammonium persulfate and 35 μl TEMED. Mix gently, and pour into clamped plates (0.4-mm spacers). Allow to polymerize at least 45 min, room temperature.

The experimental protocol outlined here for the AFLP assay is derived in large part from the method described by Zabeau and Vos (1993), with only a few slight modifications. The AFLP assay requires a number of DNA manipulations, most of which can be done on large numbers of samples simultaneously. First, high-quality genomic DNA (0.2–2.5 μg per individual) is digested to completion with a single- or double-restriction enzyme combination. Although it need not be cesium chloride purified, the DNA must be of sufficient purity to allow efficient restriction; DNA that may be slightly degraded, but otherwise clean, can be used. Two methods we find suitable are the CsCl purification method (or its miniprep variation) of Murray and Thompson (1980) and the miniprep method described by Chen and Dellaporta (1994). Although any of several different restriction enzymes can be used, the following example utilizes *Taq*I and *Hind*III. Currently, many laboratories consistently use *Mse*I and *Eco*RI. All of these enzymes are methylation insensitive and seem to have sites that are well distributed in plant genomes (M. Zabeau, personal communication). Between 0.2 and 2.5 μg of high-molecular-weight genomic DNA is digested with 5 units/μg of *Taq*I in a 50-μl volume at 65°C for approximately 3 hr in $1\times$ RL buffer, then is digested further in the same buffer with 5 units/ μg of *Hind*III at 37°C for 3 hr. If using the *Mse*I + *Eco*RI combination, the two enzymes can be digested simultaneously in the same buffer at 37°C. The digestion products generated from each input genomic DNA are a mixture of symmetric fragments, in this example bordered at both ends by either *Taq*I or *Hind*III sites, and asymmetric fragments flanked by both a *Taq*I and a *Hind*III site.

Next, double-stranded adaptors are ligated to the ends of the restriction fragments. It is possible to perform the digestion and ligation reactions simultaneously if all the enzymes are compatible with a 37°C reaction temperature and if the adaptor sequences are designed with a single base alteration within the restriction site, to eliminate continual recutting of the ligated products. Whether the ligation is combined in the same step, or is done subsequent to the restriction digestion, the following is added to the 50 μl restriction mixture: 1 μl $10\times$ RL buffer, 1 μl each double-stranded adaptor (see below), 1.2 μl 10 mM ATP, 1 μl T4 DNA ligase (at 1 unit/μl), and 4.8 μl H_2O. The ligation reaction is incubated at 37°C for

3 hr. Double-stranded adaptors are generated by slowly annealing equimolar amounts of the two partially complementary single-stranded component oligonucleotides of each pair. Each adaptor contains a restriction enzyme 1/2-site that can ligate to the corresponding sites on the genomic DNA fragments. For this example, the double-stranded *Taq*I adaptor (Taq-Ad) at 50 pmol/µl is produced by combining 5000 pmol each of Taq.AdF and Taq.AdR single-stranded oligonucleotides with H_2O to a final 100-µl volume. For the 5 pmol/µl *Hind*III adaptor (biotin-Hnd-Ad), 500 pmol each of the corresponding single stranded oligonucleotides are combined in a final volume of 100 µl. To generate the double-stranded molecules, these mixtures are incubated at sequentially decreasing temperatures: 65°C for 15 min, 37°C for 15 min, room temperature for 15 min, then 4°C. For this example, appropriate adaptors are:

```
                                                         AAGCTT
  biotin-Hnd-Ad  Hnd-Ad-F  5'-biotin-CTCGTAGACTGCGTACC-3'
                 Hnd-Ad-R              3'-CTGACGCATGGTCGA-5'

                                                    TCGA
  Taq-Ad         Taq-Ad-F            5'-GACGATGAGTCCTGAC-3'
                 Taq-Ad-R            3'-TACTCAGGACTGGC-5'
```

A nonbiotinylated version of the Hnd-Ad-F oligonucleotide may be used if biotin selection is not done in the subsequent step; the alternative preamplification method does not require a biotinylated adaptor (see below). For best results, these constituent oligonucleotides are all nonphosphorylated to prevent adaptor-to-adaptor ligation. The overhanging ends for the restriction sites are intact, but a single-base site mutation elsewhere in the site prevents recutting at the site after ligation. The forward oligonucleotide in the adaptor for the 6-bp cutter can contain a biotin moiety at its 5', nonligatable end. This biotinylated adaptor is required for performing one of two different methods for reducing the complexity of the template DNA mixture in preparation for the PCR amplification. The biotin-requiring method involves biotin–streptavidin selection of only those genomic fragments bordered at one end with the 6-bp site adaptor. After ligation, fragments carrying this adaptor (fragments bordered by either two *Hind*III sites or one *Hind*III site and one *Taq*I site) are selected from the total pool of ligation products using streptavidin-coated paramagnetic beads. All fragments bordered by two *Taq*I sites, therefore, are discarded. Briefly, to each ligation is added 150 µl of a suspension containing 10 µl paramagnetic beads in STEX buffer. Following gentle mixing for 30–60 min at room temperature to allow streptavidin–biotin binding, the beads are purified using a magnetic rack, and then washed extensively with the same buffer. After five successive washes, the beads are resuspended in a volume of $TE_{0.1}$ corresponding to 1 µl/10 ng genomic DNA initially used for the digestion. This selected subset of genomic restriction fragments, known as template-DNA, is used directly (with beads present) in the subsequent amplification.

The second method for reducing the complexity of the template DNA mixture prior to PCR amplification is to perform a selective preamplification. This preamplification is performed "cold," using unlabeled primers that carry a low level of selectivity. There are four possible +1 preselective primers for each restriction site adaptor; the four primers each carry sequence corresponding to the adaptor/reconstructed restriction site plus a single arbitrary nucleotide at the 3'-end. For the *Hind*III and *Taq*I adaptors in this example, appropriate +1 primers would be, respectively:

5'-GACTGCGTACCAGCTTN-3'
5'-GATGAGTCCTGACCGAN-3'

where N is any base. In a single preamplification reaction, any one of the four +1 selective primers corresponding to the 6-bp site is combined with any one of the four +1 selective primers for the 4-bp site, making for 4 × 4 = 16 possible amplifications for each original DNA sample. Each of the preamplification product submixtures, then, contains a unique subset of the genomic fragments that can serve as template for the subsequent labeled AFLP reaction that uses primers with two or three selective nucleotides. The DNA fragments in each subset should be mutually exclusive of the fragments amplified in the other subsets. Surprisingly, this amplification selects against "symmetrical" restriction fragments (i.e., those carrying the same restriction site at each end), and effectively enriches the asymmetric fragments, bordered at both ends by different sites. The preamplifications are performed with the following conditions: 5 µl of a 1:10 dilution (in $TE_{0.1}$) of the restriction–ligation mixture is combined with 5 µl of 10× PCR buffer, 1 µl of dNTP mix, 2 µl of each +1 preselective primer (each at 50 ng/µl), 35 µl H_2O, and 0.2 µl AmpliTaq polymerase (5 U/µl). Thermocycle conditions for this preamplification requires 20 cycles of 94°C for 30 sec, 56°C for 60 sec, 72°C for 60 sec. The products of each preamplification are diluted 1:50 in $TE_{0.1}$ and then used directly as substrate in the AFLP PCR reaction.

The PCR amplification employs pairs of adaptor/enzyme site-specific oligonucleotide primers, typically 17–18-mers. One of each pair, the one corresponding to the 4-bp site (*Taq*I in this example), is first 5'-end-labeled with either ^{32}P or [33]P. Labeled amplification products will be generated, and detected, only from the asymmetrical template-DNA fragments con-

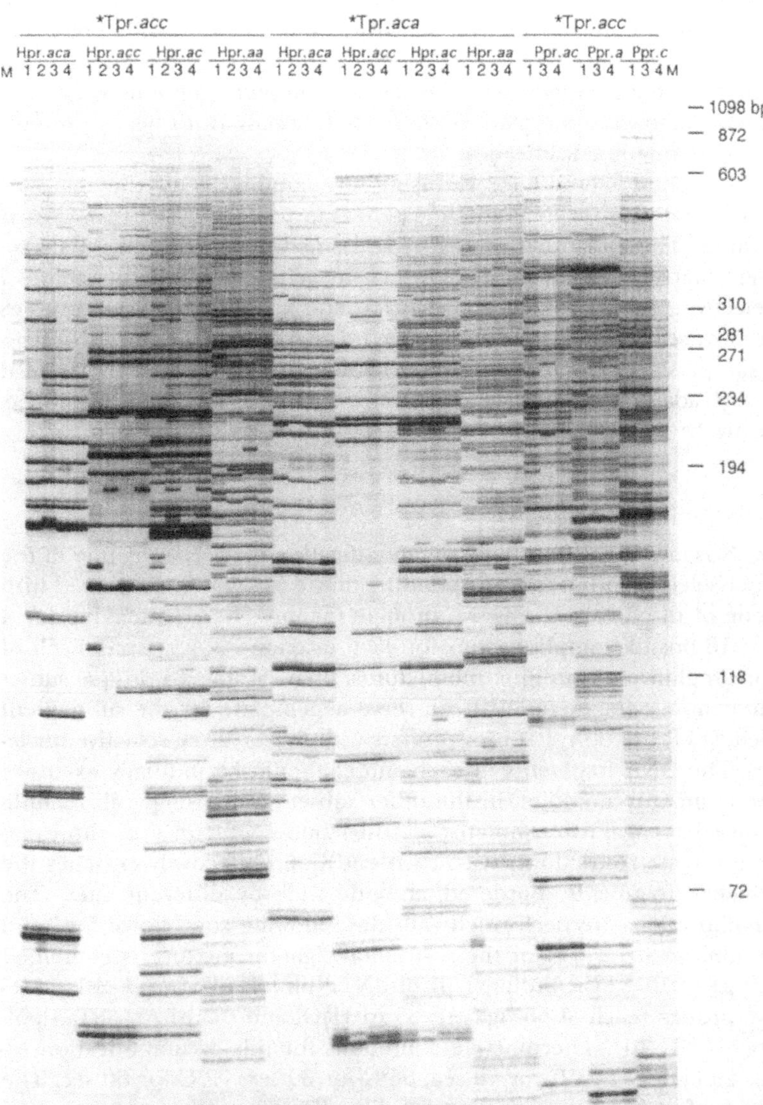

Figure 4.7 Amplified fragment length polymorphism analysis of four diverse *Glycine* cultivars, *G. soja* PI81.762, *G. max* N85-2176, *G. max* Noir-1, and *G. max* Wolverine (lanes 1, 2, 3, 4, respectively). Template DNAs were prepared as described in the text, using *Hind*III + *Taq*I or *Pst*I + *Taq*I restriction enzymes in combination with the appropriate adaptors (see text). PCR amplification from each template DNA was performed with the indicated combinations of primers. In all cases, only the *Taq*I site-specific primer is radiolabeled at its 5'-end with [33]P (indicated by *). The identity of the 3'-variable nucleotide extension (e.g., *acc*) on each PCR

taining both a 6-bp and a 4-bp site. To label enough primer for 30 amplification reactions, 150 ng primer oligonucleotide (~25 pmol of a 17–18mer) is combined with 50 μCi[γ-^{33}P]ATP (~17 pmol) and 5 units T4 polynucleotide kinase in 1$\times$ kinase buffer in a 30-μl reaction volume, and incubated at 37°C for 30–60 min.

The key to this method lies in the design of the primer sequences and in the choice of combinatorial pairs of the different primers for a given amplification. The target site of each primer is anchored at the 5'-end by the synthetic adaptor and restriction site, common on all fragments in the biotin-selected fragment pool. However, the 3'-most ends of these primers are designed to be variable in both length and sequence, such that an individual primer may anneal to only the subset of restriction fragments that carry the precise, but randomly chosen, sequence immediately internal to the restriction site. The length of the variable 3'-sequence on the primers influences the multiplex ratio of the amplification reaction; each primer combination coamplifies a different subset of the total population of template-DNA fragments. Typically, these primers carry two or three such 3' selective nucleotides. Lengths of 3' sequence greater than three nucleotides are no longer completely selective, and thus should not be used. Suitable primers corresponding to the *Hind*III and *Taq*I ends are described in the legend to Fig. 4.7. This is not a comprehensive list. Eight combinations of primer pairs are possible from this list, and each combination will amplify a different subset of template-DNA fragments, illustrated in Fig. 4.7. In general, if 3 selective nucleotides are used on each adaptor-directed primer, then $(4)^3 \times (4)^3 = 4096$ different individual primer combinations can be used against any particular template preparation. In reality, only a much more manageable subset of these combinations is appropriate for each target genome (M. Zabeau, personal communication).

Each AFLP amplification reaction is set up at room temperature. When several template-DNAs are to be amplified with the same primer set, a cocktail of the common reaction components can be made. These can be done as standard "cold-start" reactions, providing the first dena-

---

primer is indicated. For any combination of PCR primers, differentially amplified products likely to result from both dominant and codominant polymorphisms among the four soybean lines are visible. The primers used carry 1, 2, or 3 selective 3'-nucleotides: *Hpr.aa*, 5'-CTGCGTACCAGCTTaa-3'; *Hpr.ac*, 5'-CTGCGTACCAGCTTac-3'; *Hpr.acc*, 5'-CTGCGTAC-CAGCTTacc-3'; *Hpr.aca*, 5'-CTGCGTACCAGCTTaca-3'; *Ppr.ac*, 5'-GACTGCGTACATGCAG-ac-3'; *Ppr.a*, 5'-GACTGCGTACATGCAGa-3'; *Ppr.c*, 5'-GACTGCGTACATGCAGc3'; *Tpr.acc*, 5'; -TGAGTCCTGACCGAacc-3'; *Tpr.aca*, TGAGTCCTACCGAacc-3'. The 5'-portion of each corresponds to adaptor sequence, the underlined portion to the restriction site (with the altered base in bold), and the lowercase portion to the 3'-selective extension.

turation is performed as soon as possible after reaction set-up. If a hot-start technique (Erlich *et al.*, 1991) is used, it is important to allow the template-DNA to incubate for a few minutes in the presence of buffer, dNTPs, and polymerase, before the first denaturation. For each amplification reaction (20 μl volume), 2 μl of the appropriate template DNA is combined with 1 μl of ^{33}P- or ^{32}P-labeled primer (5 ng) and 0.5 μl of a 50 ng/μl stock (25 ng) of the unlabeled version of the same primer corresponding to the 4-bp restriction site adaptor. The template DNA is either from a biotin-selected mixture, or from a 1:10–1:50 dilution of a preamplified reaction mixture. Also added are 0.6 μl of a 50 ng/μl stock of unlabeled primer corresponding to the 6-bp site adaptor, 2 μl 10× PCR buffer, 0.8 μl 5 m*M* dNTP mixture, 0.5 units of Amplitaq DNA polymerase, and 13 μl H$_2$O. PCR amplification is initiated immediately.

To ensure template discrimination through the selective 3'-ends of the primers, it is important to maintain stringent annealing conditions for the amplification reaction. Thus, these amplifications are best performed using "touchdown" conditions (Don *et al.*, 1991). This amplification scheme, which begins cycling with high annealing temperatures with the subsequent cycles annealing at incrementally lower temperatures, provides good specificity in the earliest amplifications when inefficient template discrimination can lead to erroneous results. Because of the precise temperature shifts required, AFLP amplifications are best achieved using a thermocycler capable of rapid and extremely accurate temperature equilibration. For the PCR primers listed above, all with calculated T_m (in 50 m*M* Na) of 40–45°C, the following amplification conditions are used with a Perkin–Elmer 9600 thermocycler; if a different thermocycler model is used, the times for each programmed step likely need to be optimized and any variation in sample well temperatures and well-to-well temperature variation within the block must be addressed.

Denature	94.0°C	30 sec	
Anneal	65.0°C	30 sec	
Extend	72.0°C	60 sec	1 cycle
Denature	94.0°C	30 sec	
Anneal	64.3°C	30 sec	
and decrease by 0.7°C per cycle			
Extend	72.0°C	60 sec	11 cycles
Denature	94.0°C	30 sec	
Anneal	56.0°C	30 sec	
Extend	72.0°C	60 sec	20–25 cycles

Although the amplification products can be viewed on agarose gels, the resolution is poor and it is difficult to distinguish polymorphic bands, especially in a highly multiplexed assay. Instead, the radioactively tagged, polymorphic fragments are best visualized by autoradiographic detection from high-resolution denaturing polyacrylamide gels. An equal volume (20 µl) of formamide loading buffer containing bromophenol blue and xylene cyanol dyes is added to each completed reaction, the samples are heated to 94°C for 3 min, and are then loaded onto a 4–6% polyacrylamide/7 M urea/1X TBE gel (40 cm). After electrophoresis at constant power setting of 60–70 W for 1.5–2 hr in 1X TBE, the gel is dried onto a filter paper support, and then exposed to X-ray film at −80°C. The exact length of the electrophoretic run depends on the range of products to be resolved, but usually the bromophenol blue dye should migrate to the very bottom of the gel. Figure 4.7 shows a typical autoradiographic image of a gel containing AFLP products of both *Hind*III + *Taq*I and *Pst*I + *Taq*I reactions.

This protocol has outlined the method for performing the AFLP assay using radioisotopic detection on high-resolution manual polyacrylamide gels. Very recently, a reagent kit has become available (BRL Life Technologies) which utilizes the *Eco*RI/*Mse*I restriction enzyme combination and radioisotopic detection of the amplification products. This kit may be ideal for investigators new to the AFLP method. Currently, there are two alternatives to the use of radioisotopes. The first is the detection of unlabeled AFLP products on polyacrylamide gels by silver staining (T. Vantoai, personal communication). The second alternative now available is to use fluorescent detection of the amplification products, currently best carried out using a fluorescent AFLP kit now available commercially (Applied Biosystems, Inc.) and an automated sequencer.

2. Advantages and Other Considerations

Although the AFLP assay is potentially very informative for genome analysis and offers the opportunity for a high multiplex ratio, this method involves multiple steps, requires a fair amount of technical skill, and is somewhat more cumbersome and laborious than other assays that have been discussed in this chapter. In addition, several cautions are in order. First, the AFLP assay requires a fairly large amount of genomic DNA, although a single batch of biotin-selected template-DNA can serve for 10–100 individual PCR amplifications and preamplified template DNA can be perpetually generated from a single batch of starting material. In addition, non-Mendelian polymorphisms, resulting for example from incomplete digestion or DNA base methylation, can lead to artifactual polymorphic bands between genomes. Therefore, the genomic DNAs used for AFLP assays must be of the highest quality, free of all restriction enzyme-inhibiting contaminants, and the enzymes chosen should not be methylation-sensitive. Despite all these precautions, however, non-Mendelian polymorphisms still may be produced. The best precaution against becoming fooled by such artifactual polymorphism is to always perform duplicate amplifications on every template and to make templates from duplicate DNA preparations for each genome to be tested. This assay is best performed with radioactively labeled primers, which requires special laboratory procedures. However, it is possible that a safer, nonradioactive version of the assay using fluorescently labeled primers or silver staining can be devised. An agarose gel and ethidium bromide or pico-green detection can be used for reaction products having a low multiplex ratio and for which the polymorphisms being tracked are sufficiently different in size. Finally, the cost of this assay is relatively high, with a rough estimate of consumable costs at ~$1–2 per individual amplification reaction. Much of this cost stems from the expense of the radioisotope and the synthetic oligonucleotides.

Despite all these cautions, the AFLP assay is attractive for plant genome analysis for several reasons. First, AFLP assays are expected to reveal relatively high levels of polymorphism between genomes, higher than those observed with standard RFLP analysis, because of the multiplicative influence of moderately polymorphic restriction sites and the likelihood for random sequence differences adjacent to these sites. These polymorphisms may be codominant, because of fragment length differences, or they may be dominant, due to the presence or absence of individual restriction sites between genomes. Both types of polymorphism are detectable; however, the allelic relationships between codominant polymorphic bands may not be immediately obvious. Second, the AFLP reactions can be easily adjusted to identify from just a few loci to several hundred or more. This is accomplished largely by differentially choosing restriction enzymes with varying site frequencies in the genome, and by varying the length of the variable nucleotide extension at the 3′-end of the primers. At a low multiplex ratio, single loci can be tracked within a population; the DNA sequences at the ends of a single amplified fragment can be deduced by cataloging the successful amplifications from a collection of PCR primers differing only at their 3′ nucleotide(s). On the other hand, the high multiplex ratios achievable with AFLP offer a great advantage for identifying and tracking large numbers of loci, as well as for DNA fingerprinting and for whole-genome comparisons, such as for recurrent parent analysis. Third, although the AFLP assay relies on discrimination among potential target sites via the selective 3′-ends on the primers, it differs from allele-specific PCR in that AFLP requires no prior knowledge about allele-specific sequences.

Certainly, one drawback to AFLP is the inability to immediately discern the allelic relationships among the polymorphic fragments observed between individual genomes. However, the DNA sequence of these amplified fragments can be determined in two ways. The amplifications can be performed with successive primers, each with known, but differing, 3′-end sequences, to reveal, one nucleotide at a time, the DNA sequence at the end of the fragment. Alternatively, a fragment of interest can be isolated from the gel and its DNA sequence determined by standard methods, from which locus-specific primers then can be designed. Thus, it is possible to convert the AFLP method into a sequence-based assay, allowing the tracking of individual polymorphic loci. Any AFLP band, whether from a highly multiplexed reaction or from a reaction amplifying just one band, can be mapped by performing the AFLP reactions on individuals from a mapping population. These bands (loci) localize to seemingly random sites in the genome. In addition, it is possible to identify and isolate polymorphic AFLP bands that cosegregate with a phenotypic trait of interest. This can be done utilizing a bulked segregant approach or by examining each individual in a population. If the gene conferring the trait is not yet cloned, these tightly linked AFLP markers may help to facilitate gene isolation.

B. Interrepeat Amplifications

Simultaneous analysis of DNA sequence variation at multiple loci increases the information content of any genome assay. Conserved, highly repetitive sequences in the genome can be used as markers for such a multiplexed assay. Oligonucleotide primers that correspond to known, repeated sequences can be used singly or in various combinations in a PCR reaction to amplify the single-copy DNA flanked by these repeats (Fig. 4.8) It is likely that such a strategy will reveal polymorphisms resulting both from length variation between conserved repeat sites (codominant loci) and from individual repeat sites whose presence or absence differs (dominant

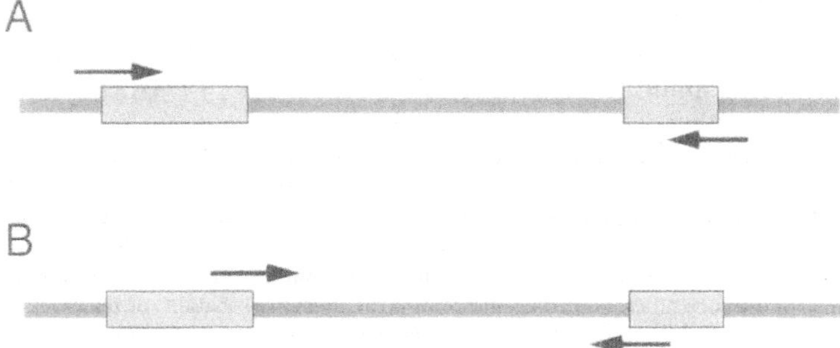

Figure 4.8 Arrangement of PCR primers for Inter-SSR Amplification (ISA). Gray rectangles represent copies of a dispersed repeat, for example SSRs. (A) 5′-Anchored primers, (B) 3′-anchored primers.

loci) between genomes. Ideally, these repetitive sequences are randomly dispersed along all chromosomes, allowing a truly random assessment of the genome. In addition, it is important that the number of repeats be great enough to ensure a high probability that pairs of repeats will lie within an amplifiable distance (50–5000 bp) of one another. As an example, a hypothetical repeat sequence present in 20,000 copies in the soybean genome (2C, $\sim 2 \times 10^9$ bp) (Arumuganathan and Earle, 1991) will, on average, reside within 2 kb of another member of the same repeat family at approximately 400 positions in the genome.[8] In the larger maize genome (2C, $\sim 4.5 \times 10^9$ bp), such closely spaced repeats will occur only 200 times. Nevertheless, both genomes have the potential for the simultaneous coamplification of a large number of individual genetic loci. This strategy has been exploited to great advantage in mammalian genomes using the high copy number ($\sim 500,000$) and well-conserved Alu-type interspersed repeats (Sinnett *et al.*, 1990; Zietkiewicz *et al.*, 1992).

Repeat families in plants are now just beginning to become well characterized. In some cases, such as for rDNA (Flavell, 1980) and for telomeric satellite sequences (Ganal *et al.*, 1988; Wu and Tanksley, 1993b), these repeats are not well dispersed, and therefore are not good candidates for whole-genome analysis. To date, no dispersed repeat analogous to the Alu family of repeats of mammals has yet been identified in plant genomes. Currently, the best candidate for interrepeat analysis in plants is the simple sequence repeat (see preceding section). In the human genome, $(CA)_n$ repeats (where $n \geq 10$) are estimated to be present at a haploid copy number of $\sim 50,000–100,000$ (Weber and May, 1989). Such copy number estimates are not so well established in plants (Lagercrantz *et al.*, 1993). Nevertheless, it is likely that $(AT)_n$, $(GA)_n$, and $(CA)_n$ repeats are well represented in plant genomes, including soybean, maize, rice, and *Arabidopsis* (Akkaya *et al.*, 1992; Morgante and Olivieri, 1993; Wu and Tanksley, 1993a).

SSR sequences in plants have been used successfully as target sites for interrepeat PCR amplifications (Fig. 4.9, Zietkiewicz *et al.*, 1994). Discrete,

[8]The likely number of repeats occuring within a certain distance of each other in the genome can be estimated as a function of the repeat number, genome size, and required minimum and maximum distance between adjacent repeats, based on the following assumptions (M. Hanafey, private communication): If N events occur randomly and independently along a line of length S, the number of events in any subinterval has a Poisson distribution. The probability of distance X between adjacent events is the Poisson probability of zero events in an interval, which is $e^{-X*N/S}$. The probability of distances less than or equal to X is obtained by integrating $e^{-X*N/S}$ with respect to X, and normalizing the result so that in the limit as X approaches infinity, the integral approaches 1. The result is $(1-e^{-X*N/S})$. The probability of a distance falling between two values X_1 and X_2 (where $X_2 > X_1$) is just $(1 - e^{-X_2^*N/S}) - (1 - e^{-X_1^*N/S})$. The expected number of segments with a distance between X_1 and X_2 is $(e^{-X_1^*N/S} - e^{-X_2^*N/S}) * (N - 1)$.

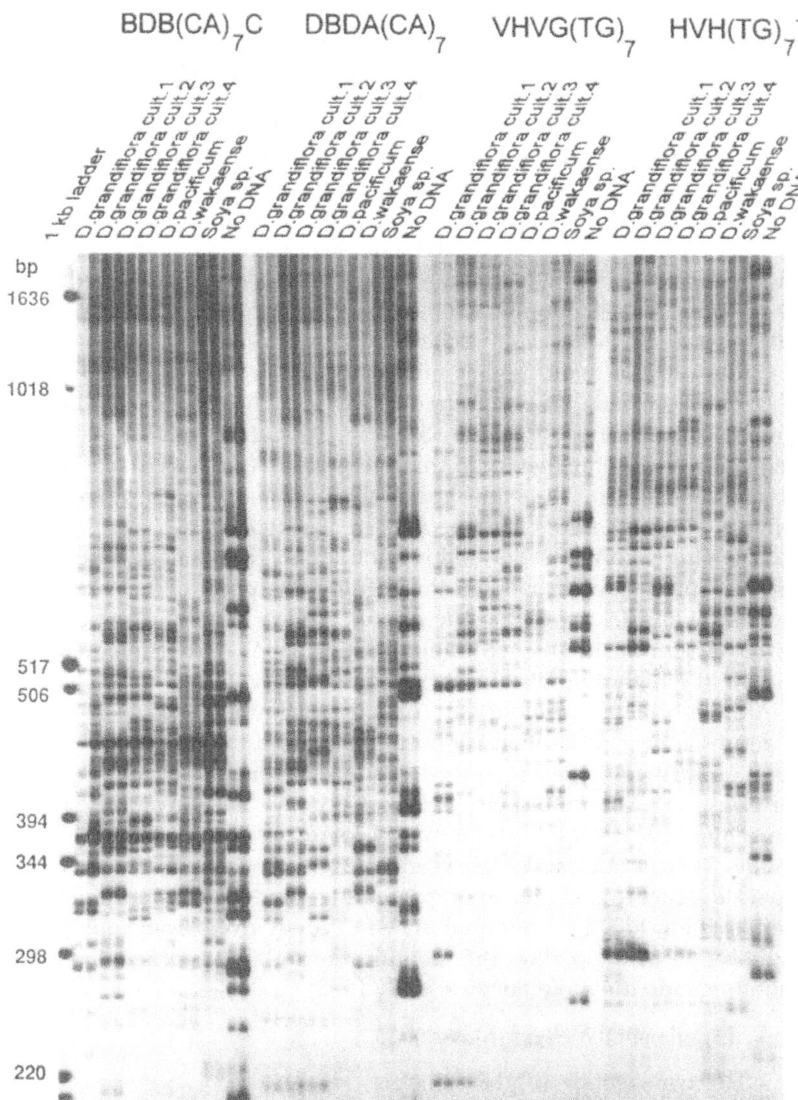

Figure 4.9 Example of ISA: amplification of inter-SSR regions from several species and cultivars of chrysanthemum and soybean *(Glycine max)* cv. Bonus using 5′-anchored simple sequence repeat-specific primers indicated. Each amplification was performed in duplicate to demonstrate reproducibility. For details see text and Zietkiewicz *et al.* (1994). Figure courtesy of E. Zietkiewicz and K. Wolff.

resolvable amplification products are produced only when two target sequences are spaced relatively closely (<5 kb) in the genome. For this strategy to be successful, it is important to "anchor" the SSR-specific PCR primer with unique sequences flanking one side of the SSR (Fig. 4.8), since a nonanchored primer will anneal at random positions within the repeating units at a target locus and produce falsely heterogeneous amplification products. A $3'$-anchored SSR primer, for example, might contain 7–9 tandemly repeating units at its $5'$-end, followed by 2–4 bp of random sequence at the extreme $3'$ priming end; this random sequence either can be a single, nondegenerate sequence, or can harbor some level of degeneracy. In either case, such a primer would be expected to anneal to a subset of the total number of SSR loci in the genome, and prime only from those loci that match the random sequence at the $3'$-end of the primer. Similarly, a $5'$-anchored SSR primer would contain random sequences at its extreme $5'$-end, so that the primer is anchored to a unique flanking sequence on the end opposite that priming the polymerization. Unlike the $3'$-anchored situation, allelic polymorphisms in the products of a $5'$-anchored amplification will result not only from length differences in the unique sequence between the SSR locus target sites, but also from length variation in the SSR sequences themselves. Therefore, the $5'$-anchored version of the inter-SSR amplification (ISA) assay is predicted to generate the highest levels of polymorphism.

Although these techniques are now just beginning to be performed, the results appear promising. High multiplex ratios can be obtained, and alteration of the anchor sequence leads to different subsets of amplification products. As more dispersed repeats become better characterized in plants, interrepeat amplifications may become more generalized for plant genome analysis. Certainly, segregation analysis of amplified, randomly dispersed repetitive sequences representing random polymorphic loci, whether these loci are identified via AFLP or derived from interrepeat amplifications, should allow the identification of DNA markers linked to desired genetic traits (Zietkiewicz *et al.*, 1992).

1. Experimental Method: Materials

Deoxynucleotide (dNTP) mixture
5 mM each of dATP, dCTP, dGTP, dTTP, pH 7.0 in sterile, ultrapure H_2O
AmpliTaq DNA polymerase
From Perkin–Elmer/Hoffman La Roche
Concentration: 5 units/μl
Use at 0.1 unit/μl final reaction volume
10$\times$ PCR buffer

10× concentration	To prepare 10 ml
100 mM Tris-HCl, pH 8.5	1.0 ml 1 M stock
15 mM MgCl$_2$	0.15 ml 1 M stock
500 mM KCl	5.0 ml 1 M stock
	Sterile, ultrapure H$_2$O to 10 ml

Use at a 1× final concentration

The primers used by Zietkiewicz *et al.* (1994) should provide a good starting point for interrepeat amplification experiments. These degenerate primers are: (CA)$_8$RY, (CA)$_8$RG, (CA)$_7$RTCY, BDB(CA)$_7$C, DBDA(CA)$_7$, VHVG(TG)$_7$, HVH(TG)$_7$T, where R is a purine, Y is pyrimidine, B is every base except A, D is every base except C, H is every base except G, and V is every base except T. These primers were designed to provide 5′-end or 3′-end anchoring of the primer at the junction of the repeated sequence with surrounding nonrepeat sequences (Fig. 4.8). It is likely that the 5′-anchored primers also provide some undesirable unanchored priming throughout the SSR repeats (J. Vogel, unpublished observations). Therefore, refinement of the primer design for better anchoring is likely to further improve the results. The protocol shown here follows the experimental conditions of Zietkiewicz *et al.* (1994). The primers are ^{32}P-labeled using polynucleotide kinase and high specific activity [γ-^{32}P]ATP. Amplification from 25 ng of high-molecular-weight genomic DNA is performed in 1× PCR buffer, 0.01% gelatin, 0.01% Triton X-100, 0.2 mM dNTPs, 2% formamide (deionized), and 1 μM primer (diluted 1:4 with unlabeled primer), in 20 μl total volume. The reaction mixture is denatured in the thermocycler for 7 min at 94°C and rapidly cooled to 4°C, and 1 unit AmpliTaq DNA polymerase is added. Twenty-eight cycles of 30 sec at 94°C, 45 sec at 52°C, 120 sec at 72°C are performed in a thermocycler (e.g., Perkin–Elmer 9600), followed by final extension at 72°C for 7 min. The reaction products are mixed with sucrose-containing loading buffer and separated without prior denaturation on a 50-cm-long, 0.4 mm 6% acrylamide-N,N′-methylenebisacrylamide (30:1) gels containing 3 M urea and 1x TBE buffer. The electrophoresis is performed at room temperature at 14V/cm for 10–13 hr. The gels are vacuum-dried and autoradiographed using standard conditions.

X. Choosing Appropriate Technology

Which of the multitude of DNA polymorphism analysis techniques should one choose? The answer depends first of all on the nature of the project and the needs dictated by the specific application, and also on the facilities and skills that are available. Simple Sequence Repeat markers are codominant and have the highest information content of all marker types. The very high cost of their development, however, may not be justified for many laboratories. In some species, such as *Arabidopsis* and soybean, enough primer sequences soon may become available to satisfy most needs. We expect that in the next few years such primer sets will become available for the major experimental organisms. The inter-SSR amplification technique allows the presence and degree of polymorphism of various SSRs in any genome to be assessed without any prior development (i.e., SSR locus characterization) costs. If the objective is to characterize (fingerprint) genomes or to perform a study of genetic diversity in a population, then multiplex methods such as interrepeat amplification, AFLP, or multiplexed RAPDs are appropriate. AFLP also allows for the extremely rapid identification of new markers (primarily dominant), either spread throughout the whole genome or confined to specific genetic regions. RAPD technology provides a very easy entry into mapping, even for a worker with no experience in molecular methods. Consequently RAPD analysis is frequently chosen by those beginning to study a novel genetic system, but having minimal amounts of DNA or other resources at their disposal. Although the reproducibility of RAPD patterns between laboratories remains a concern, appropriate precautions and care with the assay will eliminate most problems (see discussion above). The dominant nature of RAPD markers also may limit the utility of RAPDs for some mapping needs. RFLP technology offers codominant markers, and is easy and convenient if existing collections of RFLP probes are available. Concerns regarding the use of radioactive materials can be avoided if nonradioactive detection systems (e.g., chemiluminescence) are used. RFLP analysis is also the fastest way to place cloned genes or other genomic fragments on an existing map. This is now straightforward in maize, where standard mapping populations exist, and a common mapping data set is maintained (Burr *et al.*, 1988). Once informative markers have been identified, and sequence information on individual alleles obtained, one of the many sequence-based polymorphism assays can be chosen to devise a diagnostic test that can be used for the large-scale population analyses required in many plant breeding applications (Rafalski and Tingey, 1993b).

Acknowledgments

We thank our colleagues at DuPont and in the genome mapping community for their contributions to the methods and ideas discussed in this chapter. We thank Barbara Mazur for helping improve the manuscript.

References

Akkaya, M. S., Bhagwat, A. A., and Cregan, P. B. (1992). Length polymorphism of simple sequence repeat DNA in soybean. *Genetics* **132**(4), 1131–1139.

Allard, R. W. (1956). Formulas and tables to facilitate the calculation of recombination values in heredity. *Hilgardia* **24**, 235–278.

Anonymous (1992). "The Genius System User's Guide for Filter Hybridization." Boehringer-Mannheim, IN.

Arnheim, N., Strange, C., and Erlich, H. (1985). Use of pooled DNA samples to detect linkage disequilibrium of polymorphic restriction fragments and human disease: Studies of the *HLA* class II loci. *Proc. Natl. Acad. Sci. U.S.A.* **82**, 6970–6974.

Arumuganathan, K., and Earle, E. D. (1991). Nuclear DNA content of some important plant species. *Plant Mol. Biol. Rep.* **9**(3), 208–218.

Ausubel, F. M., Brent, R., Kingston, R. E., Moore, D. D., Seidman, J. G., Jmith, J. A., and Struhl, K. (1987–1993). "Current Protocols in Molecular Biology," Vols. 1 and 2. Greene/J. Wiley,

Bárány, F. (1991). Genetic disease detection and DNA amplification using cloned thermostable ligase. *Proc. Natl. Acad. Sci. U.S.A.* **88**, 189–193.

Barua, U. M., Chalmers, K. J., Hackett, C. A., Thomas, W. T. B., Powell, W., and Waugh, R. (1993). Identification of RAPD markers linked to *Rhynchosporium secalis* resistance locus in barley using near isogenic lines and bulked segregant analysis. *Heredity* **71**, 177–184.

Beckmann, J. S., and Soller, M. (1983). Restriction fragment length polymorphisms in genetic improvement: Methodologies, mapping and costs. *Theor. Appl. Genet.* **67**, 35–43.

Beckmann, J. S., and Weber, J. L. (1992). Survey of human and rat microsatellites. *Genomics* **12**, 627–631.

Beier, D. R., Dushkin, H., and Sussman, D. J. (1992). Mapping genes in the mouse using single-strand conformation polymorphism analysis of recombinant inbred strains and interspecific crosses. *Proc. Natl. Acad. Sci. U.S.A.* **89**, 9102–9106.

Bell, C. J., and Ecker, J. R. (1994). Assignment of thirty microsatellite loci to the linkage map of *Arabidopsis*. *Genomics* **19**, 137–144.

Botstein, D., White, R. L., Skolnick, M. H., and Davis, R. W. (1980). Construction of a genetic map in man using restriction fragment length polymorphisms. *Am. J. Hum. Genet.* **32**, 314–331.

Burr, B., Burr, F. A., Thompson, K. H., Albertsen, M. C., and Stuber, C. (1988). Gene mapping with recombinant inbreds in maize. *Genetics* **118**, 519–526.

Caetano-Anolles, G. (1993). Amplifying DNA with arbitrary oligonucleotide primers. *PCR Methods Appl.* **3**(2), 85–94.

Caetano-Anolles, G., Bassam, B. J., and Gresshoff, P. M. (1991). High resolution DNA amplification fingerprinting using very short arbitrary oligonucleotide primers. *Bio/Technology* **9**, 553–557.

Carland, F. M., and Staskawicz, B. J. (1993). Genetic characterization of the *pto* locus of tomato—semi-dominance and co-segregation of resistance to *Pseudomonas syringae* pathovar tomato and sensitivity to the insecticide Fenthion. *Mol. Gen. Genet.* **239**, 17–27.

Chalmers, K. J., Barua, U. M., Hackett, C. A., Thomas, W. T. B., Waugh, R., and Powell, W. (1993). Identification of RAPD markers linked to genetic factors controlling the milling energy requirement of barley. *Theor. Appl. Genet.* **87**, 314–320.

Chen, J., and Dellaporta, S. (1994). Urea-based plant DNA miniprep. *In* "Maize Handbook" (M. Freeling and V. Walbot, eds.), pp. 526–527. Springer-Verlag, New York.

Dellaporta, S. L., Wood, J., and Hicks, J. B. (1985). Maize DNA miniprep. *In* "Molecular Biology of Plants" (R. Malmberg, J. Messing, and I. Sussex, eds.), pp. 36–37. Cold Spring Harbor Lab. Press, Cold Spring Harbor, NY.

Dietrich, W., Katz, H., Lincoln, S. E., Shin, H.-S., Friedman, J., Dracopoli, N. C., and Lander, E. S. (1992). A genetic map of the mouse suitable for typing intraspecific crosses. *Genetics* **131**, 423–447.

Don, R. H., Cox, P. T., Wainwright, B. J., Baker, K., and Mattick, J. S. (1991). Touchdown PCR to circumvent spurious priming during gene amplification. *Nucleic Acids Res.* **19**(14), 4008.

Erlich, H. A., Gelfand, D., and Sninsky, J. J. (1991). Recent advances in the polymerase chain reaction. *Science* **252**, 1643–1651.

Ermak, G. Z., Prosnyak, M. I., Vecher, A. A., and Kartel, N. A. (1990). AT repeats in barley genome. *FEBS Lett.* **272**(1–2), 193–196.

Feinberg, A. P., and Vogelstein, B. (1983). A technique for labelling DNA restriction endonuclease fragments to high specific activity. *Anal. Biochem.* **132**, 6–13.

Flavell, R. (1980). The molecular characterization and organization of plant chromosomal DNA sequences. *Annu. Rev. Plant Physiol.* **31**, 569–596.

Ganal, M. W., Lapitan, N. L. V., and Tanksley, S. D. (1988). A molecular and cytogenetic survey of major repeated DNA sequences in tomato *(Lycopersicon esculentum)*. *Mol. Gen. Genet.* **213**, 262–268.

Giovannoni, J. J., Wing, R. A., Ganal, M. W., and Tanksley, S. D. (1991). Isolation of molecular markers from specific chromosomal intervals using DNA pools from existing mapping populations. *Nucleic Acids Res.* **19**, 6553–6558.

Greaves, D. R., and Patient, R. K. (1985). $(AT)_n$ is an interspersed repeat in the *Xenopus* genome. *EMBO J.* **4**(10), 2617–2626.

Hadrys, H., Balick, M., and Schierwater, B. (1992). Applications of randomly amplified polymorphic DNA (RAPD) in molecular ecology. *Mol. Ecol.* **1**, 55–63.

Haley, S. D., Miklas, P. N., Stavely, J. R., Byrum, J., and Kelly, J. D. (1993). Identification of RAPD markers linked to a major rust resistance gene block in common bean. *Theor. Appl. Genet.* **86**, 505–512.

Huang, M.-M., Arnheim, N., and Goodman, M. F. (1992). Extension of base mispairs by Taq DNA polymerase: Implications for single nucleotide discrimination in PCR. *Nucleic Acids Res.* **20**(17), 4567–4573.

Innis, M. A., Gelfand, D. H., Sninsky, J. J., and White, T. J., eds. (1990). "PCR Protocols: A Guide to Methods and Applications." Academic Press, San Diego, CA.

Ito, T., Smith, C. L., and Cantor, C. R. (1992). Sequence-specific DNA purification by triplex affinty capture. *Proc. Natl. Acad. Sci. U.S.A.* **89**, 495–498.

Kandpal, R. P., Kandpal, G., and Weissman, S. M. (1994). Construction of libraries enriched for sequence repeats and jumping clones, and hybridization selection for region-specific markers. *Proc. Natl. Acad. Sci. U.S.A.* **91**, 88–92.

Karagyozov, L., Kalcheva, I., and Chapman, V. M. (1993). Construction of random small-insert genomic libraries highly enriched for simple sequence repeats. *Nucleic Acids Res.* **21**(16), 3911–3912.

Konieczny, A., and Ausubel, F. M. (1993). A procedure for mapping Arabidopsis mutations using co-dominant ecotype-specific PCR-based markers. *Plant J.* **4**, 403–410.

Kuppuswamy, M. N., Hoffman, J. W., Kasper, C. K., Spitzer, S. G., Groce, S. L., and Baja, S. P. (1991). Single nucleotide primer extension to detect genetic disease: Experimental application to hemophilia B (factor IX) and cystic fibrosis genes. *Proc. Natl. Acad. Sci. U.S.A.* **88**, 1143–1147.

Kwok, S., Kellogg, D. E., McKinney, N., Spasic, D., Goda, L., Levenson, C., and Sninsky, J. J. (1990). Effects of primer—template mismatches on the polymerase chain reaction: Human immunodeficiency virus type-1 model studies. *Nucleic Acids Res.* **18**(4), 999–1005.

Lagercrantz, U., Ellegren, H., and Andersson, L. (1993). The abundance of various polymorphic microsatellite motifs differs between plants and animals. *Nucleic Acids Res.* **21**(5), 1111–1115.

Landegren, U., Kaiser, R., Sanders, J., and Hood, L. (1988). A Ligase-mediated gene detection technique. *Science* **241**, 1077–1080.

Lander, E. S., Green, P., Abrahamson, J., Barlow, A., Daly, M. J., Lincoln, S. E., and Newburg, L. (1987). Mapmaker: An interactive computer package for constructing primary linkage maps of experimental and natural populations. *Genomics* **1**, 174–181.

Landry, B. S., Kesseli, R., Leung, H., and Michelmore, R. W. (1987). Comparison of restriction endonucleases and sources of probes for their efficiency in detecting restriction fragment length polymorphisms in lettuce (*Lactuca sative* L.). *Theor. Appl. Genet.* **74**, 646–653.

Lawler, F. C., Stoffel, S., Saiki, R. K., Chang, S.-Y., Landre, P. A., Abramson, R. D., and Gelfand, D. H. (1993). High-level expression, purification, and enzymatic characterization of fulllength thermus aquaticus DNA polymerase and a truncated form deficient in 5′ to 3′ exonuclease activity. *PCR Methods Appl.* **2**(4), 275–287.

Levinson, G., and Gutman, G. A. (1987). Slipped-strand mispairing: A major mechanism for DNA sequence evolution. *Mol. Biol. Evol.* **4**, 203–221.

Lincoln, S. E., Daly, M. J., and Lander, E. S. (1991). "Primer" software, available from E. Lander, Whitehead Institute, Cambridge, MA.

Litt, M., and Luty, J. A. (1989). A hypervariable microsatellite revealed by in vitro amplification of a dinucleotide repeat within the cardiac muscle action gene. *Am. J. Hum. Genet.* **44**, 397–401.

Litt, M., Hauge, X., and Sharma, V. (1993). Shadow bands seen when typing polymorphic dinucleotide repeats: Some causes and cures. *BioTechniques* **15**(2), 280–284.

Livak, K. J., and Hainer, J. W. (1993). A microtiter plate assay for determining apolipoprotein E genotype and discovery of a rare allele. *Hum. Mutat.* **3**, 379–385.

Lu, Z., Douthitt, M. P., Taffs, R. E., Ran, Y., Norwood, L. P., and Chumakov, K. M. (1993). Quantitative aspects of the mutant analysis by PCR and restriction enzyme cleavage (MAPREC). *PCR Methods Appl.* **3**, 176–180.

Lyall, J. E. W., Brown, G. M., Furlong, R. A., Ferguson-Smith, M. A., and Affara, N. A. (1993). A method for creating chromosome-specific plasmid libraries enriched in clones containing $(CA)_n$ microsatellite repeat sequences directly from flow-sorted chromosomes. *Nucleic Acids Res.* **21**(19), 4641–4642.

Lynn Senior, M., and Heun, M. (1993). Mapping maize microsatellites and polymerse chain reaction confirmation of the targeted repeats using a CT primer. *Genome* **36**, 884–889.

Mansfield, E. S., and Kronick, M. N. (1993). Alternative labeling techniques for automated fluorescence based analysis of PCR products. *BioTechniques* **15**(2), 274–279.

Marino, M. A., Turni, L. A., Del Rio, S. A., and Williams, P. E. (1994). Molecular size determinations of DNA restriction fragments and polymerase chain reaction products using capillary gel electrophoresis. *J. Chromatogr. A* **676**, 185–189.

Martin, G. B., Williams, J. G. K., and Tanksley, S. D. (1991). Rapid identification of markers linked to *Pseudomonas* resistance gene in tomato by using random primers and near isogenic lines. *Proc. Natl. Acad. Sci. U.S.A.* **88**, 2336–2340.

Martin, G. B., Devicente, M. C., and Tanksley, S. D. (1993). High resolution linkage analysis and physical characterization of the *pto* bacterial resistance locus in tomato. *Mol. Plant-Microbe Interact.* **6**, 26–34.

Martin, R., Hoover, C., Grimme, S., Grogan, C., Holtke, J., and Kessler, C. (1990). A highly sensitive, non-radioactive DNA labelling and detection system. *BioTechniques* **9**(6), 762–768.

Mellersh, C., and Sampson, J. (1993). Simplifying detection of microsatellite length polymorphisms. *BioTechniques* **15**(4): 582–584.

Michelmore, R. W., Paran, I., and Kesseli, R. V. (1991). Identification of markers linked to disease resistance genes by bulked segregant analysis: A rapid method to detect markers in specific genomic regions using segregating populations. *Proc. Natl. Acad. Sci. U.S.A.* **88**, 9828–9832.

Miklas, P. N., Stavely, J. R., and Kelly, J. D. (1993). Identification and potential use of a molecular marker for rust resistance in common bean. *Theor. Appl. Genet.* **85**, 745–749.

Mohabeer, A., Hiti, A., and Martin, W. J. (1991). Non-radioactive single strand conformational polymorphism (SSCP) using Pharmacia "PhastSystem." *Nucleic Acids Res.* **19**(11), 3154.

Morgante, M., and Olivieri, A. M. (1993). PCR-amplified microsatellites as markers in plant genetics. *Plant J.* **3**(1), 175–182.

Morgante, M., Rafalski, J. A., Biddle, P., Tingey, S., and Olivieri, A. M. (1994). Genetic mapping and variability of seven soybean simple sequence repeat loci. *Genome* **37**, 763–769.

Mullis, K., Faloona, S., Scharf, S., Saiki, R., Horn, G., and Erlich, H. (1986). Specific enzymatic amplification of DNA in vitro: The polymerase chain reaction. *Cold Spring Harbor Symp. Quant. Biol.* **51**, 263–273.

Murray, M. G., and Thompson, W. F. (1980). Rapid isolation of high molecular weight plant DNA. *Nucleic Acids Res.* **8**, 4321–4325.

Nickerson, D. A., Kaiser, R., Lappin, S., Stewart, J., Hood, L., and Landegren, U. (1990). Automated DNA diagnostics using an ELISA-based oligonucleotide ligation assay. *Proc. Natl. Acad. Sci. U.S.A.* **87**, 8923–8927.

Odelberg, S. J., and White, R. (1993). A method for accurate amplification of polymorphic CA repeat sequences. *PCR Methods Appl.* **3**, 7–12.

Orita, M., Suzuki, Y., Sekiya, T., and Hayashi, K. (1989). Rapid and sensitive detection of point mutations and DNA polymorphisms using the polymerase chain reaction. *Genomics* **5**, 874–879.

Ostrander, E. A., Jong, P. M., Rine, J., and Duyk, G. (1992). Construction of small-insert genomic DNA libraries highly enriched for microsatellite repeat sequences. *Proc. Natl. Acad. Sci. U.S.A.* **89**, 3419–3423.

Paran, I., and Michelmore, R. W. (1993). Development of reliable PCR-based markers linked to downy mildew resistance genes in lettuce. *Theor. Appl. Genet.* **85**(8), 985–993.

Paran, I., Kesseli, R., and Michelmore, R. (1991). Identification of restriction-fragment-length-polymorphism and random amplified polymorphic DNA markers linked to downy mildew resistance genes in lettuce, using near-isogenic lines. *Genome* **34**(6), 1021–1027.

Penner, G., Chong, J., Levesque, M., Molnar, S., and Fedak, G. (1993). Identification of a RAPD marker linked to the oat stem rust gene Pg3. *Theor. Appl. Genet.* **85**, 702–705.

Rafalski, J. A., and Tingey, S. (1993a). RFLP map of soybean *(Glycine max)* 2N = 40. *In* "Genetic Maps: Locus Maps of Complex Genomes" (S. J. O'Brien, ed.) 6th ed., Book 6. pp. 149–156. Cold Spring Harbor Lab. Press, Cold Spring Harbor, NY.

Rafalski, J. A., and Tingey, S. V. (1993b). Genetic diagnostics in plant breeding: RAPDs, microsatellites and machines. *Trends Genetics.* **9**(8), 275–280.

Rafalski, J. A., Tingey, S. V., and Williams, J. G. K. (1991). RAPD markers—a new technology for genetic mapping and plant breeding. *AgBiotech News Inf.* **3**(4): 645–648.

Reiter, R. S., Williams, J. G. K., Feldmann, K. A., Rafalski, J. A., Tingey, S. V., and Scolnik, P. A. (1992). Global and local genome mapping in *Arabidopsis thaliana* by using recombinant inbred lines and random amplified polymorphic DNAs. *Proc. Natl. Acad. Sci. U.S.A.* **89**(4), 1477–1481.

Rigas, B., Welcher, A. A., Ward, D. C., and Weissman, S. M. (1986). Rapid plasmid library screening using RecA-coated biotinylated probes. *Proc. Natl. Acad. Sci. U.S.A.* **83**, 9591–9595.

Ronald, P. C., Albano, B., Tabien, R., Abenes, L., Wu, K. S., McCouch, S., and Tanksley, S. D. (1992). Genetic and physical analysis of the rice bacterial blight disease resistance locus, XA21. *Mol. Gen. Genet.* **236**, 113–120.

Sambrook, J., Fritsch, E. F., and Maniatis, T. (1989). "Molecular Cloning: A Laboratory Manual," 2nd ed. Cold Spring Harbor Lab., Cold Spring Harbor, NY.

Serikawa, T., Kuramoto, T., Hilbert, P., Mori, M., Yamada, J., Dubay, C. J., Lindpainter, K., Ganten, D., Guenet, J.-L., Lathrop, G. M., and Beckman, J. S. (1992). Rat gene mapping using PCR-analyzed microsatellites. *Genetics* **131**, 701–721.

Shattuck-Eidens, D. M., Bell, R. N., Neuhausen, S. L., and Helentjaris, T. (1990). DNA sequence variation within maize and melon: Observations from polymerase chain reaction amplification and direct sequencing. *Genetics* **126**, 207–217.

Shibata, D., Kato, T., and Tanaka, K. (1991). Nucleotide sequence of a soybean lipoxygenase gene and the short intergenic region between an upstream lipoxygenase gene. *Plant Mol. Biol.* **16**, 353–359.

Sinnett, D., Deragon, J.-M., Simard, L. R., and Labuda, D. (1990). Alumorphs—Human DNA polymorphisms detected by polymerase chain reaction using Alu-specific primers. *Genomics* **7**, 331–334.

Sokolov, B. P. (1990). Primer extension technique for the detection of single nucleotide in genomic DNA. *Nucleic Acids Res.* **18**(12), 3671.

Syvanen, A.-C., Aalto-Setala, K., Harju, L., Kontula, K., and Soderlund, H. (1990). A primer-guided nucleotide incorporation assay in the genotyping of apolipoprotein E. *Genomics* **8**, 684–692.

Tautz, D. (1989). Hypervariability of simple sequences as a general source for polymorphic DNA markers. *Nucleic Acids Res.* **17**(16), 6463–6471.

Taylor, G. R., Haward, S., Noble, J. S., and Murday, V. (1992). Isolation and sequencing of CA/GT repeat microsatellites from chromosomal libraries without subcloning. *Anal. Biochem.* **200**, 125–129.

Thomas, M. R., and Scott, N. S. (1993). Microsatellite repeats in grapevine reveal DNA polymorphisms when analysed as sequence-tagged sites (STSs). *Theor. Appl. Genet.* **86**, 985–990.

Tingey, S. V., and del Tufo, J. P. (1993). Genetic analysis with RAPD marklers. *Plant Physiol.* **101**, 349–352.

Vignal, A., Gyapay, G., Hazan, J., Nguyen, S., Dupraz, C., Cheron, N., Becuwe, N., Tranchant, M., and Weissenbach, J. (1993). A nonradioactive multiplex procedure for genotypeing of microsatellite markers. *In* "Methods in Molecular Genetics" (K. W. Adolph, ed.), Vol. 1, pp. 211–221. Academic Press, San Diego, CA.

Waugh, R., and Powell, W. (1992). Using RAPD markers for crop improvement. *Trends Biotechnol.* **10**, 186–191.

Weber, J. L., and May, P. E. (1989). Abundant class of human DNA polymorphisms which can be typed using the polymerase chain reaction. *Am. J. Hum. Genet.* **44**, 388–396.

Weber, J. L. (1990). Informativeness of human (dC-dA)$_n$ · (dG-dT)$_n$ polymorphisms. *Genomics* **7**, 524–530.

Weissenbach, J., ed. (1992). "The Genethon Microsatellite Map Catalogue." Genethon, Evry, France.

Welsh, J., and McClelland, M. (1990). Fingerprinting genomes using PCR with arbitrary primers. *Nucleic Acids Res.* **18,** 7213–7218.

Welsh, J., Petersen, C., and McClelland, M. (1991). Polymorphisms generated by arbitrarily primed PCR in the mouse: Application to strain identification and genetic mapping. *Nucleic Acids Res.* **19**(2), 303–306.

Williams, J. G. K., Kubelik, A. R., Livak, K. J., Rafalski, J. A., and Tingey, S. V. (1990). DNA polymorphisms amplified by arbitrary primers are useful as genetic markers. *Nucleic Acids Res.* **18**(22), 6531–6535.

Williams, J. G. K., Rafalski, J. A., and Tingey, S. V. (1993a). Genetic analysis using RAPD markers. *In* "Methods in Enzymology" (R. Wu, ed.), Vol. 218, pp. 704–740. Academic Press, San Diego, CA.

Williams, J. G. K., Reiter, R. S., Young, R. M., and Scolnik, P. A. (1993b). Genetic mapping of mutations using phenotypic pools and mapped RAPD markers. *Nucleic Acids Res.* **21**(11), 2697–2702.

Wu, D. Y., Ugozzoli, L., Pal, B. K., and Wallace, R. B. (1989). Allelespecific enzymatic amplification of β-globin genomic DNA for diagnosis of sickle cell anemia. *Proc. Natl. Acad. Sci. U.S.A.* **86,** 2757–2760.

Wu, K.-S., and Tanksley, S. D. (1993a). Abundance, polymorphism and genetic mapping of microsatellites in rice. *Mol. Gen. Genet.* **241**(1–2), 225–235.

Wu, K.-S., and Tanksley, S. D. (1993b). Genetic and physical mapping of telomeres and macrosatellites of rice. *Plant Mol. Biol.* **22,** 861–872.

Zabeau, M., and Voss, P. (1993). Selective restriction fragment amplification: A general method for DNA fingerprinting. *Eur. Pat. App.* **92402629.7** (Publ. Number 0 534 858 A1).

Zhao, X., and Kochert, G. (1993). Phylogenetic distribution and genetic mapping of a (GGC)n microsatellite from rice (*Oryza sativa* L.). *Plant Mol. Biol.* **21,** 607–614.

Zietkiewicz, E., Labuda, M., Sinnett, D., Glorieux, F. H., and Labuda, D. (1992). Linkage mapping by simultaneous screening of multiple polymorphic loci using Alu oligonucleotide-directed PCR. *Proc. Natl. Acad. Sci. U.S.A.* **89,** 8448–8451.

Zietkiewicz, E., Rafalski, J. A., and Labuda, D. (1994). Genome fingerprinting by simple sequence repeats (SSR)-anchored polymerase chain reaction amplification. *Genomics* **20,** 176–183.

5

Genome Mapping of Protozoan Parasites by Linking Clones

S. P. Morzaria

I. Introduction

Protozoa are unicellular eukaryotic organisms, many of which cause diseases in man and animals. The majority of the economically important medical and veterinary parasitic protozoa belong to the genera *Trypanosoma*, *Leishmania*, *Toxoplasma*, *Eimeria*, *Entamoeba*, *Giardia*, *Plasmodium*, *Babesia*, and *Theileria*. Understanding the basic biology of these protozoa is fundamental to the development and planning of rational disease control strategies. While modern technologies have been beneficial in cloning, sequencing, and characterizing genes encoding protozoan proteins, it is now becoming increasingly clear that the complex processes involved in gene expression and function cannot be fully understood without a basic knowledge of the structure and organization of genomes.

Unlike yeast and bacteria, experimental systems to study classical genetics for most of the parasitic protozoa are either nonexistent or in the initial stages of development. Therefore, with the exception of *Toxoplasma gondii* (Sibley *et al.*, 1993), even rudimentary genetic maps of most protozoan parasites are lacking. Consequently most of the mapping efforts have focused on generating physical maps of protozoan parasites. The single most significant technology in this regard has been the application of pulsed-field gel electrophoresis (PFGE, Schwartz and Cantor, 1984), which has enabled separation and resolution of parasite chromosomes. The majority of the mapping projects have used a "top-down" approach with the aim of producing a long-range restriction map of a genome. However, construction of physical maps is facilitated by the use of unique DNA sequences as hybridization probes. These probes provide strategic landmarks along the genome and aid in bridging gaps between genetic and physical maps. Linking clones, which span rare restriction sites and therefore can identify pairs of immediately adjacent restriction fragments, are particularly useful as hybridization probes (Smith *et al.*, 1986). Thus various levels of low-resolution physical maps, ranging from simple molecular karyotypes to complete restriction maps of many species of parasitic protozoa including *Trypanosoma* (Van der Ploeg *et al.*, 1984), *Leishmania* (Spithill and Samaras, 1985), *Plasmodium* (Sinnis and Wellems, 1988; Triglia *et al.*, 1992), *Toxoplasma* (Sibley *et al.*, 1993), and *Theileria* (Morzaria and Young, 1992) are available.

To generate higher resolution maps, "bottom-up" mapping approaches exploiting the newer "large DNA" cloning technologies, such as yeast artificial chromosomes (YACs), cosmids, and P1-based cloning vectors, are being used (Melville *et al.*, 1993; Triglia and Kemp, 1991). The bottom-up approaches to mapping depend on the generation and assembly of an overlapping set of contiguous DNA clones, or "contigs." Although contig maps are difficult to complete, they are useful because a large number of DNA markers along specified regions of a genome are generated.

A top-down approach utilizing linking clones (Smith *et al.*, 1986) has been used to construct a complete low-resolution physical map of the genome of *Theileria parva* (Morzaria and Young, 1992). This approach can be used for many relatively simple protozoan parasite genomes, including those of *Theileria, Babesia,* and *Plasmodium,* as will as to map regions of complex protozoan parasites, such as *Trypanosoma, Leishmania,* and *Toxoplasma* species. The technical details of the various methods used to derive such a map, some of the problems of using such an approach, and ways of overcoming these are described and discussed below.

II. Strategy

The mapping strategy involves two major steps. This approach is illustrated in Fig. 5.1 and the technical steps required to derive such a map are outlined below. The first step is to divide the genome into its structural

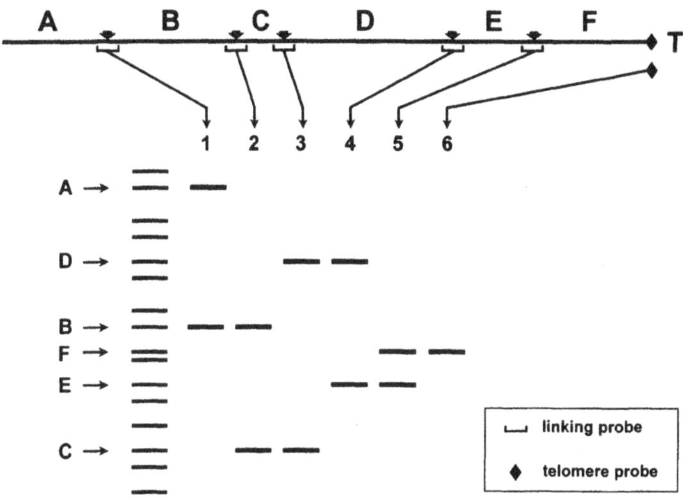

Figure 5.1 Linking clone approach to physical mapping of a chromosome. The longitudinal bar represents a part of a chromosome that is divided into six macrorestriction fragments (A–F) that are separated in PFGE (below the bar) according to their sizes. Clones of small DNA fragments bearing the restriction sites (1–5) that cut the chromosome into six fragments are selected from a library and used as linking clones to hybridize to the Southern blot of the PFGE-separated fragments. Each clone should hybridize with two restriction fragments, revealing the attachment of each fragment and the order of the fragments. Fragment 6 bearing the telomere should hybridize to linking clone 5 and the telomere probe 6. This provides a complete long-range restriction map of a chromosome. Reprinted with permission by Morzaria and Young. "Genome Analysis of *Theileria parva*," *Parasitology Today* **9** (1993).

elements, such as chromosomes and subchromosomal macrorestriction fragments, using PFGE. This is relatively easy since the sizes of the haploid nuclear genomes of most parasitic protozoa vary between 6 and 80 Mb. The chromosomes do not condense during cell division and it is possible to separate DNA up to 7 Mb. For small parasite genomes (between 5 and 9×10^6 bp) containing a small number (four) of chromosomes, such as those of *T. parva*, *Theileria annulata*, *Theileria mutans*, *Babesia bigemina*, and *Babesia bovis*, digestion with a rare-cutting restriction enzyme can be performed directly without separating the chromosomes. For larger, more complex genomes, chromosomes may need to be separated before generating macrorestriction fragments for each chromosome separately. The second step is to generate linking clones. These linking clones can then be used as probes to hybridize to Southern blots of restriction enzyme-digested genomic DNA to establish DNA fragments that are adjacent, or linked, to each other.

III. Materials

A. Stock Solutions

Stock solutions routinely used in the various methods described in this chapter are as follows:

(1) Tris-HCl (1 M, pH 7.4, 7.6, 8.0). Dissolve 121.1 g of Tris base in 800 ml of distilled H_2O. Adjust the pH with concentrated HCl and make up the volume to 1 liter. Autoclave and store at room temperature.

(2) EDTA (0.5 M, pH 8.0). Mix 186.1 g of disodium EDTA to 800 ml of distilled H_2O using a magnetic stirrer. Adjust pH with NaOH, make up the volume to 1 liter, autoclave, and store at room temperature.

(3) NaCl (5 M). Dissolve 292.2 g of NaCl in 800 ml of distilled H_2O, autoclave and store at room temperature.

(4) SSC (20×). Dissolve 175.3 g of NaCl and 88.2 g of sodium citrate in 800 ml of distilled H_2O. Adjust pH with HCl, make up volume to 1 liter, autoclave, and store at room temperature.

(5) Denhardt's solution (50×). Dissolve 5 g Ficoll, 5 g polyvinylpyrrolidone, and 5 g BSA Fraction V in 500 ml of distilled H_2O. Filter through a 0.22-μm filter, dispense in 25-ml aliquots, and store at −20°C.

(6) Proteinase K 20 mg/ml. Dissolve in sterile distilled H_2O, aliquot in 500-μl, volumes and store at −20°C.

(7) SDS 20%. Dissolve 20 g of SDS in distilled H_2O and store at room temperature.

(8) 50× TAE. Mix and dissolve 242 g of Tris-HCl, 57.1 ml of glacial acetic acid, and 100 ml of 0.5 M EDTA in 1 liter of distilled H_2O. Store at room temperature.

B. Buffers

(1) Alsever's solution

Final concentration	Per liter
113.8 mM Glucose	20.5 g
71.9 mM NaCl	4.2 g
27.2 mM Na$_3$ citrate · 2H$_2$O	8.0 g
2.61 mM Citric acid	0.55 g

Mix in 800 ml distilled H_2O. Adjust volume to 1 liter. The pH should be 6.1 and must be checked frequently.

(2) TEN

Final concentration	Per liter	
50 mM Tris-HCl, pH 7.5	50 ml	ex 1 M stock
0.1 M NaCl	20 ml	ex 5 M stock
0.15 M EDTA, pH 8.0	300 ml	ex 0.5 M stock

(3) PSG buffer

Final concentration	Per liter	
75 mM NaPO$_4$, pH 8.0	75 ml	ex 1 M stock
65 mM NaCl	13 ml	ex 5 M stock
10% Glucose, pH 8.0	100 g	

Autoclave and store at 4°C.

(4) Lysing buffer for preparing DNA from LMP embedded parasites

Final concentration	Per 100 ml	
0.25 M EDTA, pH 8.0	50 ml	
1% Sodium N-lauroyl sarcosine	1 g	
or 0.5% SDS	2.5 ml	ex 20% stock
10 mM Tris-HCl	100 μl	ex 1 M stock
1 mg/ml proteinase K	5 ml	
1 mM PMSF*	1 ml	

*Phenylmethylsulpfonyl fluoride needs to be prepared fresh as a 0.1 M solution by adding 17.5 mg of PMSF in 1 ml of isopropanol at room temperature with gentle shaking.

(5) T.E

Final concentration	Per liter
10 mM Tris-HCl, pH 7.4	10 ml
0.1 mM EDTA, pH 8.0	200 μl

(6) TBE (10×)

90 mM Tris-HCl	109 g
90 mM Boric acid	55.6 g
2.5 mM EDTA	9.3 g

(7) Hybridization solution

Final concentration	Per litre	
6× SSC	300 ml	ex 20× stock
5× Denhardt's solution	100 ml	ex 50× stock
0.5% SDS	25 ml	ex 20% stock
Sonicated denatured salmon sperm DNA	100 μg/ml	

(8) Lysing buffer for making DNA from parasite pellets

Final concentration

0.25 M EDTA, pH 8.0
1% Sodium N-lauroylsarcosine or 0.5% SDS
10 mM Tris · HCl
1 mg/ml Proteinase K

(9) Linear sucrose gradient
Top 10% sucrose
Bottom 40% sucrose
Both in 25 mM Tris-HCl, pH 8.0, 5 mM EDTA
Pour using a two-chamber gradient mixer

(10) 10× Ligase buffer

Final 1× concentration

0.5 M Tris-HCl, pH 7.8
0.1 M MgCl$_2$
0.1 M dithiothreitol
10 mM ATP
25 μg/ml BSA

(11) Incomplete SOB
20 g bacto-tryptose
5 g bacto-yeast extract
10 mM MgCl$_2$
10 mM MgSO$_4$

Dissolve in 1 liter of distilled H$_2$O, divide into 10 aliquots of 100 ml and sterilize by autoclaving.

(12) Complete SOB
When ready to use add 250 μl of 1 M KCl and 1 ml of MgCl$_2$ per 100 ml of incomplete SOB.

(13) SOC
Add 2 ml of 1 M glucose per 100 ml of complete SOB.

(14) TFB (Hanahan, 1983; Sambrook *et al.*, 1989)

Final concentration	Per liter	
10 mM MES	10 ml	ex stock 1 M MES
45 mM MnCL$_2$ · 4H$_2$O	8.91 g	
10 mM CaCl$_2$ · 2H$_2$O	1.47 g	
100 mM KCl	7.46 g	
3 mM Hexamine cobalt chloride	0.80 g	

Filter through 0.45-μm filter, divide into 10-ml aliquots, and store at −20°C.

(15) NZYCM/ampicillin plates

10 g NZ amine

5 g NaCl

5 g Bacto-yeast extract

1 g Casamino acids

2 g $MgSO_4 \cdot 7H_2O$

Mix the above chemicals in 800 ml of distilled H_2O and make up the volume to 1 liter. Autoclave, cool to 55°C, add 100 μg/ml ampicillin, mix well, and pour plates as required.

IV. Procedures

A. Purification of Parasites

For successful physical mapping it is important that pure, chromosome-sized DNA be prepared. This is relatively easy for the protozoan parasites that spend part of their life-cycle extracellularly and/or can be cultured *in vitro*. Examples of these are *Trypanosoma, Leishmania, Toxoplasma, Eimeria, Entamoeba,* and *Giardia* species. Some of the intracellular parasitic protozoa present problems because an additional purification step is required to separate parasites from the host nuclear DNA. For *Plasmodium, Babesia,* and *Theileria* parasites this is not a major problem because they multiply in host erythrocytes which are nonnucleated. Additionally, the former two parasites can also be grown in culture. Two separate protocols are described, one for intraerythrocytic parasites of the genera *Theileria* and *Babesia* (Conrad *et al.,* 1987) and the other for extracellular parasites of the genus *Trypanosoma* (Lanham and Godfrey, 1970). The protocols described below can be adapted for almost all other protozoan parasites.

1. *Theileria* and *Babesia* Species

(a) Collect 2 liters of whole blood from cattle (piroplasm parasitaemia between 10–50%) in an equal volume of ice-cold Alsever's solution supplemented with 50 IU/ml heparin, and centrifuge in 500-ml aliquots at 2500 g for 30 min at 4°C.

(b) Aspirate and discard the supernatant and the top 5% of the cells at the interface of the plasma and erythrocytes. This depletes the sample of leukocytes.

(c) Resuspend the packed erythrocytes in an equal volume of Alsever's solution and centrifuge as in (a) above. This constitutes a wash. Repeat these washes two more times in Alsever's solution, then perform the last wash in TEN. Estimate the volume of packed erythrocytes after the final wash. This should be between 400 and 500 ml. The sample can be left overnight at 4°C in Alsever's solution before the last wash in TEN.

(d) Prepare a fresh solution of saponin (1 mg/ml) in distilled water and warm the erythrocyte pellet and saponin separately at 37°C for 30 min.

(e) Add an equal volume of prewarmed saponin to the erythrocyte pellet and mix for 10 sec. The lysis is immediate and the solution should be clear.

(f) Add to the mixture 4 times its volume of TEN buffer, mix thoroughly by inverting the container 10 to 20 times, and centrifuge at $2500g$ for 20 min at 4°C to pellet most of the ruptured erythrocytic stroma.

(g) Aspirate and save the supernatant which contains free parasites and discard the pellet.

(h) Centrifuge the supernatant at $7000g$ at 4°C for 30 min, aspirate the supernatant carefully, as the pellet is soft and loose, and discard the supernatant. Estimate the pellet volume, which should be between 100 and 150 ml.

(i) Wash pellet by resuspending in an equal volume of ice-cold TEN, and centrifuge at $7000g$ at 4°C for 30 min. Repeat washing until the pellet is free of hemoglobin and is almost translucent white. This takes about three to five more washes. It is important to perform these washes to remove residual hemoglobin in the sample which may interfere with certain enzyme manipulations.

(j) After the last wash leave the parasite pellet on ice and embed within 30 min.

(k) Approximately 90% loss of parasites can occur during the above purification procedure. However, attempts to reduce this high level of loss results in contamination with leukocytes.

2. *Trypanosoma* species

(a) Equilibrate 500 g of DEAE cellulose (DE52, Whatman) in 2 liters of PSG buffer. Adjust the pH to 8.0 with 1:10 dilution of phosphoric acid and leave for 30 min. Decant supernatant and resuspend in PSG (pH 8.0) for another 30 min. Repeat two more times and then store the washed DE52 at 4°C.

(b) Place a disc of filter paper (Whatman 41) at the bottom of a syringe barrel and pour the prewashed DE52 in the barrel. Allow to pack and keep adding DE52 until the final volume of the packed column is about 8 times the volume of the blood to be separated. For larger volumes of blood a sintered glass column can be used.

(c) Place a PSG-soaked filter disc on top of the packed column and pour gently over the column ice-cold blood previously diluted in an equal volume of PSG. When the blood has passed into the column add PSG equal to the diluted blood volume over the column.

(d) Check eluate regularly for trypanosomes under a light microscope and start collecting when the parasites are observed in a sterile tube kept on ice. Check eluate at regular intervals and stop collecting when no parasites are observed.

(e) Estimate the number of trypanosomes present by counting a drop under a hemocytometer. Centrifuge the sample at 1000g for 15 min at 4°C to pellet and resuspend trypanosomes at a concentration of 2 × 10^9/ml. Leave at 4°C and use as soon as possible.

The approximate final concentration of various protozoan parasites in agarose for preparation of HMW DNA is given below.

Theileria parva, T. mutans, T. annulata, and *T. buffeli:* 1–2 × 10^{10} parasites/ml.
Babesia bigemina and *B. bovis:* 1–2 × 10^{10} parasites/ml.
Trypanosoma brucei, T. vivax, and *T. congolense:* 1 × 10^9 parasites/ml.
Leishmania spp.: 0.5–1 × 10^9 parasites/ml.
Plasmodium spp.: 1–2 × 10^9 parasites/ml.
Giardia lamblia: 1–2 × 10^8 parasites/ml.
Toxoplasma gondii: 0.5–1 × 10^9 parasites/ml.

For optimum separation of chromosomes and large restriction fragments, use 5 × 4 × 1 mm (20 µl) agarose blocks containing DNA at a concentration of 100 µ/ml.

B. Preparation of Agarose-Embedded HMW DNA

(1) Melt 1.5% low-melting-point (LMP) (e.g., InCert, FMC, Rockland, ME) agarose and keep at 50°C. Other grades of LMP agarose can be used provided that they have been tested for purity, since the contaminants may interfere with subsequent enzyme manipulation.

(2) Warm parasite suspension briefly to bring to 42°C.

(3) Mix equal volumes of liquefied LMP agarose cooled to 45°C and prewarmed parasite suspension gently.

(4) Distribute with a pipette into a mould, avoiding air bubbles. Two glass slides clamped together with 1-mm gap, premoulded templates purchased commercially (e.g., Pharmacia-LKB) or "home-made" plastic moulds can be used.

(5) Leave the mould on ice until the mixture solidifies (approximately 15 min).

(6) Transfer the solidified agarose slabs into at least 3 times the volume of lysing buffer.

(7) Incubate at 50°C for 48 hr with gentle shaking. Change lysing buffer at least twice during this procedure. The blocks will turn clear, indicating that lysis and proteolysis are complete.

(8) Wash the slabs with 10 mM EDTA (pH 8.0) four times, each time for 30 min at room temperature with gentle shaking.

(9) Store blocks at 4°C. If long-term storage (>3 months) is desired store in lysis buffer that does not contain SDS or in 100 mM EDTA. This prevents degradation of the DNA.

C. Identification of Appropriate Rare-Cutting Restriction Enzymes

The aim is to select an enzyme that produces approximately 30–50 DNA fragments larger than 50 kb, each of which can be resolved clearly by PFGE. The estimated G + C content of a genome may aid in determining rare-cutting restriction enzymes that might be of use in generating a small number of macrorestriction fragments. However, in practice, it is better to test a number of enzymes. A list of enzymes that can produce large DNA fragments is shown in Chapter 1. Of these enzymes, the two 8-base cutters, *Not*I and *Sfi*I, have been used widely in many mapping projects. For *T. parva*, *T. annulata*, *T. taurotragi*, *T. buffeli*, *B. bigemina*, and *B. bovis*, *Sfi*I has been found to be the most appropriate enzyme. *Not*I produces only 8 fragments for *T. parva*, too small a number to enable construction of a useful physical map. On the other hand, in the highly GC-rich *T. mutans* genome, *Sfi*I and *Not*I produce numerous DNA fragments below 50 kb, which are harder to resolve clearly in a gel.

1. Day 1

(1) Cut out a 5 × 4 × 1-mm agarose block (containing approximately 2 μg DNA) and wash in 2 ml of T.E containing 1 mM phenylmethylsulfonyl fluoride (PMSF) (needs to be prepared fresh as a 0.1 M solution in isopropanol) for 1 hr at room temperature with gentle shaking. PMSF is used to inactivate the proteinase K.

(2) Repeat the above washing step five times, using T.E (lacking PMSF) at room temperature with gentle shaking. Change T.E every 30 min.

(3) Transfer the dialyzed block into a 1.5-ml Eppendorf tube containing 200 μl of 1× restriction enzyme buffer without BSA and equilibrate at 4°C for 1 hr.

(4) Aspirate the buffer and add 100 μl of fresh restriction enzyme buffer with the appropriate concentration of BSA and 40 units of restriction enzyme (20 units/μg) as recommended (Smith *et al.*, 1988).

(5) Incubate at the appropriate temperature (e.g., *Not*I at 37°C, *Sfi*I at 50°C) for 2 hr to overnight.

2. Day 2

(6) Aspirate the buffer and the enzyme and store agarose blocks at 4°C in lysis buffer without SDS or in 100 mM EDTA. If the block is to be used immediately, add 500 μl of the buffer in which the PFGE is to be run.

D. Separation of Chromosomes and Macrorestriction Fragments by PFGE

The PFGE techniques are described in Chapter 1. It is important to emphasize that the ability to separate large-sized DNA molecules by PFGE is crucial to a mapping project using the linking clone strategy. At ILRI we routinely use the contour-clamped homogenous electric-field electrophoresis (CHEF) system (Chu *et al.*, 1986) and the double inhomogenous configuration (Smith *et al.*, 1986) used in the Pharmacia-LKB Pulsaphor apparatus.

The PFGE conditions for optimum resolution of heterogeneously sized chromosomes and macrorestriction fragments vary greatly between different protozoa. The parameters used for separating chromosomes and *Sf*I and *Not*I fragments for some of the protozoan parasites are described below. A combination and modifications of these conditions can be adapted easily to resolve any fragment size ranging from 25 kb to 7 Mb. It is important to note that the conditions described below were obtained after numerous experiments exploring a combination of several factors (see Chapter 1). A useful guide to separating different-sized DNA molecules for CHEF and the Pulsaphor apparatus are given in Vollarth and Davis (1987) and Smith *et al.* (1988), respectively.

1. Size Markers

A wide range of size markers need to be employed to determine the sizes of DNA molecules separated with PFGE systems. For most protozoan parasites a combination of bacteriophage *lambda* DNA concatemers and *Saccharomyces cerevisiae* and *Schizosaccharomyces pombe* chromosomes cover a broad range of DNA sizes from approximately 50 kb to 6 Mb. These size markers are available commercially and methods for their preparation have been described (Watebury and Lane, 1987; Vollarth and Davis, 1987; Schwartz and Cantor, 1984).

2. Separation of *Sf*I Fragments of *Theileria* and *Babesia* Species

All the *Sf*I restriction fragments of *T. parva*, *T. annulata*, *T. taurotragi*, *B. bigemina*, and *B. bovis* can be separated using CHEF. Electrophoresis is performed in 10.5 × 14-cm agarose gels prepared in 0.38× TBE. The buffer is circulated at 12°C during electrophoresis. It is obviously desirable to obtain discrete separation of all the macrorestriction fragments on a gel at one run. However, this is not always possible due to comigration of some similarly sized DNA molecule. For the above parasites, 10-sec pulses for 16 hr and 40-sec pulses for 5 hr at 10 V/cm enable resolution of almost all the fragments in a gel. Comigrating fragments in the size ranges 50–100 and 350–600 kb can be further resolved using 3-sec pulses for 15 hr and 40-sec pulses for 10 hr, respectively, using the same voltage gradient.

3. Separation of *Theileria* and *Babesia* Chromosomes

The *Theileria* species *(T. parva; T. mutans; T. taurotragi,* and *T. annulata)* and stocks of *Babesia* species *(B. bigemina* and *B. bovis)* examined so far (S. P. Morzaria, unpublished results) have four chromosomes each. Using a home-made CHEF apparatus, we have easily separated the *T. parva* chromosome 1 of 3.2 Mb but the remaining three chromosomes, whose sizes vary between 2.3 and 2.1 Mb have been difficult to resolve despite numerous attempts using a combination of different parameters. However, using the Pulsaphor-LKB point electrode system we had little difficulty in separating these three chromosomes. Using this system we have also successfully resolved chromosomes of all the parasites mentioned above. The separation is performed in 1% agarose, prepared in 1× TBE, using point electrodes (North and West walls, one anode on each wall in position 60, East and South walls, three cathodes on each wall in positions 30, 110, and 210) and pulses of 900 sec for 24 hr, 600 sec for 24 hr, 480 sec for 24 hr, and 400 sec for 24 hr at 3 V/cm. The buffer is circulated at 12°C. The point electrode configuration described above is known as the double inhomogeneous mode. Because of the inhomogeneous field the separation of DNA fragments does not occur in a straight line. However, with the availability of several new systems with sophisticated programmers (see Chapter 1), all providing homogeneous electric fields, it should be possible to separate chromosomes of all the parasites mentioned above.

4. Separation of *Trypanosoma* Chromosomes

Van der Ploeg *et al.* (1984) first described the separation of *Trypanosoma* chromosomes. For the parasites of this genus we use a protocol developed at ILRI (N. B. Murphy, personal communication) for the Pulsaphor-LKB HEX electrode configuration. Chromosomes from 50 kb up to 2 Mb can be separated in 1.5% agarose with 1× TBE using a voltage gradient of 8 V/cm and pulses of 250 sec for 30 hr, 180 sec for 24 hr, and 120 sec for 10 hr. To increase the separation up to 3 Mb, use 800-, 600-, 250-, 180-, and 120-sec pulses for 30 hr each at 4–4.5 V/cm. To separate the largest size classes of chromosomes (i.e., around 5 Mb), use pulses of 2700 sec for 50 hr at 60 V, 1800 sec for 50 hr at 80 V, 1200 sec for 35 hr at 2–3 V/ cm. For the start of such a run the DNA is first run into the gel at 120-sec pulses for 2 hr at 160 V before the use of longer pulse times as described above. All gels are cooled to approximately 8°C by setting the circulating water at 5°C.

5. Separation of *Toxoplasma gondii* Chromosomes

Sibley and Boothroyd (1992) have described the separation of *T. gondii* chromosomes using the transverse alternating field electrophoresis (TAFE) gels (Gardiner *et al.*, 1986). The Beckman Geneline System I was used to resolve 9 chromosomes from 2 to >6 Mb in 0.5% chromosome grade agarose. They used four defined steps of voltage, switch time, and duration as follows: 40 V, 30 min, 36 hr; 50 V, 25 min, 24 hr; 60 V, 22 min, 24 hr; 70 V, 15 min, 36 hr in 0.25× TAE buffer cooled at 12°C.

6. Separation of *Plasmodium falciparum* Chromosomes

All 14 *Plasmodium falciparum* chromosomes were first separated by Kemp *et al.* (1987). Using 1% agarose gels in 0.5× TBE and a three-step separation (210-sec pulses for 48 hr at 8.7 V/cm, 420-sec pulses for 72 hr at 6.9 V/cm, and 740-sec pulses for 72 hr at 5.0 V/cm), all the chromosomes can be separated.

7. Southern Blotting and Hybridization

Both acid depurination and UV nicking can be used for breaking large DNA fragments into smaller pieces. Acid depurination works well for fragment sizes between 1 Mb and 50 kb but produces variable results when the target DNA fragments are >1 Mb. UV nicking is more reliable and is used routinely for larger DNA fragments. Both methods are described briefly.

1. Southern Blotting

(a) Stain the gel in distilled water with ethidium bromide (0.5 μg/ml) for 1 hr and then destain for 1 hr in fresh distilled water with gentle shaking.

(b) Photograph using a UV transilluminator. Note the total time exposed to UV during visualization and photography.

(c) Depurinate by immersing the gel for two washes of 15 min each in 500 ml of 0.25 *M* HCl.

(d) UV nicking is performed as described in Chapter 1.

(e) Denature, neutralize, and Southern blot onto a nylon filter as described in Sambrook *et al.* (1989) for 24–36 hr in 15× SSC.

(f) Fix DNA on to the membrane either by baking in an oven at 80°C or by UV irradiation of 120 mJ in a UV oven.

2. Hybridization

(a) Preincubate filters in hybridization solution for at least 5 hr at 65°C.

(b) Remove filter and incubate in the hybridization solution containing radiolabeled probe ($1–5 \times 10^8$ cpm/μg) for 6 hr at 65°C.

(c) Wash filter twice in 0.2× SSC/0.1% SDS at 65°C for 1 hr before autoradiographic exposure.

(d) Thirty to 40 hr exposure is usually sufficient to obtain a signal.

E. Identification of Telomeres

Linear chromosomes of parasitic protozoa contain conserved telomeric sequences at either end and have DNA sequence arrangements similar to

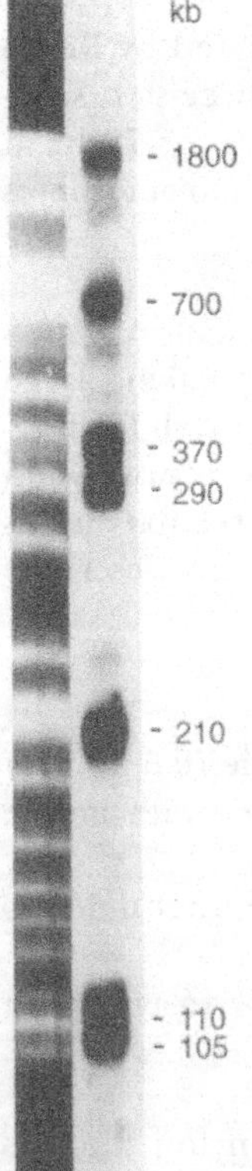

Figure 5.2 Identification of macrorestriction fragments bearing telomeres. Hybridization of an oligonucleotide telomere probe from *P. berghei* to an *Sfi*I-digested blot of *T. parva* genomic DNA reveals eight (two comigrating bands at 110 kb) fragments bearing telomeres.

the telomeres of other eukaryotic organisms (Blackburn and Szostak, 1984). Telomeric sequences of many protozoan parasites (e.g., *Plasmodium, Theileria, Trypanosoma,* and *Leishmania*) have been cloned and sequenced. These contain simple repeats and, usually, are devoid of restriction enzyme sites. The telomeric repeat sequence $(CCCTGAACCTAAA)_2$ of *P. berghei* has been found to hybridize with other *Plasmodium, Theileria, Babesia,* and *Trypanosoma* species (Morzaria *et al.,* 1990). Thus, this oligonucleotide can be used as a probe to identify macrorestriction fragments bearing telomeres as shown for *T. parva* (Fig. 5.2).

F. Construction of a Linking Library

A linking library consists of an enriched population of clones of parasite DNA containing a site for a specific rare-cutting restriction enzyme. Preparation of a library of linking clones requires first that a representative library be constructed from total genomic DNA; from this, clones containing the desired rare restriction site are selected. The scheme for the preparation of a *Sau*3AI library, enrichment for clones bearing *Sfi*I sites, and their use as probes is outlined in Fig. 5.3. In constructing a representative *Sau*3AI plasmid library of the parasite DNA, care is taken to avoid ligation of multiple genomic fragments, which would compromise the use of the library as a source of linking fragments. This is achieved by partial filling-in of the ends of both the vector and the target DNA. Appropriate nucleotides are used to generate compatible "sticky" ends. This approach almost completely abolishes ligation of target molecules to each other, or self-ligation of the vector. It is recommended that the library be constructed from small-sized DNA fragments (4–6 kb) to minimize the potential problem associated with cloning repetitive sequences in the linking clones.

1. Partial Digestion of *Theileria parva* Genomic DNA by *Sau*3AI

(a) For the preparation of genomic DNA use purified piroplasms as described above for *Theileria* and *Babesia.* Prepare genomic DNA by suspending 5 ml of purified pellet in an equal volume of lysing buffer containing EDTA (pH 8.0), proteinase K, and SDS to a final concentration of 0.5 *M*, 100 µg/ml, and 0.5%, respectively. Leave at 60°C for 10–20 hr, gently swirling the solution periodically. Then follow the standard protocol for the preparation of high-molecular-weight DNA as described in Sambrook *et al.,* (1989). Keep DNA in T.E at a final concentration of between 200 and 500 µg/ml.

(b) Set up a partial digestion reaction with an appropriate enzyme. For *T. parva* DNA, *Sau*3AI was used as follows:

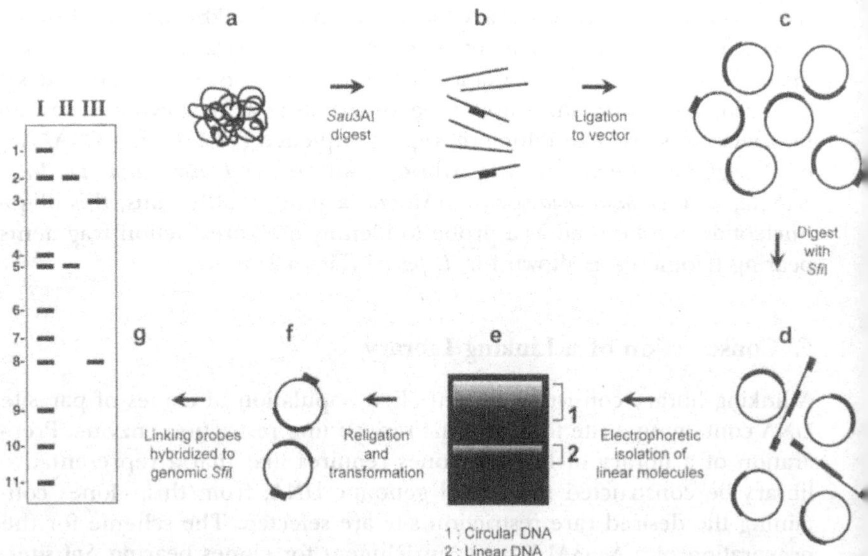

Figure 5.3 Generation of linking clones and their use in mapping. A representative library of *Sau*3AI partially digested fragments (b) is prepared from parasite genomic DNA (a) in a plasmid vector. Only a small minority of the library consists of DNA fragments bearing rare-cutting restriction sites (c, closed rectangles). These are linking clones, which are selected by first linearizing them with the appropriate restriction enzyme (d) and then physically separating them from the bulk of the circular DNA molecules by electrophoresis using SeaPlaque agarose (e2). The linearized plasmids are recircularized and recovered after reintroduction in *E. coli* (f) and then selected unique linking clones are used as probes in Southern blots of PFGE-separated macrorestriction fragments. Each clone will hybridize to two restriction fragments (g, lanes II and III). Reprinted with permission by Morzaria and Young. "Genome Analysis of *Theileria parva*," *Parasitology Today* **9** (1993).

(i) Set up five reactions, each containing 7 μg of *T. parva* DNA. Use a different amount of restriction enzyme for each tube, i.e., 0.8, 1.2, 2.0, 2.8, or 3.6 units of *Sau*3A, and digest in a final volume of 50 μl at 37°C for 1 hr.

(ii) Stop reaction by chilling each tube on ice and adding EDTA to a final concentration of 20 m*M* (e.g., 4 μl of 0.25 *M* EDTA to 50-μl reaction).

(iii) Mix the contents of all five tubes, extract once with an equal volume of buffered phenol/chloroform, once with chloroform. Precipitate the DNA by adding and mixing Na acetate to a final concentration of 0.3 *M* (i.e., 20 μl of 3 *M* Na acetate in 200 μl of DNA solution) and 2 vol of ice-cold ethanol. Chill DNA on ice for 1 hr, pellet the DNA by centrifugation at 12,000*g* for 5 min, rinse the pellet in 70% ethanol.

(iv) Dissolve DNA in T.E and size-separate on a linear sucrose gradient (top 10% sucrose and bottom 40% sucrose, both in 25 m*M* Tris-HCl, pH 8.0, 5 m*M* EDTA), poured using a two-chamber gradient mixer. The sample over the gradient is centrifuged in SW60 Ti rotor at 55,000 rpm for 12 hr at 20°C. The sample is fractionated by upward displacement with 50% sucrose using an ISCO gradient fractionator.

(v) Analyze DNA fractions by electrophoresis on agarose gels to identify those carrying fragments of 4.5–5.5 kb.

(vi) Combine these and extract with an equal volume of buffered phenol/chloroform and then with chloroform, ethanol precipitate and resuspend in T.E at a concentration of between 50 and 100 μg/ml.

2. Partial Filling-in Reaction To Generate Compatible Sticky Ends in the Vector and Target DNAs

(a) Digest to completion 5 μg of the pBluescript SK M13 plasmid vector (Strategene, San Diego, CA) with 5–10 units of *Sal*I restriction enzyme in a 50-μl reaction volume.

(b) Check that the digestion is complete by electrophoresing a sample (0.5 μg) on a 1% minigel.

(c) Set up filling-in reactions separately for the vector (pBluescript) and target *T. parva* DNAs as follows:

| | Filling-in: | |
Reagents	vector *Sal*I	Target *Sau*3AI
1 m*M* dATP	—	1.0 μl
1 m*M* dCTP	1.0 μl	—
1 m*M* dGTP	—	1.0 μl
1 m*M* dTTP	1.0 μl	—
10× ligase buffer	1.0 μl	1.0 μl
dH₂O	3.5 μl	3.0 μl
Vector (50 μg/ml)	2.5 μl	—
Target (80 μg/ml)	—	3.0 μl (240 ng)
Klenow (4 units/μl)	1.0 μl	1.0 μl

(d) Mix and incubate at 25°C for 20 min.

(e) Transfer to a 68°C water bath for 20 min.

(f) Leave at 4°C until ready for ligation. The above reaction fills the vector and target with T + C and G + A, respectively.

3. Ligation of Partially Filled Vector and Target DNAs

(a) The ligation reaction is set up as follows:

Reagents	Control	Vector/target ligation
T + C filled		
*Sal*I-digested vector	2.0 μl (25 ng)	8.0 μl (100 ng)
G + C filled *Sau*3A-digested *T. parva* target	—	10.0 μl (240 ng)
10× ligase buffer	2.0 μl	10.0 μl
5 mM ATP	2.0 μl	10.0 μl
dH$_2$O	13.5 μl	61.0 μl
Ligase, 4 units/μl	0.5 μl	1.0 μl

(b) Incubate for 3–5 hr at 16°C.

4. Transformation of *Escherichia coli* Strain JM83

A standard protocol for the transformation of *E. coli* cells with plasmids (Hanahan, 1983) is used. A brief description of the preparation of fresh competent cells and their transformation is given below. The detailed protocol for the preparation of SOB, SOC, and TFB can be found in Hanahan (1983).

(a) Pick four or five fresh individual colonies of *E. coli* JM83 from an agar plate and suspend in 100 ml of SOB in a 1-liter flask.

(b) Mix in a 37°C orbital shaker (250 rpm/min) incubator until O.A.$_{600}$ is 0.4–0.5.

(c) Centrifuge at 2500g for 10 min at 4°C.

(d) Pour off the supernatant and resuspend the cell pellet by gentle pipetting in 30 ml of ice-cold TFB in a 50-ml tube.

(e) Centrifuge as in (c).

(f) Pour off the supernatant and resuspend the cell pellet by gentle pipetting in 8 ml of TFB. Immediately add 280 μl DMSO (ultrapure), mix, and leave on ice for 20 min.

(g) Add an additional 280 μl of DMSO, mix, and leave on ice for 20 min.

(h) Aliquot 8 × 1 ml and 1 × 200 μl of the cell suspension in 15-ml round-bottom polystyrene tubes on ice.

(i) Add, to each of the eight separate tubes containing 1 ml cells, 10 μl (~30 ng) of the ligation mix containing vector and target DNAs.

(j) Add to the tube containing 200 μl of cells 3 μl (3 ng) of the control ligation mix containing only the vector.

(k) Leave on ice for 30 min.

(l) Heat pulse for 90 sec at 42°C.

(m) Return tubes to ice and add 4 ml of ice-cold SOC to the eight tubes containing 1 ml of competent cells and 800 μl of SOC to the control tube containing 200 μl of competent cells.

(n) Transfer the tube to a 37°C shaker water bath for 30 min.

(o) Pool cells from all eight tubes (32 ml) in 50-ml tube. Centrifuge the tube containing the pooled cells and the control tube at 2500g for 10 min at room temperature. Resuspend each pellet in one-third its original volume in SOC (i.e., the pooled cells in 10.6 ml and the control cells in 330 μl of SOC). Plate 200 μl cell suspension per 90-mm^2 NZYCM/ampicillin plates spread with 170 μl each of 20 mg/ml X-gal and 2% IPTG solutions.

(p) Leave the plates at room temperature until the liquid has been absorbed and then leave inverted at 37°C. Examine for colonies after 12–16 hr. The control plate should have either very few blue colonies (2–4 per plate) or no colonies. Others should have approximately 2500 colonies per plate and over 97% of these should be recombinants, i.e., white colonies.

(q) To isolate recombinant plasmid DNA from these bacteria add 4 ml of SOB to each plate and scrape off bacteria with a sterile L-shaped glass rod. Pool the cell suspension and centrifuge at 2500*g* for 10 min at 4°C. Prepare plasmid DNA from the pellet by alkaline lysis method and caesium chloride–ethidum bromide centrifugation (Sambrook *et al.*, 1989).

5. Enrichment of Linking Clones

The representative genomic library contains only a small proportion of fragments containing the rare-restriction site. In the *T. parva Sau*3AI genomic library it was estimated that only 1/60 of the total DNA fragments contained *Sfi*I sites, since there were only 33 *Sfi*I fragments in the total genome of 10^7 bp. Therefore, the critical step in obtaining linking clones is the ability to separate the very small numbers of clones bearing the appropriate restriction site from the bulk of the DNA. A variety of methods has been described for the preparation of libraries containing enriched populations of linking clones (Pohl *et al.*, 1988; Collins, 1988; Wallace *et al.*, 1989; Zabarovsky *et al.*, 1990; Smith *et al.*, 1987; Hayashizaki *et al.*, 1992). In most of these methods, linking fragments are normally cloned by circularization with or without a selectable marker, digestion with the appropriate enzyme, and ligation into a vector containing the relevant restriction site. However, this is not possible for linking clones carrying *Sfi*I sites because different *Sfi*I sites generally have different cohesive ends. This problem can be overcome by physically separating the linking clones by first linearizing the clones with *Sfi*I and then separating the linear molecules from the bulk of circular plasmids using standard electrophoresis through high-percentage gels of SeaPlaque agarose or pulsed-field–polyacrylamide-gel electrophoresis (Young and Morzaria, 1991; Ito and Sakaki, 1988). The SeaPlaque agarose approach has been found to be simple and efficient and has provided a 25% enrichment of linking clones expected to be present at a frequency of 1/60 in the original library of 4- to 6-kb fragments of the genomic DNA of *T. parva.*

(a) Digest DNA (5 μg) from the plasmid library with *Sfi*I (10 units/μg of DNA) in a 50-μl reaction volume at 50°C for 3 hr. Stop reaction by transferring the tube on ice and add 10 μl of 5× loading buffer prepared in TAE (see below).

(b) Use the same DNA-undigested (1 μg) and *Not*I-digested, completely linearized DNA molecules (1 μg), respectively, as controls (the vector has a *Not*I site).

(c) Prepare a 2.5% SeaPlaque agarose gel in 1× TAE. Electrophorese 4 × 15 μl of the *Sfi*I-digested DNA, one uncut control, and one *Not*I-linearized control at 15 V/cm for 5 hr.

(d) Following electrophoresis, visualize the DNA by staining with ethidium bromide. The separation should be as shown in Fig. 5.4.

(e) Recover the linearized *Sfi*I fragments from the gel by the NaI/Glassmilk method (GeneClean, BIO 101, La Jolla, CA) and religate as follows:

*Sfi*I-linearized DNA	40.0 μl (~60 ng)
5 m*M* ATP	14.0 μl
Ligase, 4 units/μl	2.0 μl
10× ligase buffer	15.0 μl
H₂0	79.0 μl
Total	150.0 μl

Incubate at 8°C for 30 hr.

(f) Transform *E. coli* JM83 as described above. Briefly, set up 14 tubes, each containing 300 μl of competent cells, and to each add 10 μl of the religated DNA. Heat-shock, add 1.2 ml of SOC per tube, leave at 37°C for 30 min, pool samples, pellet cells, and resuspend in 4 ml of SOC. Plate out 200 μl of cell suspension per 90-mm^2 NZYCM/ampicillin plate. Incubate at 37°C for 18 hr and examine for colonies. Approximately 100 colonies per plate will be observed, of which 85% should be white.

(g) From 1.5 ml of overnight cultures of single white recombinant colonies prepare minipreps (Sambrook *et al.,* 1989) for the identification of *Sfi*I linking clones.

G. Identification and Characterization of Linking Clones

1. Identification of *Sfi*I Linking Clones

(a) Run approximately 500 ng of uncut miniprep DNAs from putative recombinants on 0.8% agarose in 1× TAE gel and visualize after ethidium

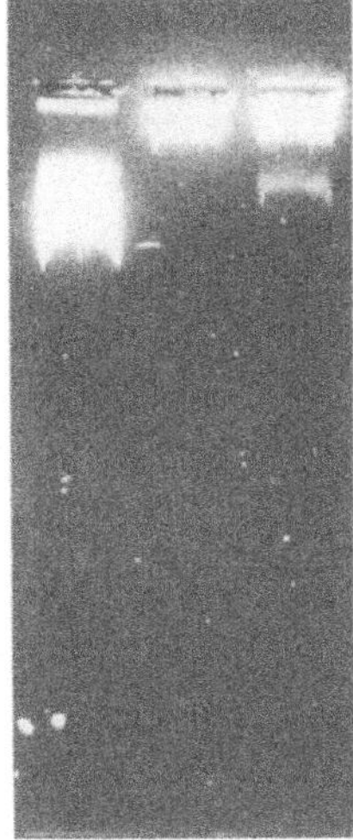

Figure 5.4 Enrichment of linking clones. A 2.5% SeaPlaque agarose gel was loaded with three samples of plasmid DNA library, *Not*I digested (N), uncut (U), and *Sfi*I cut (S). Following electrophoresis for 5 hr at 15 V/cm a small proportion of the linearized library was separated as shown in lane S, leaving behind, in or near the well, the majority of the circular DNA molecules. Similarly, the undigested plasmid library remained in or just migrated out of the well (U) while the completely linearized library moved away much farther from the well (N). Reprinted by permission of the publisher from "Agarose Entrapment Method for Production of *Sfi*I, Linking Library for *Theileria parva*," J. R. Young and S. P. Morzaria, *GATA*, Vol. 8, No. 5, pp. 148–150. Copyright 1991 by Elsevier Science Inc.

bromide staining. Use uncut pBluescript SK M13 plasmid vector DNA as control.

(b) Select the plasmids containing inserts of the expected size (4–6 kb). These are now putative linking clones.

2. Characterization of *Sfi*I Linking Clones

The redundancy of the initial library means that many different linking clones selected as above will contain the same *Sfi*I site. In order to select linking clones containing unique *Sfi*I sites, comparative restriction maps using electrophoretic analysis of *Sau*3AI and *Sau*3AI/*Sfi*I double-digests are made. The principle for selection of unique linking clones is illustrated in Fig. 5.5. Initial electrophoretic analysis of putative linking clones is performed in 0.8% agarose gel. This provides resolution of DNA fragments up to 200 bp. For smaller fragments, end-labeled DNA fragments following *Sau*3AI and *Sau*3AI/*Sfi*I digestion need to be separated on acrylamide gels. The method for separating end-labeled small DNA fragments in polyacrylamide gel is described below.

Forty-eight lanes can be run on a polyacrylamide gel as described below. This enables characterization of 24 linking clones, each clone digested with *Sau*3AI and *Sau*3AI/*Sfi*I.

(a) Mix on ice the restriction digestion buffers and restriction enzymes as follows:

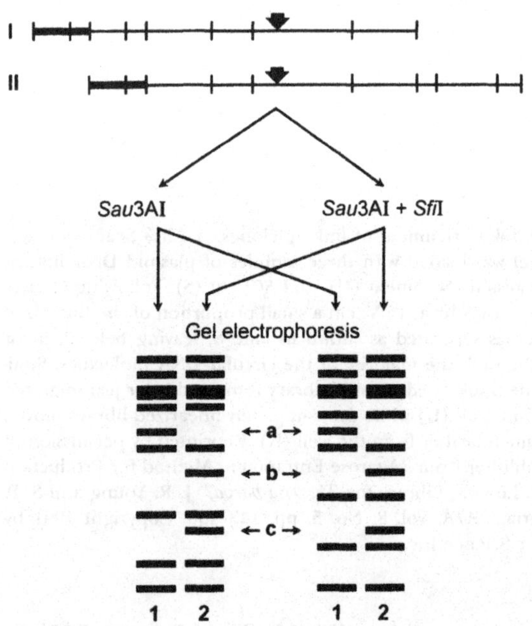

Figure 5.5 Characterization of linking clones. The enriched linking clone library contains many clones, most of which will consist of several vector and parasite *Sau*3AI fragments (horizontal lines interrupted with vertical short bars). Two clones bearing the same *Sfi*I site (vertical arrow) will always have the same *Sau*3AI fragment (a) that will cut into the same two smaller *Sfi*I fragments (b and c in lanes 1 and 2). Groups of such overlapping clones can be identified by electrophoretic comparison of *Sau*3AI and *Sau*3AI/*Sfi*I digests. Reprinted with permission by Morzaria and Young, "Genome Analysis of *Theileria parva*," *Parsitology Today* **9** (1993).

	Digestion with:	
Reagents	*Sau*3AI	*Sau*3AI/*Sfi*I
10× *Sfi*I buffer	12.5 μl	12.5 μl
dH₂O	96.0 μl	89.0 μl
*Sfi*I (10 units/μl)	—	7.0 μl
*Sau*3AI(5 units/μl)	4.0 μl	4.0 μl
Total	112.5 μl	112.5 μl

(b) Aliquot 2× 0.5 μl (~100 ng) of each of the plasmid minipreps from 24 linking clones to be characterized in 48 × 500-μl tubes. To one half of the 24 aliquots add 4.5 μl of the above *Sau*3AI mix and to the other half add 4.5 μl of the *Sau*3AI/*Sfi*I mix.

(c) Incubate for 2–3 hr at 37°C. *Sfi*I is active at this temperature for DNA in solution.

(d) Prepare the gel solution containing 5.0 g acrylamide and 0.25 g bis-acrylamide in 1× TBE. Add 200 μl of 10% ammonium persulpfate (made fresh) and 20 μl TEMED and mix immediately.

(e) Pour the solution gently between two glass plates clamped together with 0.4-mm spacers, using a 25-ml pipette.

(f) Place a shark's-tooth comb inverted between top of the plates to a depth of about 3–4 mm to give a flat top to the gel. Allow to polymerize.

(g) Remove clamps and spacers and invert the shark's-tooth comb. Fill the bottom space with 1× TBE and load into the tank. The gel is now ready for loading the samples.

(h) Set up a mix for end-labeling 48 restriction digests as follows:

10× *Sfi*I buffer	6.0 μl
1 m*M* dCTP	10.0 μl
1 m*M* dGTP	10.0 μl
1 m*M* dTTP	10.0 μl
H$_2$O	17.0 μl
Klenow	3.0 μl
[α−^{35}S]dATP (10 mCi/ml)	4.0 μl
Total	60.0 μl

(i) Mix everything and then aliquot 1.0 μl into each of 48 wells of a round-bottom microtitre plate.

(j) Add to these wells 2.0 μl of each of the restriction digests from (b) above and leave at room temperature for 10 min.

(k) Add 1.0 μl of TBE–dye–glycerol loading buffer to each tube and mix.

(l) Load 1.5 μl of the above into each track using a Hamilton syringe and ensure that the sample is evenly layered.

(m) Allow 10 min for the samples to settle and then run the gel at 300 V until the bromophenol blue reaches the bottom of the gel (approximately 2.5 hr).

(n) Remove the gel from the tank, separate the glass plates, and soak the gel attached to one of the plates in 10% methanol/10% acetic acid for 15–20 min. Place dry 3MM paper onto the gel and press down firmly.

(o) Lift off the paper with the gel attached. Dry on gel dryer for 30 min.

(p) Expose the dried gel to X-ray film. A typical result is shown in Fig. 5.6. *Sau*3AI fragments bearing *Sfi*I sites can be easily identified.

H. Ordering of Macrorestriction Fragments Using Linking Clones as Probes

Unique linking clones can be identified using the above two methods, i.e., single- (*Sau*3AI) and double-digestion (*Sau*3AI/*Sfi*I) of linking clones and

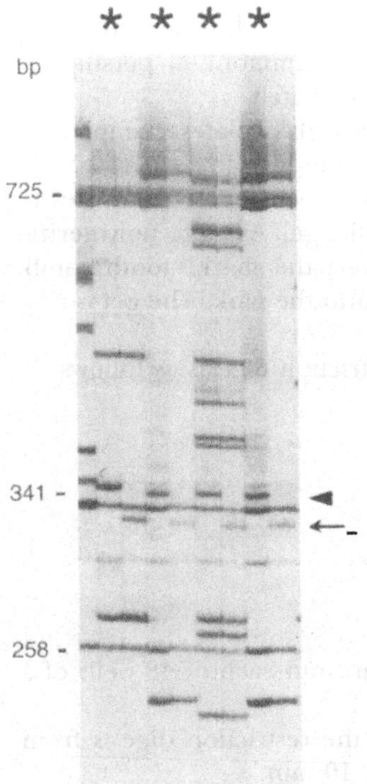

Figure 5.6 Linking clone characterization following polyacrylamide gel electrophoresis of end-labeled *Sau*3AI and *Sau*3AI/*Sfi*I double-digested plasmids. Similar clones are depicted by asterisks. Note that the parasite *Sau*3AI 350-bp fragment (arrow head) bearing *Sfi*I site disappears from the lane, resulting into two smaller 330-bp (arrow) and 20-bp (not shown in the gel) fragments.

visualizing the DNA fragments on either ethidium bromide-stained gels or on autoradiographs following end-labeling and separating in polyacrylamide gels. Each linking clone thus identified should hybridize to two unique *Sfi*I fragments of the genomic *T. parva* DNA. The linking clones, as minipreps, can be used directly as radiolabeled probes on Southern blots of *Sfi*I restriction enzyme-digested genomic DNA separated by PFGE. Random priming and nick translation labeling methods are both suitable. The procedure used for obtaining several strips of filter paper, on which *Sfi*I-digested genomic fragments of *T. parva* are transferred, is described below.

(1) Digest 12 agarose blocks (4 × 5 × 1 mm) containing *T. parva* genomic DNA with *Sfi*I.

(2) Pour a 1.5% agarose gel made in 0.38× TBE in a tray (10.5 × 14 cm) with a preparative comb (well former).

(3) Insert the agarose blocks in the well and place them directly beside each other leaving no space in the well.

(4) Seal the well with 1.5% agarose, place the gel in CHEF apparatus and run at appropriate voltage and pulse times.

(5) Remove the gel, stain as described, and photograph.

(6) Enlarge the photograph and scale to match exactly the original dimension of the gel. An example of a gel with *Sfi*I-resolved fragments is shown in Fig. 5.7.

(7) Southern-blot the gel onto Nylon filter and then cut-out twelve 8- to 10-mm-wide longitudnal strips. Usually several gels are run this way so that between 35 and 50 good strips can be obtained.

(8) Probe each strip with a unique linking clone. Each linking clone should hybridize to two *Sfi*I fragments.

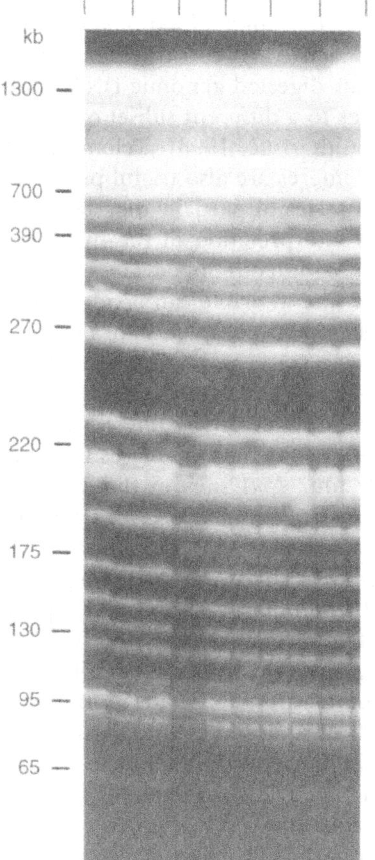

Figure 5.7 Agarose gel showing separation of *Theileria parva Sfi*I fragments from the genomic DNA. The fragments were separated as described under Procedures. Note that several comigrating fragments in the region of 800, 200, and 100 kb required different PFGE conditions for complete resolution of all the fragments.

I. Completion and Confirmation of the Map Using Other Techniques

The probability of obtaining a set of unique linking clones spanning all the rare-cutting restriction enzyme sites in a genome is low. For example, in attempting to construct an *Sfi*I restriction map of *T. parva*, it was estimated that it would be necessary to analyze 200 random linking clones to have a greater than 95% probability of obtaining clones spanning each of 29 *Sfi*I sites. Examination of 199 linking clones yielded 21 unique groups of linking clones enabling the establishment of 21 links between *Sfi*I fragments. Although further searching using the above strategy would have yielded a few more linking clones, the return for the effort expended would have been low. Therefore, it is usually necessary to resort to alternative approaches to complete and confirm a map.

Half of each linking clone, obtained by digesting a linking clone with *Sfi*I, separating on a gel, and purifying each fragment, can be used to probe the hybridization pattern of partially digested genomic DNA. Each half of a linking clone fragment hybridizes to a different subset of partial digest products as shown in Fig. 5.8 (Smith *et al.*, 1993). Telomeric sequences identifying particular ends of telomeres are also useful probes in similar analysis as they will hybridize to a series of partially digested fragments in one direction only. This makes interpretation of the hybridization patterns simpler. In this way, the ambiguous unlinked macrorestriction fragments for which no linking clones are available can be identified. These two approaches should theoretically enable completion of the map.

A number of approaches can be used to confirm data obtained from the application of linking clones and to resolve any ambiguities. Double-digestion of macrorestriction fragments with two rare-cutting restriction enzymes allows further subdivision of the genome and increases the map resolution. For example, there are only four *Not*I sites in the *T. parva* genome. *Sfi*I-digested and *Sfi*I/*Not*I-digested genomic DNA, separated by PFGE, enables the exact location of *Not*I sites in *Sfi*I fragments. Two-dimensional PFGE, involving first the separation of *Not*I fragments followed by digestion of each of the *Not*I fragments with *Sfi*I and separation of *Sfi*I fragments, provides an unambiguous way of assigning the fragments to subchromosomal genomic regions. This, combined with two-dimensional PFGE in which chromosomes are separated in the first dimension followed by digestion of each chromosome with *Sfi*I or *Not*I alone and separation of these fragments in the second dimension, enables identification and assignment of the macrorestriction fragments to chromosomes (see Chapter 6). Further confirmation of the assigned *Sfi*I fragments can be obtained by using PFG-purified macrorestriction fragments

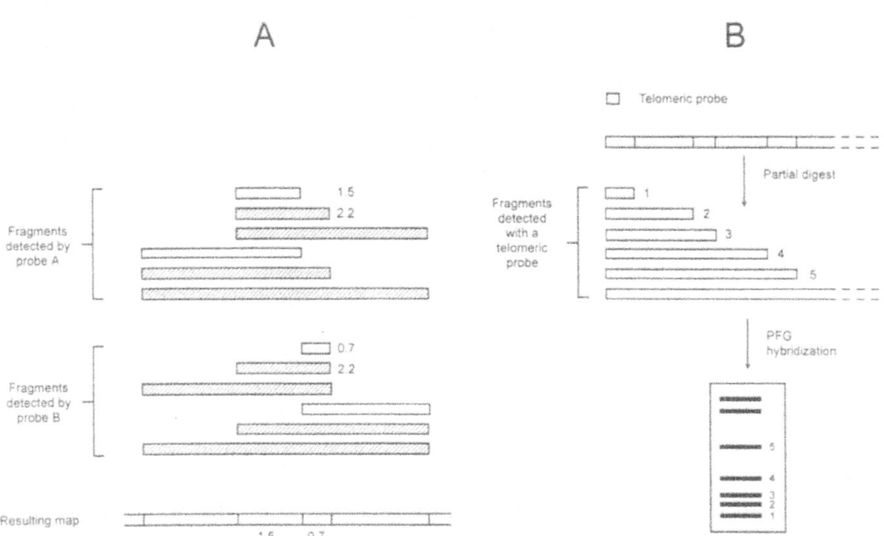

Figure 5.8 Use of partial restriction enzyme digestion to construct physical maps (A). Single probes (A or B) used in hybridization experiments to partially digested DNA detect bidirectional partial digestion products. Probes on adjacent fragments detect common (shaded) and unique (unshaded) bands. (B) Indirect-end labeling experiment with telomeric sequences reveals the restriction sites close to the end of the chromosome. Partial digestion experiments are easy to interpret with telomeric probes because fragments extend in only one direction.

as probes to hybridize to separated chromosomes. Similarly, linking clones can be used as probes to hybridize to chromosomes to provide additional confirmation of the assigned megarestriction fragments to chromosomes. A combination of the above approaches is usually adequate in enabling the completion of a restriction map.

V. General Remarks and Conclusions

The approaches and protocols described here can be used to construct low-resolution physical maps of a number of parasitic protozoa. Such maps are useful in understanding the complexity of genome structures and also provide a framework for more detailed analysis of genomes. The approach depends principally on the use of PFGE and linking clones to construct a map, but other techniques are necessary to provide a complete map. The linking clone approach is simple for genomes that have a small number of chromosomes and a small number of megarestriction fragments.

A number of problems can occur using the approach. Partial digestion of rare-cutting restriction sites can complicate the interpretation of Southern blot data with linking clones. This was particularly true with *Sfi*I sites in the *T. parva* genome and, despite the use of large amounts of the enzyme, some sites were resistant to complete digestion. If *Sfi*I sites have these characteristics then use of alternative restriction enzymes should be explored. The presence of repetitive sequences in linking clones could present a problem in interpreting the hybridization data. This can be prevented best by selecting short genomic fragments for preparing the linking library. Fortunately, most protozoan parasites do not contain highly dispersed repetitive sequences, unlike higher eukaryotic genomes.

The protocols described here enabled the successful construction of an *Sfi*I restriction map of the *T. parva* genome. A number of variant protocols have been described, especially for the selection and analysis of linking clones. The final choice of approach and use of protocols will depend greatly on the physical characteristics of the genome to be mapped.

Acknowledgments

Drs. John Young, Vish Nene, and Noel Murphy are thanked for their helpful advice in the preparation of the manuscript. This is ILRI Publication No. 1279.

References

Blackburn, E. H., and Szostak, J. (1984). The molecular structure of centromeres and telomeres. *Annu. Rev. Biochem.* **53,** 163–194.

Chu, G., Vollrath, D., and Davis, R. W. (1986). Separation of large DNA molecules by contour-clamped homogenous electric fields. *Science* **234,** 1582–1585.

Collins, F. S. (1988). Chromosome jumping. *In* "Genome Analysis: A Practical Approach" (K. E. Davies, ed.), pp. 73–94. IRL Press, Oxford.

Conrad, P. A., Iams, K. P., Brown, W. C., Sohanpal, B., and Ole-MoiYoi, O. K. (1987). DNA probes detect genomic diversity in *Theileria parva* stocks. *Mol. Biochem. Parasitol.* **25,** 213–226.

Gardiner, K., Laas, W., and Patterson, D. (1986). Fractionation of large mammalian DNA restriction fragments using vertical pulsed-field gel electrophoresis. *Somatic Cell Mol. Genet.* **12,** 185–195.

Hanahan, D. (1983). Studies on transformation of *Escherichia coli* with plasmids. *J. Mol. Biol.* **166,** 557–580.

Hayashizaki, Y., Hirotsune, S., Hatada, I., Tamatsukuri, S., Miyamoto, C., Furuichi, Y., and Mukai, T. (1992). A new method for constructing *Not*I linking and boundary libraries using a restriction trapper. *Genomics* **14,** 733–739.

Ito, T., and Sakaki, Y. (1988). A novel procedure for selective cloning of *Not*I linking fragments from mammalian genome. *Nucleic Acids Res.* **16,** 9177–9184.

Kemp, D. J., Thompson, J. K., Walliker, D., and Corcoran, L. M. (1987). Molecular karyotype of *Plasmodium falciparum:* Conserved linkage map and expandable histidine-rich protein genes. *Proc. Natl. Acad. Sci. U.S.A.* **84,** 7672–7676.

Lanham, S. M., and Godfrey, D. G. (1970). Isolation of salivarian trypanosomes from man and other mammals using DEAE-cellulose. *Exp. Parasitol.* **28,** 521–534.

Melville, S. E., Barrett, M. P., Sweetman, J. P., Ajioka, J. W., and Le Page, R. W. F. (1993). Chromosome polymorphism in *Trypanosoma brucei brucei* and the selection of chromosome-specific markers. *In* "Genome Analysis of Protozoan Parasites" (S. P. Morzaria, ed.), pp. 123–132. ILRAD, Kenya.

Morzaria, S. P., and Young, J. R. (1992). Restriction mapping of the genome of the protozoan parasite *Theileria parva. Proc. Natl. Acad. Sci., U.S.A.* **89,** 5241–5245.

Morzaria, S. P., Spooner, P. R., Bishop, R. P., Musoke, A. J., and Young, J. R. (1990). *Sf*I and *Not*I polymorphisms in *Theileria* stocks detected by pulsed field gel electrophoresis. *Mol. Biochem. Parasitol.* **40,** 203–212.

Pohl, T. M., Zimmer, M., Macdonald, M. E., Smith, B., Bucan, M., Poustka, A., Volinia, S., Searle, S., Zehetner, G., Wasmuth, J., Gusella, J., Lehrach, H., and Frishauf, A. M. (1988). Construction of a *Not*I linking library and isolation of new markers close to the Huntington's disease gene. *Nucleic Acids Res.* **16,** 9185–9198.

Sambrook, J., Fritsch, E. F., and Maniatis, T. (1989). "Molecular Cloning: A Laboratory Manual," 2nd ed. Cold Spring Harbor Lab., Press, Cold Spring Harbor, NY.

Schwartz, D. C., and Cantor, C. R. (1984). Separation of yeast chromosome-sized DNAs by pulsed-field gel electrophoresis. *Cell (Cambridge, Mass.)* **37,** 67–75.

Sibley, L. D., and Boothroyd, J. C. (1992). Construction of a molecular karyotype for *Toxoplasma gondii. Mol. Biochem. Parasitol.* **51,** 291–300.

Sibley, L. D., Pfefferkorn, E. R., and Boothroyd, J. C. (1993). Development of genetic systems for *Toxoplasma gondii. Parasitol. Today* **1,** 392–395.

Sinnis, P., and Wellems, T. E. (1988). Long-range restriction maps of *Plasmodium falciparum* chromosomes: Crossing over and size variation among geographically distant isolates. *Genomics* **3,** 287–293.

Smith, C. L., Warburton, P., Gaal, A., and Cantor, C. R. (1986). Analysis of genome organization and rearrangements by pulsed field gradient gel electrophoresis. *Genet. Eng.* **8,** 45–70.

Smith, C. L., Lawrence, S. K., Gillespie, G. A., Cantor, C. R., Weissman, S. M., and Collins, F. S. (1987). Strategies for mapping and cloning macro-regions of mammalian genomes. *In* "Methods in Enzymology" (M. Gottesman, ed.), Vol. 151, pp. 461–489. Academic Press, San Diego, CA.

Smith, C. L., Klco, S. R., and Cantor, C. R. (1988). Pulsed field gel electrophoresis and the technology of large DNA molecules. *In* "Genome Analysis: A Practical Approach" (K. Davies, ed.), pp. 41–72. IRL Press, Oxford.

Smith, C. L., Oliva, R., Wang, D., Grothues, D., and Lawrance, S. (1993). Physical strategies for the molecular dissection of genomes. *In* "Genome Analysis of Protozoan Parasites" (S. P. Morzaria, ed.), pp. 123–132. ILRAD, Kenya.

Spithill, T. W., and Samaras, N. (1985). The molecular karyotype of *Leishmania major* and mapping of Â- and β-tubulin gene families to multiple unlinked chromosomal loci. *Nucleic Acids Res.* **13,** 4155–4161.

Triglia, T., and Kemp, D. J. (1991). Large fragments of *Plasmodium falciparum* DNA can be stable intron cloned in yeast artificial chromosomes. *Mol. Biochem. Parasitol.* **44,** 207–212.

Triglia, T., Wellems, T. E., and Kemp, D. J. (1992). Towards a high-resolution map of the *Plasmodium falciparum* genome. *Parasitol. Today* **8,** 225–228.

Van der Ploeg, L. H. T., Schwartz, D. C., Cantor, C. R., and Borst, P. (1984). Antigenic variation in *Trypanosoma brucei* analysed by electrophoretic separation of chromosome-sized DNA molecules. *Cell (Cambridge, Mass.)* **37,** 77–84.

Vollrath, D., and Davis, R. W. (1987). Resolution of DNA molecules greater than 5 megabases by contour-clamped homogeneous electric fields. *Nucleic Acids Res.* **15,** 7865–7876.

Wallace, M. R., Fountain, J. W., Brereton, A., and Collins, F. S. (1989). Direct construction of a chromosome-specific *Not*I linking library from flow-sorted chromosomes. *Nucleic Acids Res.* **17,** 1665–1677.

Watebury, P. G., and Lane, M. J. (1987). Generation of *lambda* phage concatemers for use as pulsed-field electrophoresis size markers. *Nucleic Acids Res.* **15,** 3930.

Young, J. R., and Morzaria, S. P. (1991). Agarose entrapment method for the production of *Sfi*I linking library for *Theileria parva*. *Genet. Anal. Tech. Appl.* **8,** 148–150.

Zabarovsky, E. R., Boldog, F., Thompson, T., Scanlon, D., Winberg, G., Marcsek, Z., Erlandsson, R., Stanbridge, E. J., Klein, G., and Sumegi, J. (1990). Construction of a human chromosome specific *Not*I linking library using a novel cloning procedure. *Nucleic Acids Res.* **18,** 6319–6324.

6

Macrorestriction Mapping and Analysis of Bacterial Genomes

Ute Römling, Rainer Fislage, and Burkhard Tümmler

I. Introduction

The relatively small genome of 0.5–10 Mbp makes bacteria an appropriate target for the comprehensive study of genome organization by pulsed-field gel electrophoresis (PFGE) techniques. For the construction of a physical map, the bacterial chromosome is digested with a rare-cutting restriction endonuclease, the generated macrorestriction fragments are separated by PFGE, and the fragments are subsequently ordered by strategies that have previously been applied to the mapping of small plasmids. Alternatively, genomic DNA can be cloned into cosmids, phages, or bacterial or yeast artificial chromosomes, and individual clones of the library are subsequently ordered into a chromosome-spanning contig (for details of the latter approach see Chapters 5, 7, and 11). This chapter focuses on the construction of macrorestriction maps of bacterial genomes. The PFGE techniques described herein are generally applicable to any prokaryote.

II. Materials

All buffers and materials are to be autoclaved unless otherwise stated.

A. 10× TBE Buffer

Final concentration	To prepare 1 liter
0.9 M Tris 0.9 M Boric acid 0.04 M EDTA	Add 109 g Tris, 56 g boric acid, and 7.95 g Na$_2$ EDTA·2H$_2$O. Fill up to 1 liter with water. Measure pH. It should be within 8.3–8.6.

B. 0.5× TBE

Final concentration	To prepare 1 liter
45 mM Tris 45 mM Boric acid 2 mM EDTA	Add 50 ml 10× TBE buffer and adjust the volume to 1 liter with water.

C. Restriction Enzyme Buffers

Usually the recipes of the suppliers are followed. Larger amounts of buffer should be prepared for preincubation and adjustment of plugs and agarose lanes to be digested. A list of some commonly used buffers is given. For *Spe*I we use one buffer regardless of the enzyme supplier. *Pac*I and *Dpn*I give good results in the buffers mentioned below. DNA cut with restriction enzymes having recognition sites of at least 8 bp (I-*Ceu*I, I-*Sce*I, *Pac*I, *Swa*I) or cloned enzymes (*Dpn*I) sometimes requires proteinase K treatment after digestion to remove the enzyme from the DNA.

1. *Spe*I Buffer (Klenow end-labeling buffer)

Final concentration	To prepare 100 ml
10 mM Tris-HCl 50 mM NaCl 10 mM MgCl$_2$ pH 7.5 (37°C)	Add 1 ml 1 M Tris, pH 8.0, 5 ml 1 M NaCl, 1 ml 1 M MgCl$_2$. Fill up to 95 ml with water, adjust pH to 7.8 (22°C) with HCl and volume to 100 ml with water.

2. *Dpn*I Buffer

Final concentration	To prepare 100 ml
10 mM Tris-HCl 150 mM NaCl 10 mM MgCl$_2$ pH 7.5 (37°C)	Add 1 ml 1 M Tris, pH 8.0, 15 ml 1 M NaCl, 1 ml 1 M MgCl$_2$. Fill up to 95 ml with water, adjust pH to 7.8 (22°C) with HCl and volume to 100 ml with water.

3. PacI Buffer

Final concentration	To prepare 100 ml
25 mM Tris-acetate 100 mM K-glutamate 10 mM Mg-acetate	Add 2.5 ml 1 M Tris-acetate, pH 8.0, 10 ml 1 M K-glutamate, 1 ml 1 M Mg-acetate. Fill up to 95 ml with water, adjust pH to 7.6 with acetic acid and volume to 100 ml with water.

4. SwaI Buffer

Final concentration	To prepare 100 ml
50 mM Tris-HCl 100 mM NaCl 10 mM MgCl$_2$	Add 5 ml 1 M Tris-HCl, pH 8.0, 10 ml 1 M NaCl, 1 ml 1 M MgCl$_2$. Fill up to 95 ml with water, adjust pH to 7.8 (22°C) with HCl and volume to 100 ml with water.

5. PmeI Buffer

Final concentration	To prepare 100 ml
20 mM Tris-acetate 50 mM K-acetate 10 mM Mg-acetate	Add 2 ml 1 M Tris-acetate, pH 8.0, 5 ml 1 M K-acetate, 1 ml 1 M Mg-acetate. Fill up to 95 ml with water, adjust pH to 7.9 (25°C) with acetic acid and volume to 100 ml with water.

6. I-CeuI Buffer

Final concentration	To prepare 100 ml
10 mM Tris-HCl 10 mM MgCl$_2$	Add 1 ml 1 M Tris-HCl, pH 9.0, 1 ml 1 M MgCl$_2$. Fill up to 95 ml with water, adjust pH to 8.6 (25°C) with HCl and volume to 100 ml with water.

D. Tris (1 M, pH 8.0/9.0) (pH of Tris Buffers Is Strongly Temperature Dependent)

To prepare 100 ml: Add 12.1 g Tris into 80 ml water using a magnetic stirring bar. Adjust pH to 8.0/9.0 (25°C) with concentrated HCl. Adjust volume to 100 ml.

E. NaCl (1 M)

To prepare 100 ml: Dissolve 5.84 g NaCl in 80 ml water using a magnetic stirring bar. Adjust volume to 100 ml.

F. MgCl$_2$ (1 M)

To prepare 100 ml: Mix 20.3 g MgCl$_2$ · 6H$_2$O into 70 ml water using a magnetic stirring bar. Adjust volume to 100 ml.

G. K-Glutamate (1 *M*)

To prepare 100 ml: Dissolve 20.3 g K-glutamate · H_2O in 80 ml water. Adjust volume to 100 ml.

H. Mg-Acetate (1 *M*)

To prepare 100 ml: Dissolve 21.45 g Mg-acetate · $4H_2O$ in 70 ml water. Adjust volume to 100 ml.

I. K-Acetate (1 *M*)

To prepare 100 ml: Add 9.8 g K-acetate to 80 ml water. Mix and adjust volume to 100 ml.

J. DTT (0.5 *M*)

To prepare 10 ml: Mix 0.77 g dithiothreitol (DTT) with 9 ml autoclaved water in a sterile tube. Adjust volume to 10 ml and distribute 1 ml into each sterile Eppendorf cup. Store at −20°C.

K. BSA (10 mg/ml)

To prepare 10 ml: Add 100 μg BSA to 9 ml autoclaved water in a sterile tube. Adjust volume to 10 ml, distribute 1 ml into each sterile Eppendorf cup, and store at −20°C.

L. β-Mercaptoethanol (0.1 *M*)

To prepare 1 ml: Add 7 μl 14.4 *M* β-mercaptoethanol to 993 μl autoclaved water in a sterile Eppendorf cup. Prepare freshly before use; do not store or freeze.

M. 10× TE

Final concentration	To prepare 500 ml
100 m*M* Tris	Add 6.1 g Tris and 18.6 g Na_2EDTA · $2H_2O$ to 450 ml water.
100 m*M* EDTA	Adjust pH to 7.4 while stirring using HCl or NaOH
pH 7.4	

N. 1× TE

Final concentration	To prepare 500 ml

10 mM Tris	Add 50 ml 10× TE to 450 ml water. Mix and autoclave.
10 mM EDTA	
pH 7.4	

III. Topology of Bacterial Genomes

The number, size, and topology of genetic entities are the fundamental characteristics of a genome. The bacterial genome has long been considered to consist of one circular chromosome and one or more circular plasmids. PFGE studies invalidated this classical view. Bacteria show a variety of genomic configurations (Hinnebusch and Tilly, 1993). Several members of the α-group Protobacteriaceae have more than one chromosome (Suwanto and Kaplan, 1989) (examples are displayed in Fig. 6.1); linear chromosomes were discovered in *Streptomyces* species (Lin *et al.*, 1993), the spirochaete *Borrelia burgdorferi* (Ferdows and Barbour, 1989), and *Agrobacterium tumefaciens* (Allardet-Servent *et al.*, 1993).

To construct a physical genome map, one should start by resolving the genomic configuration of the organism to be mapped. PFGE per-

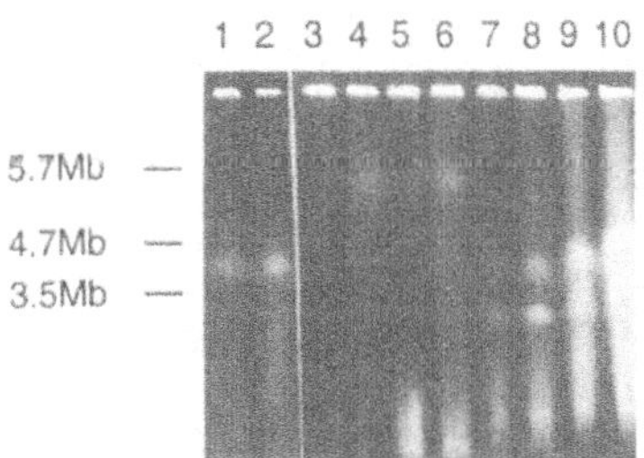

Figure 6.1 Comparison of untreated and γ-irradiated plugs of bacteria. *Pseudomonas lemoignei* reveals a chromosome of 3.9 Mbp. This correlates well with the sum of restriction digest fragments (4.1 Mbp). A plasmid is also probably present but this has not been verified yet. *Herbaspirillicum seropedicae* shows a chromosome of 5.7 Mbp. *Alcaligenes eutrophus* has two megareplicons of 3.5 and 2.5 Mbp in size. Dosage was always 10 Gy. 1, *P. lemoignei* LMG 2207, untreated; 2, *P. lemoignei*, irradiated; 3, *H. seropedicae* LMG 6513, untreated; 4, LMG 6513, irradiated; 5, *H. seropedicae* LMG 6512, untreated; 6, LMG 6512, irradiated; 7, *A. eutrophus* LMG 1201, untreated; 8, LMG 1201, irradiated; 9, *A. eutrophus* LMG 1199, untreated; 10, LMG1199, irradiated. Experimental conditions: CHEF DRII; 1.3 V/cm; 0.5× TBE; 0.6% agarose; 8°C; pulse times linearly increased from 2500 to 4750 sec for 145 hr.

mits visualization of chromosomes and plasmids. Unsheared bacterial DNA is prepared by inclusion of intact bacteria into agarose blocks prior to cell lysis (for details see Chapter 1). In the initial PFGE experiments the chosen pulse times and ramps should separate linear DNA molecules of up to 10 Mb. Under these conditions linear chromosomes and plasmids, as well as supercoiled circular DNA of up to about 150 kb, enter the gel and will be resolved, whereas open circular molecules will remain trapped in the agarose block. A subsequent PFGE run with different pulse times will discriminate between the pulse-time-dependent migration characteristics of linear DNA molecules and the invariant band position of a circle. The size of linear chromosomes and linear plasmids is determined from a calibration curve that is constructed from the relative mobilities of size markers separated in adjacent lanes of the gel. A restriction digest of bacteriophage λ DNA, a ladder of λ oligomers, and the linear chromosomes of the lower eukaryotes *Saccharomyces cerevisiae*, *Candida albicans*, and *Schizosaccharomyces pombe*, which cover the whole size range found in plasmids and chromosomes of bacteria, could serve as size markers. However, erroneous size values will be obtained if the DNA concentration is too high during sample preparation, which strongly retards band migration. Embedding cells at concentrations of 1, 2.5, and 5×10^9 bacteria per milliliter agarose is recommended for visualization of all genetic entities.

Plasmids are not reliably detected by PFGE. Small plasmids (<10 kb) rapidly diffuse out of agarose plugs and larger circular plasmids may not migrate during PFGE. The isolation must be optimized on a case-to-case basis. We used a modified Eckhardt method (Eckhardt, 1978), i.e., gel electrophoresis of the supernatants of bacteria gently lysed in gel slots, in order to clearly identify plasmids of *Pseudomonas aeruginosa*. Alternatively, plasmid minipreparation (Birnboim, 1983; Lennon and DeCicco, 1991) and subsequent PFGE or conventional agarose gel electrophoresis were used to isolate the plasmids.

Determining the size of circular chromosomes requires the additional step of linearizing the DNA prior to PFGE, since large circular DNA remain trapped in the plug instead of migrating through the gel. Thus, after PFGE of embedded samples, circular chromosomes will produce an intense fluorescent signal from the agarose plug, which reflects the trapped circles, and faint bands in the Megabase range, which reflect linear fragments arising due to random breakage of a portion of the circular molecules. "Random single-hit" linearization provides information about the size and number of circular chromosomes in a bacterial genome. As described in the next paragraph, agarose-trapped DNA samples are exposed to irradiation with X- or γ-rays (Beverley, 1989) or to digestion kinetics

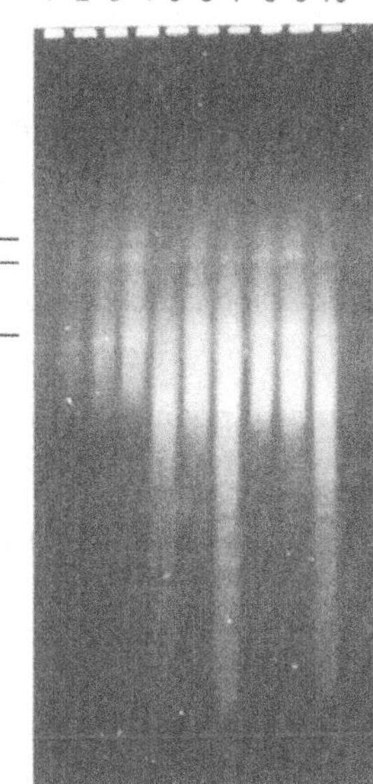

Figure 6.2 Short-time treatment of *Burkholderia cepacia* ATCC 25416 DNA with the frequent cutting restriction enzyme *Eco*Rl. Three replicons of 3.65, 3.2, and 1 Mbp are seen in the digested sample. 1, No enzyme; 2, amount of enzyme: 0.01 U, incubation time 5 min; 3, 0.05 U, 5 min; 4, 0.1 U, 5 min; 5, 0.5 U, 5 min; 6, 1 U, 5 min; 7, 5 U, 5 min; 8, 0.5 U, 1 min; 9, 1 U, 1 min; 10, 5 U, 1 min. Experimental conditions: CFGE; 1.3 V/cm; 0.25× TBE; 0.6% agarose; 8°C; pulse times linearly increased from 2000 to 4500 sec for 155 hr; angle 120°.

with a frequently cleaving restriction endonuclease. Separation by PFGE results in a smear of high-molecular-weight bands with the upper boundaries representing the size of the linearized single-cut chromosomes (Figs. 6.1 and 6.2).

A. Protocols for the Determination of the Number and Size of Replicons

1. Bacterial DNA

Some bacterial DNA will give distinct chromosomal bands without treatment of the plug (see Fig. 6.1). Use bacterial cell densities of 1, 2.5, and 5×10^9 cells/ml for three independent experiments.

2. Random Single-Hit Linearization by Radiation (Beverley, 1989) (Fig. 6.1)

(1) Take an agarose plug of $6 \times 5 \times 1$ mm in an Eppendorf cup and add 1 ml TE.

(2) Irradiate with dosages ranging from 10 to 100 Gy by using, e.g., a ^{60}Co source.

(3) Run a gel for the expected size range and use an untreated plug as standard.

Note. Referring to step 3: Bacterial chromosomes may vary in size from 1 Mb to approximately 10 Mb. Hence, pulse times between 1000 and 4000 sec at 1.3 V/cm for at least 90 hr are recommended (agarose concentration 0.6%).

3. Short-Time Exposure to Frequently Cutting Restriction Endonucleases (e.g., *Eco*RI) (Fig. 6.2)

(1) Take an agarose plug of $6 \times 5 \times 1$ mm in an Eppendorf cup.
(2) Equilibrate three times with 1 ml *Eco*RI buffer for 30 min each time at room temperature.
(3) Remove buffer, add 60 µl *Eco*RI buffer, 1 µl 0.5 M DTT, and 1 µl 10 mg/ml BSA and store the cup on ice.
(4) Add restriction enzyme (0.01–5 U) mix and leave it on ice for 30 min.
(5) Incubate 5 min at 37°C.
(6) Stop reaction by adding 1 ml TE.
(7) Run a gel for the expected size range and use an untreated plug as standard.

Note. Referring to steps 4 and 5: Enzyme amount and incubation time strongly depend on the particular lot of enzyme and the source of DNA. Therefore, the given values are not absolute but may vary over a wide range.

The aforementioned protocols will elucidate the number and topology of the elements of the bacterial genome and their approximate size and copy number. Next, one should distinguish whether each genomic entity is a chromosome or an extrachromosomal plasmid. Chromosomes encode vital cellular functions, like ribosomal RNA genes, which may be identified by hybridization of the PFGE-separated linearized genome with a ribosomal *rrn* probe. An alternative is the digestion of the agarose-embedded DNA with the intron-encoded endonuclease I-*Ceu*I (Liu *et al.*, 1993) which recognizes a 19- to 24-bp large-consensus sequence in the 23S rDNA gene that is conserved among chloroplast, mitochondrial, and prokaryotic genomes (Liu *et al.*, 1993). Hence, genomic digests with the commercially available restriction enzyme I-*Ceu*I allow an evaluation of the number of *rrn* operons in the bacterial strain. A chromosome with n *rrn* operons harboring the recognition sequence is cleaved into n I-*Ceu*I fragments if it is circular and into $n + 1$ I-*Ceu*I fragments if it is linear.

IV. Construction of Macrorestriction Maps

A. Choice of Restriction Endonucleases

For the construction of a macrorestriction map, the genomic DNA is digested with restriction endonucleases that cut only rarely and the fragments are subsequently separated by PFGE. Ordering of fragments will result in a low-resolution anonymous physical genome map that is converted into an informative genetic map by the assignment of genes (Smith *et al.*, 1987). Each restriction enzyme should cleave the chromosome into an informative, but still resolvable number of 5 to 40 fragments. Significant information about the genome organization and the map location of genes is only obtained if at least 20 sites are assigned, but of course a more refined map is desirable. This goal is achieved either by multiple enzymes, each of which cleaves at a few sites, or by two to three enzymes, each of which cleaves at multiple sites. At least two restriction enzymes should be employed for mapping in order to mutually prove the consistency of each macrorestriction map. It is often hard to establish genomic fragment order for islands of small fragments or for alternating sequences of very large and small fragments. In these cases a second or third restriction endonuclease will facilitate the mapping of regions harboring unfavorable spacings of sites for the first restriction enzyme, and vice versa.

Criteria for the selection of restriction enzymes are the GC content, degree of methylation and codon usage of the bacterial species to be typed, and the length of the recognition sequence of the enzyme. Table 6.1 provides a survey of commercially available restriction endonucleases that are especially useful for macrorestriction analysis of bacterial genomes. In general, restriction endonucleases with a 7- or 8-bp recognition sequence are expected to cleave bacterial chromosomes into only a few fragments. In the case of restriction enzymes recognizing a 6-bp sequence, enzymes which cleave a sequence made up by A or T residues should be tested for GC-rich genomes and enzymes with GC-recognition sites for AT-rich genomes. The tetranucleotide CTAG is underrepresented in most bacteria (McClelland *et al.*, 1987) and at least one of the four enzymes *Spel*, *Xbal*, *Nhel*, and *Avr*II that are harboring CTAG in their recognition sequence has normally proved to be useful for genome mapping. In many bacteria some palindromic sequences are counterselected. No general rules can be given, but the enzymes listed in section 3c of Table 6.1 may be tried first according to the criteria of GC-content and phylogeny.

B. Identification of Fragment Number

The range of separation allowing the identification of all fragments depends on the genome size and the number of restriction sites. In order to uncover double or multiple bands and to allow resolution of all fragments, the conditions for electrophoresis should be separately optimized

Table 6.1
Restriction Enzymes Used for Physical Mapping of Bacterial Genomes

Enzyme	Recognition sequence	Application (example)	Over-/under-representation[a]
1. Restriction enzymes with a 4-bp recognition sequence			
*Dpn*I	G^{m6}A↓TC	*Pseudomonas aeruginosa* PAO	
2. Restriction enzymes with a 5-bp recognition sequence			
*Nci*I	CC↓^{G}/$_c$GG	*Campylobacter jejuni*	−
3. Restriction enzymes with a 6-bp recognition sequence			
(a) Restriction enzymes for GC-rich genomes			
*Dra*I	TTT↓AAA	*Streptomyces coelicolor* M145	±
*Ssp*I	AAT↓ATT	*Thermus thermophilus*	±
*Asn*I (*Ase*I)	AT↓TAAT	*Rhodobacter sphaeroides* 1.2.4.	−
(b) Restriction enzymes for AT rich genomes			
*Sma*I	CCC↓GGG	Mollicutes	−
*Nae*I	GCC↓GGC	*Haemophilus influenzae* Rd	−
*Eag*I	C↓GGCCG	*Sulfolobus acidocaldarius*	−
*Bss*HII	G↓CGCGC	*Campylobacter jejuni*	±
*Bgl*I	GCCNNNN↓NGGC	Mollicutes	±
*Sac*II (*Sst*II)	CCGC↓GG	*Clostridium perfringens*	±
*Apa*I	GGGCC↓C	*Campylobacter jejuni*	±
(c) Cut less than expected from GC content			
GC content between 70 and 45%			
*Spe*I	A↓CTAGT	*Rhodobacter sphaeroides* 2.4.1.	−
*Xba*I	T↓CTAGA	*Thermococcus celer* Va13	−
*Nhe*I	G↓CTAGC	*Neisseria gonorrhoeae*	−
		Methanobacterium thermoautotrophicum	−
*Avr*II (*Bln*I)[b]	C↓CTAGG	*Myxococcus xanthus*	−
		Anabaena sp. strain PCC 7120	−
Archaebacteria (regardless of GC content)			
*Pvu*I	CGAT↓CG	*Methanococcus voltae*	−
*Bam*HI	G↓GATCC	*Haloferax mediterranei*	−
*Bgl*II	A↓GATCT	*Methanococcus voltae*	−
*Bcl*I	T↓GATCA	*Methanococcus voltae*	−
Bacteria with low GC content (until 35%)			
*Sal*I	G↓TCGAC	*Campylobacter jejuni*	
*Xho*I	C↓TCGAG	Mollicutes	
*Bam*HI	G↓GATCC	Mollicutes	
*Kpn*I	GGTAC↓C	Mollicutes	
*Asp*718I	G↓GTACC		
*Nru*I	TCG↓CGA	*Helicobacter pylori* VA802	−
*Mlu*I	A↓CGCGT	*Borrelia burgdorferi*	−
*Fsp*I	TGC↓GCA	*Clostridium perfringens*	−
Bacteria with high GC content (from 70%)			
*Eco*RI	G↓AATTC	*Thermus thermophilus*	−
*Mun*I	C↓AATTC	*T. thermophilus*	−

continues

for at least three size ranges: 500 to 50 kbp, 50 to 15 kbp with pulse times from 1 to 4 sec (1.5% agarose gels), and below 15 kbp (conventional 1% agarose gel). Electrophoresis conditions providing high resolution are particularly critical for separating fragments in the intermediate size range between 10 and 50 kb, because the mobility of these fragments reflects contributions of both continuous and pulsed-field gel electrophoresis. Fragment inversion can occur by the selection of different pulse times (Fig. 6.3). Consequently, the most common errors of misassignment, false size determination, and missing fragments occur in this size range. In addition, false size estimates may result from partial restriction digestion, insufficient resolution of PFGE, misinterpretation of fragment overlaps or

Table 6.1   ***continued***

Enzyme	Recognition sequence	Application (example)	Over-/under-representation[a]
*Hpa*I	GTT↓AAC	*T. thermophilus*	−
*Bfr*I	C↓TTAAG	*Streptomyces lividans* 66	−
*Nde*I	CA↓TATG	*T. thermophilus*	−
*Eco*RV	GAT↓ATC	*T. theermophilus*	−
*Bst*BI	TT↓CGAA	*T. thermophilus*	−
*Sna*BI	TAC↓GTA	*Rhodobacter sphaeroides* 2.4.1.	−
		4. Restriction enzymes with a 7-bp recognition sequence	
*Rsr*II	CG↓G^{T}/$_{A}$CCG	*Sulfolobus acidocaldarius*	±
		5. Restriction enzymes with an 8-bp recognition sequence	
*Sgr*AI[c]	C^{G}/$_{A}$↓CCGG^{G}/$_{T}$G	*Chlamydia trachomatis* L2	−
		Borrelia burgdorferi 212	
*Not*I	GC↓GGCCGC	*Escherichia coli*	−
*Sfi*I	GGCCNNNN↓NGGCC	*Bacillus subtilis*	±
*Asc*I[d]	GG↓CGCGCC	*Bacillus* sp. C-125	±
*Srf*I[c,f]	GCCC↓GGGC	No application known	
Sse 8387I[b]	CCTGCA↓GG	*Listeria monocytogenes*	±
*Pme*I[d]	GTTT↓AAAC	*Methanobacterium thermoautotrophicum*	±
*Swa*I[c]	ATTT↓AAAT	*Rhizobium melolitum* 1021	
*Pac*I[d]	TTAAT↓TAA	*Bradorhizobium japonicum* 110	−
		6. Intron-encoded endonucleases	
I-*Ceu*I[d]	TAACTATAACGGTC↑CTAA↓GGTAGCGA		
I-*Tli*I[d]	GGTTCTTTATGCGG↑ACAC↓TGACGGCTTTATG		
I-*Ppo*I[d,g]	ATGACTCTC↑TTAA↓GGTAGCCAAA		
VDE[d]	TCTATGTCGGG↑TGC↓GGAGAAAGAGGTAATG		
I-*Sce*I[c]	TAGGG↑ATAA↓CAGGGTAAT		

[a]Estimated from the G/C content of an organism; a recognition sequence was regarded as being underrepresented when the restriction enzyme cut three times less frequently than expected. Factors that lead to the underrepresentation of palindromic sequences might be (1) codon usage, (2) methylation, (3) selection against distinct palindromic sequences by the bacterial cell.

[b]Takara.

[c]Boehringer.

[d]New England Biolabs.

[e]Stratagene.

[f]Toyoko.

[g]Promega.

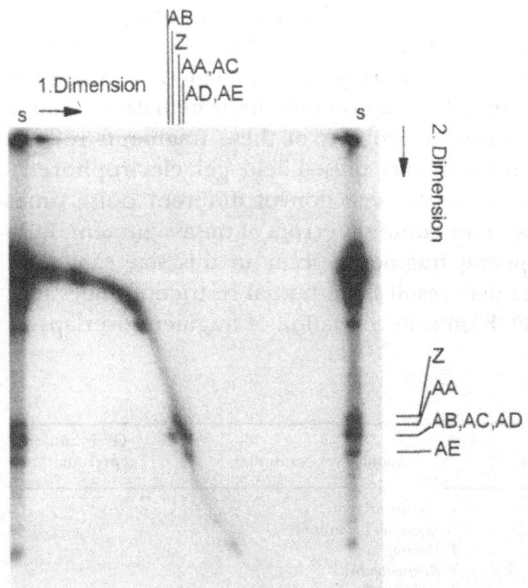

Figure 6.3 Running behavior of fragments in the size range 50–20 kbp with different pulse times. *SpeI–SpeI* two-dimensional gel of *B. cepacia* ATCC 25416 DNA. In the first dimension pulse times from 7 to 70 sec are used. In the second dimension pulse times were from 1 to 4 sec. Fragment order in the first dimension AB, Z, (AA, AC), (AD, AE), whereas in the second dimension fragments run in the order Z, AA, (AB, AC, AD), AE. Experimental conditions: CHEF; 6 V/cm; 0.5× TBE; 1.5% agarose; 8°C; pulse times first dimension: linearly increased from 7 to 30 sec for 32 hr, 35 to 40 sec for 9 hr, and 50 to 70 s for 15 hr.

plasmid content, loss of small fragments (<10 kb) by diffusion out of the agarose block during overnight incubation with restriction enzyme, or low intensity of the ethidium bromide stain, particularly in the low-molecular-weight range.

C. Mapping Strategies

Table 6.2 lists the techniques that have been applied so far to the construction of macrorestriction maps. Most maps were assembled by straightforward fragment size determinations, mainly by Southern hybridization of PFGE-separated partial, double or triple restriction digestions. These methods are labeled "A" or "C" in Table 6.2.

The most commonly applied method for mapping bacterial genomes is the hybridization of complete single or double digestions with either

Table 6.2
Macrorestriction Mapping Techniques of Bacterial Genomes

A. Southern blot hybridization of PFGE-separated macrorestriction digests
   A1. With gene probes
   A2. With linking clones of rare-cutter sites
   A3. With genomic clones
   A4. With (macro)restriction fragments
   A5. Analysis of partial digestions
   A6. Hybridization of ordered cosmid library with macrorestriction fragments
B. Anonymous two-dimensional macrorestriction mapping techniques
   B1. Reciprocal double digests
   B2. Partial–complete single digestions
   B3. B1 or B2 combined with end-labeling of terminal fragments
C. Fragment size determinations
   C1. Partial digestions
   C2. Double or multiple digestions
   C3. Restriction analysis of gel-purified fragments
D. Gene-directed mutagenesis: elimination or generation of single rare-cutter sites
E. Insertional mutagenesis
   E1. Transposons
   E2. (pro)phages
   E3. Integrative plasmids
   E4. Transformation with macrorestriction fragments
   E5. IS elements
F. Cross-protection: blockage of restriction sites by overlapping methylation with a
   methylase prior to restriction digestion
G. Comparative analysis of natural mutants (indels, inversions, etc.)

gene probes, linking clones which span rare-cutter sites or gel-purified restriction fragments obtained by cleavage with another enzyme. However, the positioning of fragments becomes equivocal if the sequence contains repeats such as IS-elements which give rise to cross-hybridization signals. This situation is encountered in all bacteria for at least the *rrn* operons which encode the ribosomal RNAs. Bacteria have 1 to 12 copies of *rrn* operons, which, moreover, contain a high density of sites for rare-cutting restriction enzymes. The correct assembly of fragment contigs with rDNA sequences at both termini may become difficult if solely based on hybridization data. An instructive example of this pitfall is the conflicting maps of *Haemophilus influenzae* Rd (Kauc and Goodgal, 1989; Lee *et al.*, 1989). Hence, the mapping of genomic regions with sequence repeats should not rely entirely on Southern hybridization.

   The combination of genetic and physical mapping is useful if linkage data or gene transfer techniques are available for the organism to be mapped. Mapping of mutants with known genomic rearrangements may be helpful in resolving ambiguous hybridization data ("G" in Table 6.2). Complementation or knock-outs of a genetically mapped locus can be utilized for physical mapping if the mutagenesis leads to changes of hybridization signals or fragment size, e.g., loss-of-function mutations in transposon mutants may be mapped by the shift of fragment size caused by the insertion of the transposon. Insertional mutagenesis ("E" in Table 6.2) is a frequently applied mapping strategy. The inserted sequence is detected by hybridization, increase of fragment size, or the loss or generation of a rare-cutter restriction site. The most recent trend in bacterial genome mapping is the combination of low-resolution macrorestriction mapping with the ordering of cloned DNA segments into contigs (see Chapter 11): Macrorestriction maps are constructed by hybridization of phage or cosmid libraries with labeled macrorestriction fragments (Charlebois *et al.*, 1991; Bukanov and Berg, 1994).

D. Two-Dimensional Mapping

The inherent problems of hybridization analysis are overcome by utilizing two-dimensional PFGE mapping techniques (label "B" in Table 6.2) (Bautsch, 1988; Römling and Tümmler, 1991). Two-dimensional PFGE mapping protocols are unbiased by fragment size and sequence repeats. They allow the construction of complete and consistent long-range restriction maps of bacterial genomes without any need for supplementary genetic data. Fragment order is established by two strategies: evaluation of partial–complete digestion gels or pairs of reciprocal double digestion gels. The two simple examples in Figs. 6.4 and 6.5 illustrate the principles of two-dimensional mapping. The mapping strategy requires optimal quality of the agarose-embedded DNA and optimal separation conditions (see the example shown in Fig. 6.6). To save money and time, one should not run more than two gels for a particular gel combination, ideally running only one.

1. Partial–Complete Mapping

In this technique, a partial restriction digest is first separated by PFGE in one dimension, then redigested to completion with the same enzyme, and subsequently resolved in the second orthogonal dimension. The genomic DNA is partially digested with one enzyme and then end-labeled in the agarose plug with ^{32}P-nucleotides by Klenow DNA polymerase (see experimental protocol below). The partial digest is separated by PFGE, the entire lane is cut out, and the gel-separated fragments in the agarose are redigested to completion with the original enzyme. The lane is oriented in the second agarose gel with the width of the lane of the first dimension becoming the height

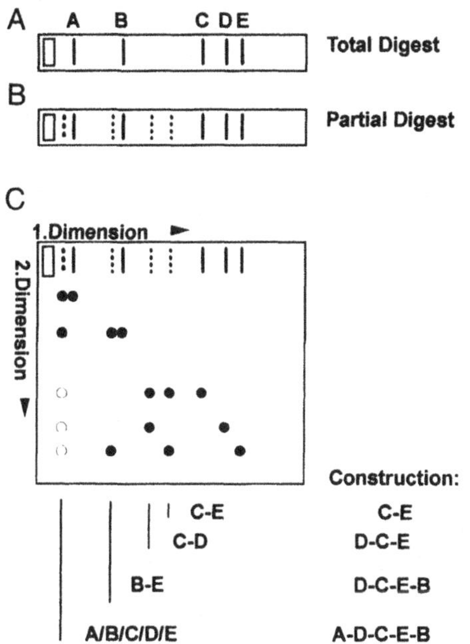

Figure 6.4 Principles of two-dimensional macrorestriction mapping by PFGE: partial/total digestion gels. (a) Complete digestion of chromosomal DNA with a rare-cutting restriction endonuclease and one-dimensional separation by PFGE. (b) Partial digestion of chromosomal DNA and one-dimensional separation by PFGE. (c) Two-dimensional partial/total digestion gel. A partial digestion is separated in the first dimension as shown in diagram (b.) The agarose lane is cut out, and the DNA in the agarose block is digested to completion with the same enzyme. A second electrophoresis perpendicular to the first run resolves the fragment composition of the partial digestion fragments. Fragments that were accidentally cleaved to completion prior to the first run are located on the diagonal. The interpretation of the spots on the two-dimensional gel is as follows: Fragments C and E as well as C and D are adjacent to each other. Thus, C must be located in the middle between D and E. B is linked with E. The position of A cannot be deduced from the gel. Fragment A may be a neighbor of D or B. End-labeling of the partial digestion with Klenow DNA polymerase resolves this ambiguity. The two fragments that are located at the ends of a partial digestion fragment incorporate radioactive nucleotides and become detectable in an autoradiogram of the two-dimensional gel. In the diagram the open circles symbolize the labeled fragments. Hence, fragments A and B constitute the terminal fragments in the largest partial digestion fragment and, consequently, A must be a neighbor of D. Thus, the fragment contig reads A-D-C-E-B.

of the second dimension. The complete digest is separated in the second dimension by PFGE, and the two-dimensional gel is stained with ethidium bromide. Subsequently, the gel is blotted for autoradiography. All fragments which are generated by complete digestion of the initial partial

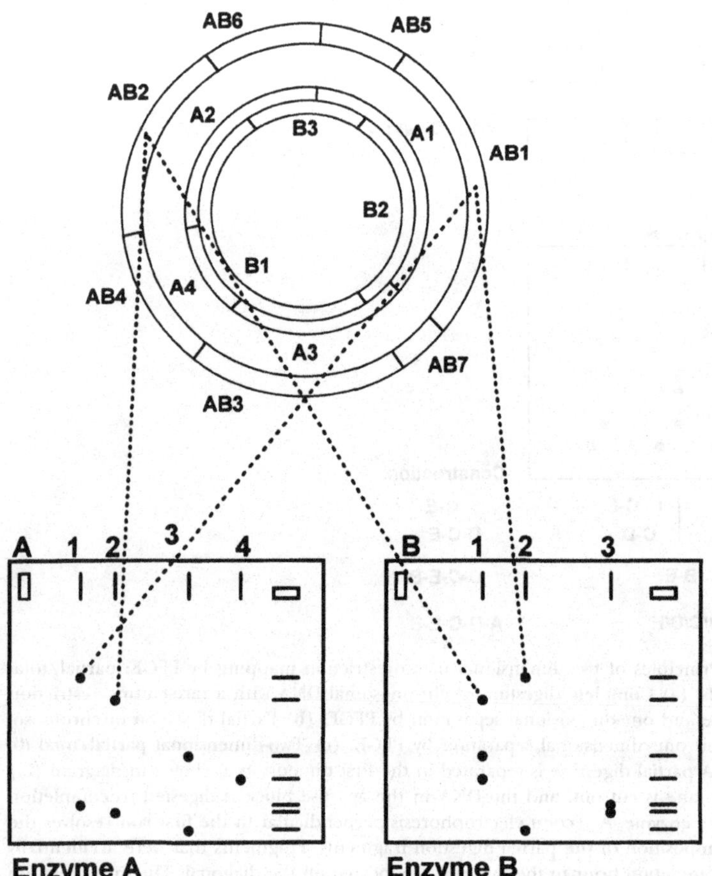

Figure 6.5 Principles of two-dimensional macrorestriction mapping by PFGE: reciprocal two-dimensional gels. A chromosome cut with two restriction enzymes, A and B, yields the same fragments independent of the succession of the digests. In this hypothetical example, cleavage with enzyme A gives fragments A1 to A4 and cutting with enzyme B fragments give B1 to B3. A double digestion results in the fragments AB1 to AB7. For reciprocal two-dimensional gels, complete digestions of either enzyme A or B are separated by PFGE, respectively. The lane is cut out and redigested with the other enzyme. An electrophoresis perpendicular to the first run separates all fragments of the double digestion. Identical spots on both two-dimensional gels are identified by fragment order. Linking fragments are present only in the double digestions. If more than two fragments, which are not cleaved by the other enzyme, are linked together their order cannot be determined from the analysis of reciprocal two-dimensional gels.

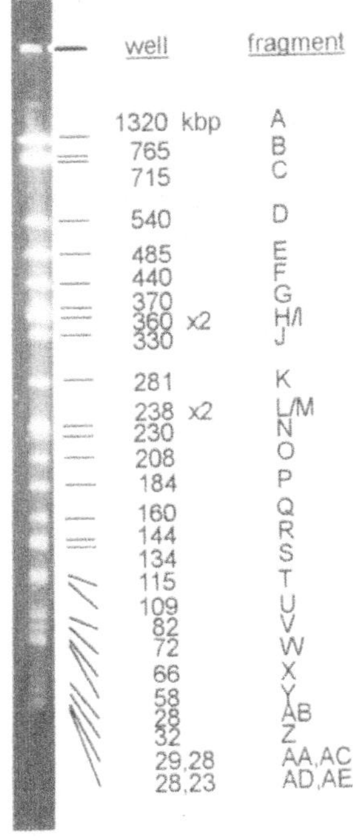

Figure 6.6 Example of an optimal separation of *B. cepacia* ATCC 25416 DNA digested with *Spe*I. This was always the first dimension when using *Spe*I in reciprocal gels. Almost all fragments between 1.32 Mb and 20 kbp are separated. Sizes are indicated. Experimental conditions: CHEF; 5.6 V/cm; 0.5× TBE; 1.5% agarose; 8°C; three linear pulse ramps from 50 to 70 sec for 25 hr, 30 to 40 sec for 10 hr, and 10 to 30 sec for 25 hr.

digestion fragments are visualized by ethidium bromide staining, with the fluorescence intensity being proportional to their length. In contrast, the autoradiogram identifies the two terminal fragments of each partial digestion fragment with comparable intensity, irrespective of size. Hence, comparative evaluation of gel stain and autoradiogram will facilitate the determination of fragment order.

Fragments that are linked to each other are directly read off from the two-dimensional gel (Römling *et al.*, 1989; Römling and Tümmler, 1991). Fragments are identified by comparison with complete digests separated in both outermost lanes of the second dimension. The molecular weight of a partial digestion fragment is determined from its position in the first dimension. Completely cleaved fragments which were produced during the partial digestion are located on the diagonal of the two-dimensional gel and serve as molecular weight markers. Should the partial digestion fragments overlap, other criteria, e.g., mass law and intensity, may be referred to for identification of the corresponding spots (see Fig. 6.11).

2. Preparation of an Appropriate Partial Digest

(1) Use a plug of $6 \times 5 \times 1$ mm containing a cell concentration of 2.5 or 5×10^9 cells/ml. Equilibrate the plug three times with 1 ml restriction enzyme buffer for 30 min each time at RT.

(2) Prepare a restriction enzyme dilution in the digest buffer. Use this dilution to prepare 60 μl restriction enzyme buffer (includes the appropriate amount of enzyme), 1 μl BSA (10 mg/ml), 1 μl DTT (0.5 *M*).

(3) Incubate at the appropriate digestion temperature (for different incubation times).

(4) Stop reaction with 1 ml TE.

(5a) When running digestion kinetics cut the plugs to a size of approximately $6 \times 2 \times 1$ mm and put them in the gel.

(5b) When running the first dimension of the final two-dimensional gel, cut the plug in half, carry out the end-labeling reaction, and put the two plugs into the gel one lane apart.

(6) Run a 1.5% agarose gel under the optimized separation conditions.

Notes. Referring to step 2: There are no robust rules of thumb for the adequate incubation conditions and the optimal amount of enzyme. Slight variations of enzyme activity may lead to irreproducible results. Enzyme activity may vary depending on the supplier and from batch to batch. It also decreases with time. Hence, it is important to keep one tube of enzyme for the partial digestion experiments and to carry out the tests speedily. Digestion conditions depend on the enzyme, the supplier, and the DNA template. Incubation time should exceed 45 min and a wide range of enzyme units should be tested. In the example below we varied the enzyme amounts between 0.5 and 3 U and the time between 2 and 8 hr. However, when we used a batch of the same enzyme from another supplier enzyme activities from 0.008 to 0.5 units and incubation times from 45 to 90 min were appropriate. Referring to step 3: Preincubation on ice is not necessary. Referring to step 5: Under the appropriate running conditions for the first dimension an upper separation limit should correspond to the sum of the molecular weight of the two largest fragments of the digest. Referring to step 6: The appropriate partial digest is the optimum compromise between resolution and information. Figure 6.7 shows an example of the step-by-step optimization for a partial–total digestion gel. Lanes 3 to 4 in Fig. 6.7A display the digest chosen, and Fig. 6.7B shows the first dimension of this digest and the expected partial digest fragments. One should bear in mind that more fragments will always show up in the two-dimensional gel than expected from the first dimension due to the multiple comigrating bands (cf. Fig. 6.7B with Fig. 6.7C).

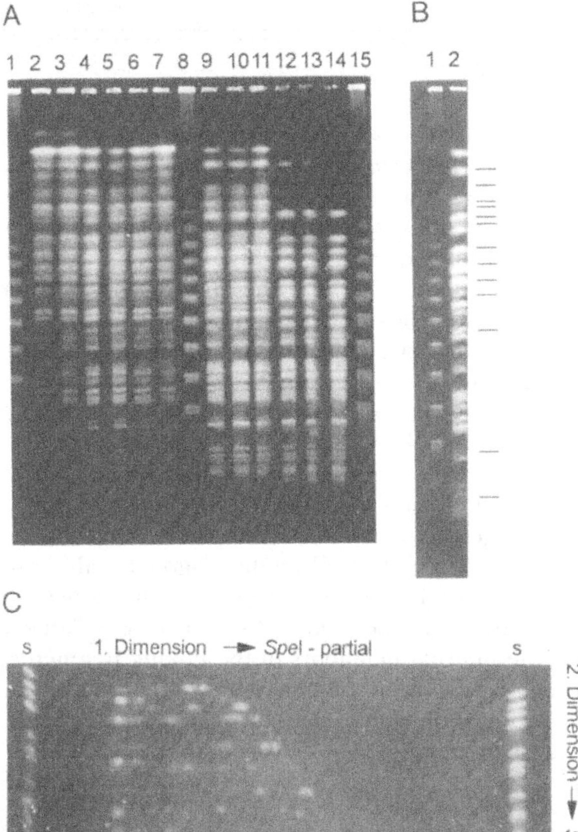

Figure 6.7 Development of a partial/total gel for *P. aeruginosa* PAO *Spe*I-digested DNA (A) Example of an optimization of *Spe*I partial digest of *P. aeruginosa* PAO DNA. An intermediate timepoint between lanes 3 and 4 was chosen for the two-dimensional gel. Fragments were resolved upto 900 kbp (SpA + SpB, 968 kbp). 1,8,15; λ-oligomer standard; 2, 0.5 U, 2 hr; 3, 0.5 U, 4 hr; 4, 0.5 U, 5.5 hr; 5, 0.5 U, 8 hr; 6, 1 U, 2 hr; 7, 1 U, 4 hr; 9, 1 U, 5.5 hr; 10, 1 U, 8 hr; 11, 3 U, 2 hr; 12, 3 U, 4 hr; 13, 3 U, 5.5 hr; 14, 3 U, 8 hr. (B) First dimension of the control lane of the two-dimensional gel. Partial digestion fragments are indicated by arrows. 1, λ-oligomer standard; 2, *Spe*I digest; 0.5 U, 4 hr 15 min. Experimental conditions as in (A). (C) Partial/total gel. Arrows indicate the direction of electrophoresis; s, *Spe*I total digest. Experimental conditions: First dimension as in A; second dimension: CHEF; 5.6 V/ cm; 0.5× TBE; 1.5% agarose; 8°C; two linear pulse ramps from 3 to 15 sec for 8 hr and from 1 to 33 sec for 30 hr.

3. Reciprocal Gels

In this technique, a complete restriction digest with enzyme A is separated in the first dimension, redigested to completion with enzyme B, and then separated in the second orthogonal direction. On a separate series of gels the order of restriction digestions is reversed. The corresponding spots on both two-dimensional gels are identified by their identical molecular weights. Linking fragments carry the recognition site for enzyme A at one end and for enzyme B at the other. The assignment of the linking fragment to its two parental fragments of the single digestions with enzyme A or B establishes the fragment overlap (see Fig. 6.5).

The linking fragments are only present in the separation of double digests and cannot be detected in single digests. The separate single digestion products from each enzyme are separated on the two outermost lanes of the second dimension of the two-dimensional gel.

Fragments of the first digest that do not contain the recognition sequence for the second enzyme will be located on the diagonal of the two-dimensional gel and can be identified by comparison with the respective single digest. On the reciprocal gel the audioradiogram of end-labeled fragments will not show the internal neighbors of the linking fragments, i.e., the fragments not digested by the second enzyme; comparison of the autoradiogram with the more complex ethidium bromide stain will again facilitate the elucidation of fragment order. To facilitate the assignment of fragments, the single digestion products from the first digestion and the double digestion products are separated on the two outermost lanes of the second dimension of the two-dimensional gel.

4. Preparation of Two-Dimensional Gels

(1) Prepare a 1.5% agarose gel (agarose: Ultra Pure, Life Technologies) in 0.5× TBE buffer.

(2) Load a $6 \times 2 \times 1$-mm agarose plug (2.5 or 5×10^9 cells/ml) in the bottom of a gel slot for the first dimension (see Fig. 6.8a).

(3) Load a second plug treated as the first one into the adjacent slot to check the quality of digestion, reproducibility of digestion in the case of a partial digest, loading, and electrophoresis run.

(4) Run the first dimension (see Figs. 6.6, 6.7B, and 6.10A as examples of running conditions). No size standard is required.

(5) Cut out the lane intended to be used in the second dimension by first marking the end of the lane with a sterile sewing thread and then using a glass plate and a scalpel to cut out the lane (Fig. 6.8b). Store this lane on a sterile glass plate at 4°C during staining and examination of the rest of the gel.

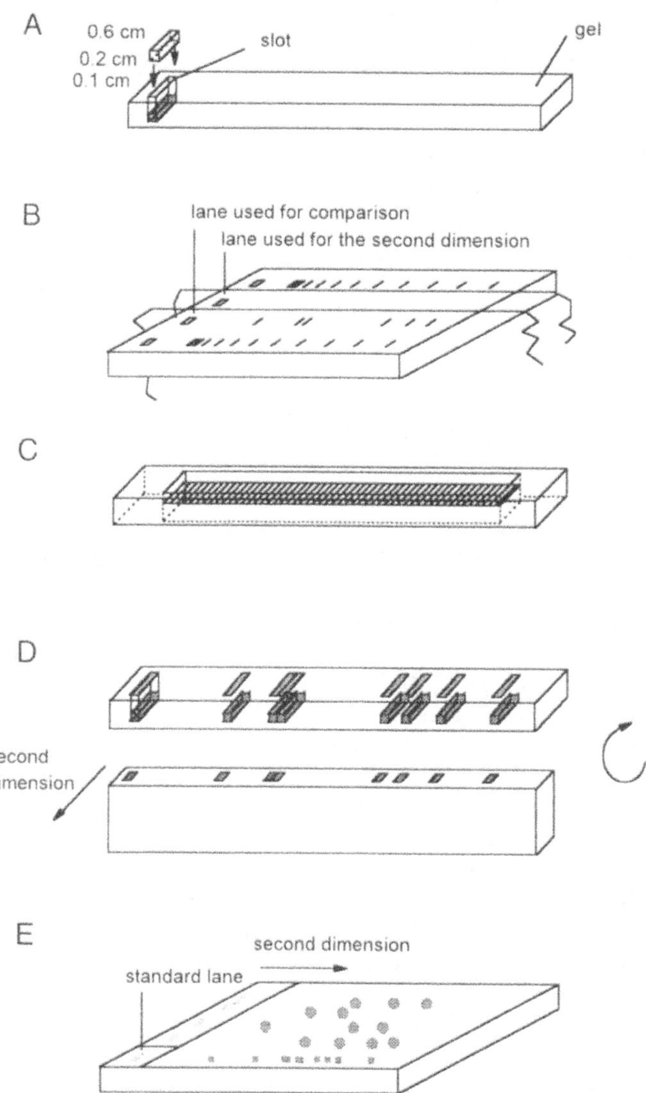

Figure 6.8 Step-by-step manual for two-dimensional gels.

(6) Stain the rest of the gel as described below. Check the quality of the control lane. If the bands are fuzzy or weak, do not use the gel.

(7) Prepare the perspex tub for incubation of the agarose lane (Fig. 6.8c). The tub consists of a whole piece of perspex with dimensions as indicated in Fig. 6.9. A perspex lid covers the tub and perspex blocks are used to adjust length and, hence, volume of buffer. Rinse the tub several times with 70% ethanol and finally soak the tub and the blocks in 70% ethanol for 5 min. Wipe the lid with 70% ethanol-soaked soft tissue. Pour the alcohol out of the tub and let it dry. Rinse with 5 ml restriction enzyme buffer.

(8) Place the outcut lane into the tub by pushing it softly with the edge of a scalpel from the glass plate. Fill the whole tub with restriction enzyme buffer. Incubate for 30 min without shaking. Repeat this procedure two times.

(9) After removing washing buffer, insert in blocks, if necessary. Fill the tub with restriction enzyme buffer until the lane is fully covered. For economic use of enzyme, no more than 3 ml restriction enzyme buffer should be used.

(10) Add DTT (33 μl of 0.5 M DTT, if 2.5 ml restriction enzyme buffer is used) and BSA (33 μl of 10 mg/ml BSA). Add restriction enzyme at a concentration of 66 U/ml. Mix thoroughly with a pasteur pipette. The concentration 66 U/ml is an average value which may be adjusted. Bear in mind that the enzyme concentration may vary according to the supplier and the batch; we used concentrations from 24 U/ml (*SpeI* (Eurogentech)) to 100 U/ml (*SwaI* (Boehringer Mannheim)).

(11) Cover the tub with the lid and incubate overnight (incubation time may be extended to 48 hr). If necessary one can add additional enzyme after 24 hr.

(12) Remove enzyme buffer with a pasteur pipette. Fill the tub to the brim with TE and leave for 30 min at room temperature to wash out enzyme. Change the buffer one or two times.

(13) Put the lane in a gel casting chamber instead of the comb. Turn the lane at an angle of 90° (at the longitudinal axis) (see Fig. 6.8d).

(14) Pour 1.5% agarose (Ultrapure, Life Technologies) around the lane in the casting chamber until a plain surface with the lane is obtained. The agarose should be close to gelling temperature.

(15) Use a scalpel to cut out two slots at the ends of the gel. Fill them with standards (see Fig. 6.8e).

(16) Run the second dimension at the desired conditions. Examples of optimal separation are given in Fig. 6.7 and 6.10.

(17) Stain the gel as described below, visualize the DNA, and take a photograph.

Notes: Referring to step 1: The gel should be as thin as possible for maximum cooling conditions during electrophoresis. No temperature gradient is generated which otherwise leads to a rocket-like running behavior of DNA fragments within the gel. To ensure results displaying clear bands use 1.5% agarose gels whenever possible, in particular when many spots are to be expected. Referring to step 2: The agarose plug should be lowered into the bottom of the mould (see Fig. 6.8a). The plug should be as high as it is wide in order to create spots in the second dimension. Avoid pressing the block since mechanical stress will cause shearing of the DNA. Referring to step 4: To separate restriction digests use the whole length of the gel. This enhances the separation and more fragments are resolved. Short running times are **not** important. Referring to step 5: Try to cut out the lane as small as possible because the width of the lane is the height of the gel in the second dimension. The thicker the gel in the second dimension the more intensity is lost. Referring to step 7: The tub for the agarose lane is just a thick perspex piece with a cut-out bowl (for dimensions see Fig. 6.9). Referring to step 9: In order to minimize the amount of enzyme used, try to use just enough restriction enzyme buffer to cover the lane for digestion. Referring to step 10: Factors that may influence the quality of the digestion include (i) BSA: acetylated BSA may inhibit the restriction enzymes. (ii) Agarose: since not all restriction enzymes work well in agarose, the restriction enzyme activity on DNA plugs encapsulated in the respective agarose should be checked. We use Ultra Pure Agarose (Life Technologies) for two-dimensional gels All tested enzymes worked well in this agarose with the exception of *Swa*I. In this case, the fragments of the first dimension were separated in 1% low-melting agarose (Sigma, type VII). Referring to step 14: Gel and lane should form an even surface. For standard two-dimensional gels we use 1.5% agarose for both directions. However, the agarose concentration for the second dimension should be the same as or higher than that for the first dimension. If this is not the case, the spots will appear distorted. Referring to step 15: In partial–total digests use a total digest as standard. In reciprocal gels use a digest of the corresponding enzyme and a double digest of the two enzymes as standards on two adjacent lanes. Referring to step 17: We stain the gel 45 min with 120 µg/ml ethidium bromide solution under shaking and destain it three times for 30 min each time with water under shaking. Even better results can be obtained by staining the gel overnight in 10 µg/ml ethidium bromide solution at 4°C without shaking and destaining as described.

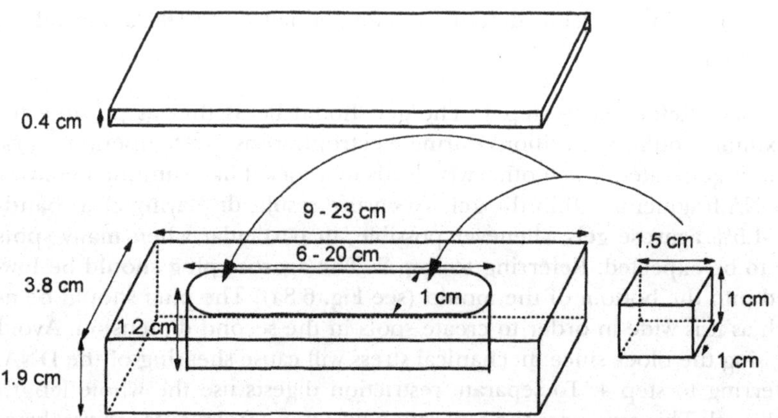

Figure 6.9 Dimension of the perspex tub used for restriction digestion of agarose lane.

5. Endlabeling of DNA in an Agarose Plug

(1) Digest the DNA (cell concentration 2.5 or 5×10^9 cells/ml) in the plug ($6 \times 5 \times 1$ mm) with the appropriate restriction enzyme in an Eppendorf cup.

(2) Remove buffer plus restriction enzyme, equilibrate twice with 1 ml TE for 30 min each time at room temperature.

(3) Adjust the plugs to the final size needed ($6 \times 2 \times 1$ mm) for the subsequent electrophoresis run and transfer it back into the Eppendorf cup.

(4) Equilibrate each plug twice again for 30 min with 1 ml *Spe*I incubation buffer at room temperature.

(5) Thoroughly remove all incubation buffer.

(6) Add 20 µl *Spe*I incubation buffer, 2 µCi α-dNTP (>3000 Ci/mmol), 1 U Klenow-polymerase. Incubate for 30 min at room temperature.

(7) Remove the buffer.

(8) Wash the plug twice with 1 ml TE for 20 min at room temperature without mixing.

(9) Check that the nucleotide is incorporated by holding the Eppendorf cup with the plug against a Geiger counter. The signal should be around 100 cps.

Notes: Referring to step 1: we use the endlabeling procedure for two purposes: (i) detection of linking fragments or end-fragments in two-dimensional gels; (ii), visualization of fragments below 5 kbp in one-dimensional gels. Klenow-polymerase incorporates nucleotides only at 5′ protruding ends. Thus, for fragment end-labeling use the dNTP complementing the first base in your restriction enzyme site. When restriction enzymes create blunt ends or 5′ recessive ends, nucleotides will be incorporated throughout the length of the fragment. This procedure is useful for increasing the sensitivity of fragment detection when compared to the ethidium bromide stain. Referring to step 6: For the success of the procedure it is strongly recommended that neither the incubation time nor the enzyme concentration be increased. If the signal intensity is too low, the amount of radioactivity should be increased. Referring to step 9: We strongly recommend the running of a one-dimensional gel before adopting the endlabeling procedure to a two-dimensional gel. After the electrophoresis run, the DNA is transferred to a nylon membrane by Southern blotting over 48 hr. Signals should be visible after 1 day exposure. For two-dimensional gels weak signals should be visible after 1 day and strong signals appear after 1 or 2 weeks. Standards should also be labeled in two-dimensional gels.

6. Construction of Physical Maps by Two-Dimensional PFGE: Practical Hints and Pitfalls

After conclusion of the experiment, an analysis of the data produced by the two-dimensional gels should be carried out. Figures 6.4 and 6.5 show the principles of interpretation. But, as Fig. 6.7 and 6.10 show, in reality the gels are much more complex. A correct interpretation of the data can only be made by referring to all of the experimental evidence given because the information hidden in such gels is of remarkable complexity. Consequently, for a fast and efficient interpretation the following guidelines should be observed.

a. Interpretation of Reciprocal Gels

(1) Identify cut and uncut fragments by comparing one-dimensional single and double digests.

(2) Identify fragment order in both reciprocal gels. Analysis based on the absolute migration distance is not possible due to the slight distortions and different running behavior of the two gels.

(3) Compare the fragment order between the gels and identify identical fragments.

(4) Identify the two linking fragments in each lane, if possible with the help of the corresponding autoradiogram.

(5) Add up fragment sizes for each original fragment in order to prove fragment size and number.

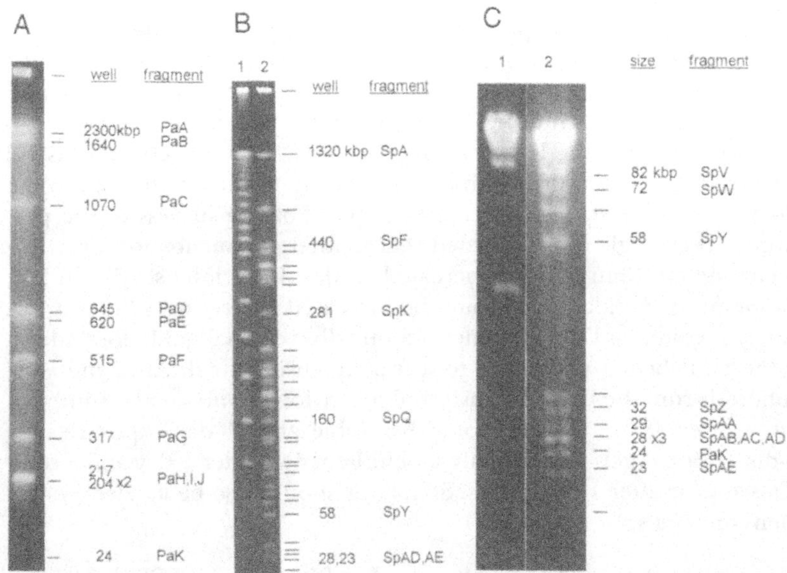

Figure 6.10 Optimized experimental conditions for a reciprocal gel. The example is a *Pad*/
Spd gel for *Burkholderia cepacia* ATCC 25416 DNA. (A) Optimized resolution of the *Pad* digest
for the first dimension. Fragments H, I, J cannot be resolved. Experimental conditions:
CHEF; 3.7 V/cm; 0.5× TBE; 1.5% agarose; 8°C; one linear pulse ramp from 80 to 350 sec
for 60 hr and one single 96-sec pulse for 20 hr. (B) Optimized resolution of the *Pad*/*Spd*
double digest over the whole size range. 1, λ-oligomer standard; 2, *Spd*/*Pad* double digest.
Experimental conditions: CHEF; 6 V/cm; 0.5× TBE; 1.5% agarose; 8°C; two linear pulse
ramps from 10 to 30 sec for 30 hr and from 50 to 70 sec for 15 hr and a single 35-sec pulse
for 3 hr. (C) Optimized resolution of the *Pad*/*Spd* double digest for fragments below 50
kbp. Since there are quite a few linking fragments below 50 kbp, an extra set of two-dimensional gels was run for this size range. 1, λ-oligomer standard; 2, *Pad*/*Spd* digest. Experimental conditions: CHEF; 6 V/cm; 0.5× TBE; 1.5% agarose; 8°C; linear pulsing from 1 to 4
sec for 36 hr. (D) Two-dimensional *Pad*/*Spd* gel for the size range between 1.3 Mbp and 20
kbp. In the first dimension the DNA was cut with *Pad* and fragments of the digest were
separated. Then the DNA was digested in the lane with *Spd* and separated in the second
dimension; s indicates *Pad*/*Spd* double digest as standard. Experimental conditions: First
dimension as in (A). Second dimension: CHEF; 6 V/cm; 0.5 TBE; 1.5% agarose; 8°C; two
linear pulse ramps from 10 to 30 sec for 30 hr and from 30 to 70 sec for 15 hr. (E) Two-
dimensional *Pad*/*Spd* gel for fragments below 50 kbp. Note that some of the small fragments
are very faint; s indicates *Pad*/*Spd* double digest as standard. Experimental conditions: First
dimension as in (A). Second dimension as in (C). (F) Autoradiogram of the gel shown in
(E). Fragments were labeled with Klenow-polymerase using 10 μCi [α-^{32}P]dCTP before the
first dimension. In the upper compression zone overexposure occurs due to the significantly
larger amount of incorporated radioactivity. Fragments down to a size of 6 kbp are seen.
Exposure time was 8 days.

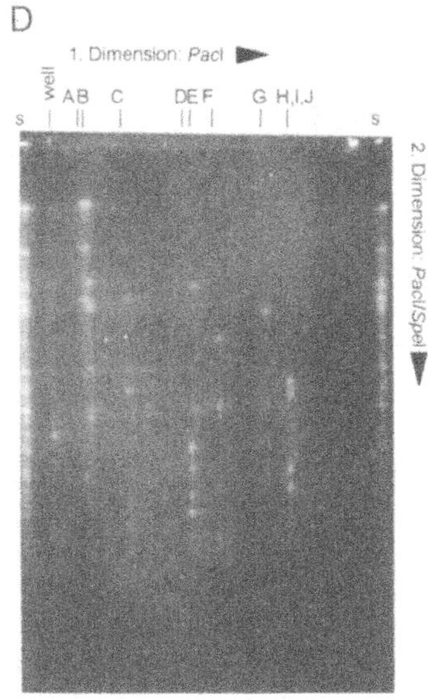

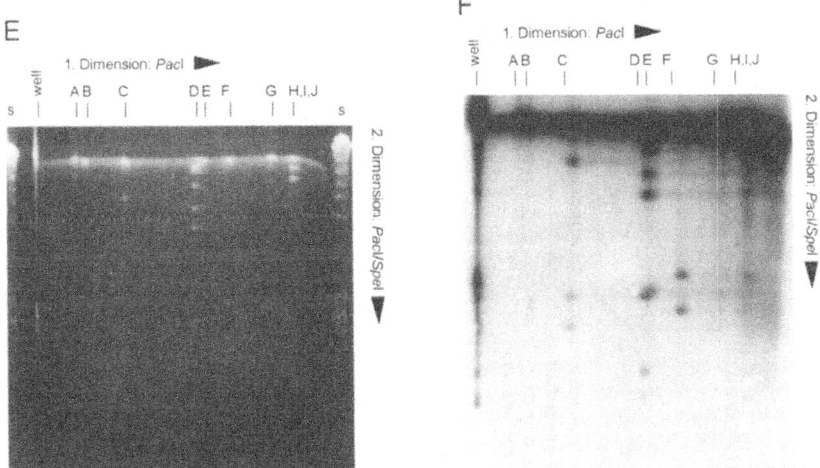

Figure 6.10 *continued*

(6) Begin construction of the fragment order by first referring to a linking fragment of one original fragment in either of the two gels and then identify the identical fragment in the second remaining gel. Then the linking fragment from the other end in the second gel is identified

Hypothetical genome map
consisting of one circular genome
and one circular plasmid:

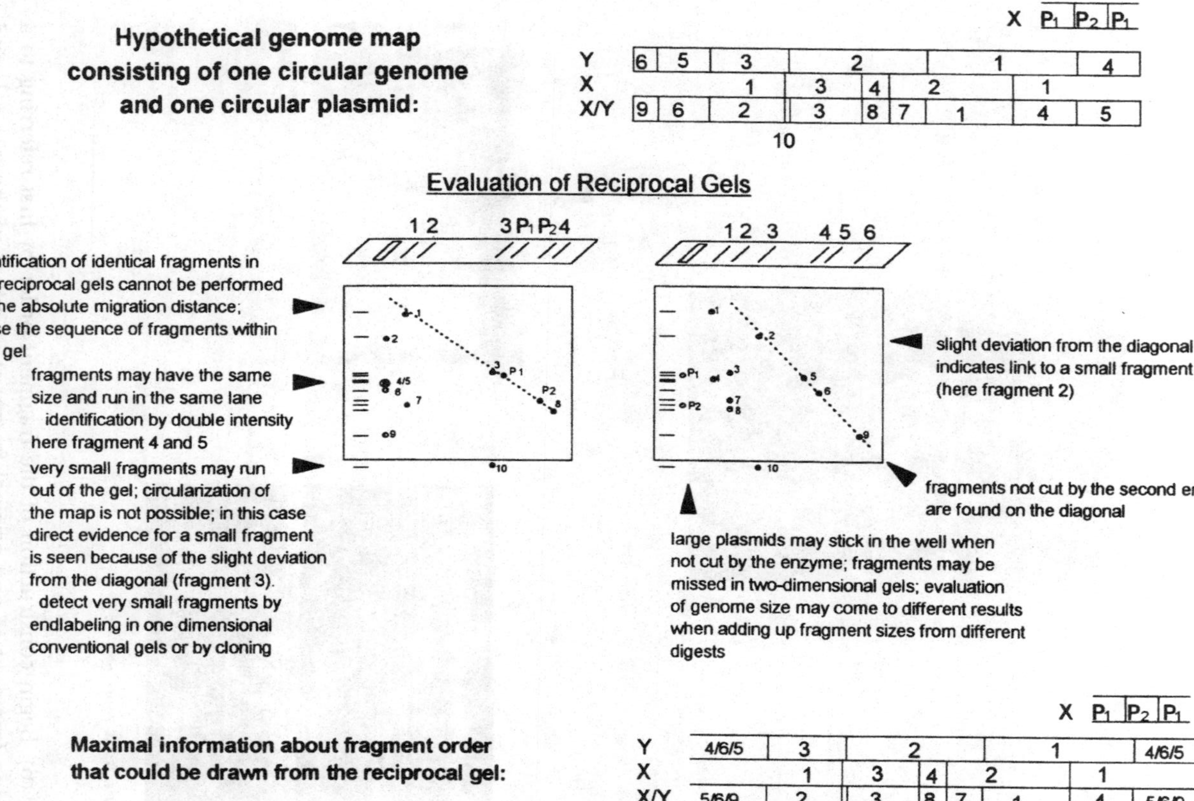

Maximal information about fragment order
that could be drawn from the reciprocal gel:

Evaluation of the Partial-Total-Gel

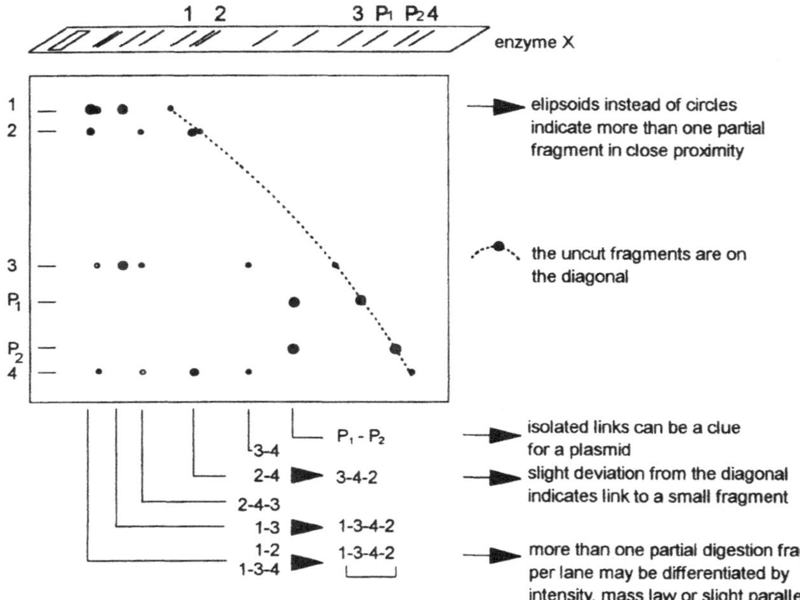

Figure 6.11 Hints for the analysis of two-dimensional gels. The hypothetical gels illustrate typical combinations of fragments that are difficult to interpret.

on the first gel, and so forth. By the end of the assignment the gaps between the linking fragments are filled with the fragments not cut by the other enzyme.

b. Interpretation of Partial/Total Gels

(1) Begin by using the smallest partial digest.

(2) Compare the sum of the fragments in one lane with the size of the original partial digestion fragment. This size may be estimated by taking the completely digested fragments from the diagonal as size standard.

(3) Compare the autoradiogram with the ethidium bromide-stained gel to aid construction of fragment order.

(4) Safe fragment orders from smaller partial digest fragments may be used to interpret a partial digestion fragment with a more complex pattern.

A hypothetical map has been constructed and interpreted in Fig. 6.11. Some of the major points which make an analysis difficult are the occurrence of double or triple fragments, small (linking) fragments, and plasmids. Experimental conditions that may be of use when interpreting data are: (i) comparison of the signals of ethidium bromide-stained gels and autoradiograms; (ii) interpretation of intensity variations of partial digestion fragments in the same lane; (iii) interpretation of deviations from the shape of a circle as two closely run fragments.

Acknowledgments

We thank Armin Bihlmaier and Philip Rodley for experimental data. The work performed in the authors' laboratory was supported by the Deutsche Forschungsgemeinschaft.

References

Allardet-Servent, A., Michaux-Charachon, S., Jurnas-Bilak, E., Karayan, L., and Ramuz, M. (1993). Presence of one linear and one circular chromosome in the *Agrobacterium tumefaciens* C58 genome. *J. Bacteriol.* **175**, 7869–7874.

Bautsch, W. (1988). Rapid physical mapping of the *Mycoplasma mobile* genome by two-dimensional field inversion gel electrophoresis techniques. *Nucleic Acids Res.* **16**, 11461–11467.

Beverley, S. M. (1989). Estimation of circular DNA size using gamma-irradiation and pulsed-field gel electrophoresis. *Anal. Biochem.* **177**, 110–114.

Birnboim, H. C. (1983). A rapid alkaline extraction method for the isolation of plasmid DNA. *In "Methods in Enzymology"* (R. Wu, L. Grossman, and K. Moldave, eds.), Vol. 100, pp. 243–255. Academic Press, New York.

Bukanov, N. O., and Berg, D. E. (1994). Ordered cosmid library and high-resolution physical-genetic map of *Helicobacter pylori* strain NCTC 11638. *Mol. Microbiol.* **11**, 509–523.

Charlebois, R. L., Schalkwyk, L. C., Hofman, J. D., and Doolittle, W. F. (1991). Detailed physical map and set of overlapping clones covering the genome of the archaebacterium *Haloferax volcanii* DS2. *J. Mol. Biol.* **222**, 509–524.

Eckhardt, T. (1978). A rapid method for the identification of plasmid desoxiribonucleic acid in bacteria. *Plasmid* **1**, 584–588.

Ferdows, M. S., and Barbour, A. G. (1989). Megabase-sized linear DNA in the bacterium *Borrelia burgdorferi*, the Lyme disease agent. *Proc. Natl. Acad. Sci. U.S.A.* **86**, 5969–5973.

Hinnebusch, J., and Tilly, K. (1993). Linear plasmids and chromosomes in bacteria. *Mol. Microbiol.* **10**, 917–922.

Kauc, L., and Goodgal, S. H. (1989). The size and a physical map of the chromosome of *Haemophilus parainfluenzae. Gene* **83**, 377–380.

Lee, J. J., Smith, H. O., and Redfield, R. J. (1989). Organization of the *Haemophilus influenzae* Rd genome. *J. Bacteriol.* **171**, 3016–3024.

Lennon, E., and DeCicco, B. T. (1991). Plasmids of *Pseudomonas cepacia* strains of diverse origin. *Appl. Environ. Microbiol.* **57**, 2345–2350.

Lin, Y.-S., Kieser, H. M., Hopwood, D. A., and Chen, C. W. (1993). The chromosomal DNA of *Streptomyces lividans* 66 is linear. *Mol. Microbiol.* **10**, 923–933.

Liu, S. L., Hessel, A., and Sanderson, K. E. (1993). Genomic mapping with I-*Ceu*I, an intron-coded endonuclease specific for genes for ribosomal RNA, in *Salmonella* spp., *Escherichia coli*, and other bacteria. *Proc. Natl. Acad. Sci. U.S.A.* **90**, 6874–6878.

McClelland, M., Jones, R., Patel, Y., and Nelson, M. (1987). Restriction endonucleases for pulsed field mapping of bacterial genomes. *Nucleic Acids Res.* **15**, 5985–6005.

Römling, U., and Tümmler, B. (1991). The impact of two-dimensional pulsed-field gel electrophoresis techniques for the consistent and complete mapping of bacterial genomes: Refined physical map of *Pseudomonas aeruginosa* PAO. *Nucleic Acids Res.* **19**, 3199–3206.

Römling, U. Grothues, D., Bautsch, W., and Tümmler, B. (1989). A physical genome map of *Pseudomonas aeruginosa* PAO. *EMBO J.* **8**, 4081–4089.

Smith, C. L., Econome, J. G., Schutt, A., Klco, S., and Cantor, C. R. (1987). A physical map of the *Escherichia coli* K12 genome. *Science* **236**, 1448–1453.

Suwanto, A., and Kaplan, S. (1989). Physical and genetic mapping of the *Rhodobacter sphaeroides* 2.4.1 genome: Presence of two unique circular chromosomes. *J. Bacteriol.* **171**, 5850–5859.

Cosmid Cloning with Small Genomes

Rainer Wenzel and Richard Herrmann

I. Introduction

Cosmid vectors are useful tools for establishing gene libraries, as has been shown for a number of different prokaryotic and eukaryotic species, like the nematode *Caenorhabditis elegans* (Coulson *et al.*, 1986) or the bacteria *Mycoplasma pneumoniae* (Wenzel and Herrmann, 1989), *Haloferax volcanii* (Charlebois *et al.*, 1991), *Mycobacterium leprae* (Eiglmeier *et al.*, 1993), and *Helicobacter pylori* (Bukanov and Berg, 1994). Their cloning capacity of up to 51 kbp keeps the number of clones required for screening at a manageable range, especially when relatively small prokaryotic genomes have to be analyzed. Since, due to their size, transformation efficiencies of cosmids are rather low, introduction of cosmids into bacterial cells is usually done by *in vitro* packaging into phage λ particles followed by infection of the host cell. Though packaging efficiency of cosmids is not as good as for phage λ-derived vectors, the cosmids do have the advantage of containing high copy number replicons which give high yields of DNA.

In the following, various strategies for the construction of an ordered set of cosmid clones representing a bacterial genome are discussed in general. As a guideline, the cloning of the entire genome of the bacterium *M. pneumoniae* in a contiguous set of cosmid clones is described (Wenzel and Herrmann, 1989). This small human pathogenic bacterium has a genome size of about 800 kbp (Wenzel *et al.*, 1992) which generally can be considered to be close to the lower limit of the coding capacity of a self-replicating cell (Morowitz, 1984).

NONMAMMALIAN GENOMIC ANALYSIS: A PRACTICAL GUIDE

This organism shall serve as a model for describing a general strategy which includes choice of the appropriate cosmid vector, construction of a cosmid library, mapping of individual clones, sorting them into contigs, filling of gaps, and, finally, giving suggestions for verifying the fidelity of an ordered set of clones.

Additional details concerning materials, reagents, and buffers which are used in the following sections can be found in the study of Sambrook *et al.* (1989).

II. General Considerations

In theory, an ideal gene bank contains all genomic sequences at the same frequency. In practice, it is hardly possible—for different reasons which shall be discussed later—to achieve this goal (Seed *et al.*, 1982).

A crucial factor is the choice of an appropriate restriction endonuclease for cloning. This enzyme should cut the target genome as randomly as possible, yielding a collection of restriction fragments with a nearly random distribution. There is usually the choice between enzymes recognizing 4- or 6-bp-long DNA sequences. Whereas 4-bp cutters produce smaller fragments of an average length of 256 (4^4, neglecting the base composition of DNA) bp in a more random manner, 6-bp cutters simplify subsequent restriction analysis of individual clones, since fewer but larger fragments of an average length of 4096 (4^6, neglecting the base composition of DNA) bp can be determined and ordered more precisely. For mapping purposes it is therefore helpful to choose a restriction endonuclease which cuts within a 6-bp DNA sequence.

As an alternative to digestion by restriction endonucleases, genomic DNA might be degraded by controlled hydrodynamic shear. For example, sonication or forcing the DNA through a narrow needle will give more random fragmentation. However, these methods do not assure a completely statistical distribution of cloned DNA fragments, since not all DNA fragments are equally well propagated in a host cell. Sometimes certain fragments cannot be cloned at all in a given system. Degradation of genomic DNA by hydrodynamic shear has several disadvantages: (i) the number of experimental steps has to be increased since the uneven overhanging ends of the sheared DNA fragments have to be repaired and then either modified with suitable adaptors for cloning or ligated with blunt ends which is less efficient; (ii) sheared fragments are not as well incorporated in cosmid libraries as DNA fragments generated by restriction endonucleases. Therefore, these methods should be applied only for special purposes, as for closing the gaps which originate during the assembly of contigs due to an uneven distribution of restriction sites in the genome (Knott *et al.*, 1988, 1989). They may also be helpful for species with genomes of unusually high or low GC contents where appropriate restriction enzymes are not available.

Another concern is the number of clones which statistically represent the complete genome. This number can be calculated according to the formula of Clarke and Carbon (1976)

$$N = \frac{\ln\ (1\text{-}P)}{\ln\ (1\text{-}f)}$$

where N is the number of clones, P is the probability for the presence of any given sequence in the library, and f represents the ratio of insert to genome size. For example, 90 cosmid clones with an average insert size of 40 kbp (representing approx. 4.5 genome equivalents) should be sufficient to cover the genome of *M. pneumoniae* at a probability of 99%, assuming a genome size of about 800 kbp.

Since virtually no restriction enzyme cuts in a purely random manner, these calculations can only give an approximate value to the number of clones needed for the construction of contiguous DNA sequences (contigs) covering the entire genome. Additional problems may arise from DNA regions which are under or even nonrepresented because they are coding for products which are toxic for the host cell or because AT-rich sequences and sequences carrying repeated DNA sequences are lost during propagation due to intramolecular recombination events. In these instances special measurements have to be taken to fill the remaining gaps. Frequently, the number of clones to be screened has to be increased or additional techniques have to be applied to fill the remaining gaps.

III. Choice of Vector

Cosmids are defined as plasmids containing one or more cohesive sites (cos sites) of bacteriophage λ DNA (Collins and Hohn, 1978). After linearization of cosmids by digestion with one or more restriction enzymes and ligation to 40- to 50-kbp DNA, among the ligation products are linear polymer DNA molecules containing cos sites spaced between 37 and 51 kbp apart. From a mixture of fragments, only linear molecules with this arrangement will be packaged efficiently *in vitro* into bacteriophage λ particles (Feiss and Siegele, 1979). After infection of an appropriate *Escherichia coli* strain with these particles the linear DNAs are injected into the host cell, where they form circles through annealing of the cohesive ends and behave thereafter like conventional plasmids. Isolation and analysis of cosmid DNA follow standard procedures with the necessary precaution to avoid shearing of such large DNAs (see Preparation of Partially Restricted High-Molecular-Weight Genomic DNA).

For mapping, choice of a well designed vector is important. An ideal cosmid for genome mapping should combine different features (see Fig. 7.1).

A specially designed multiple cloning site (MCS) contains recognition sequences for the restriction endonucleases used for cloning flanked by recognition sequences for endonucleases like *Sfi*I or *Not*I which cut rarely or not at all in the target genome (Quiang and Schildkraut, 1984). This permits linearization of the cosmid with inserted foreign DNA and facilitates restriction mapping of the cloned DNA (see below). By digestion of the cosmid DNA with these rarely cutting endonucleases, it is also possible to isolate the cloned DNA as an intact fragment. Furthermore, the cosmid can be modified by insertion of additional DNA without interrupting the cloned DNA. Alternative ways of producing unique cleavage in cosmid clones include the use of λ terminase (Wang and Wu, 1993) and the so-called Achilles cleavage (AC) as described by Ferrin and Camerini-Otero (1991), Koob and Szybalski (1990) and Koob *et al.* (1992). The enzyme λ terminase specifically cleaves the cos site, the length of which assures that this sequence would be unlikely to occur in any cloned fragment. For RecA-mediated cleavage, a specific restriction site is protected from meth-

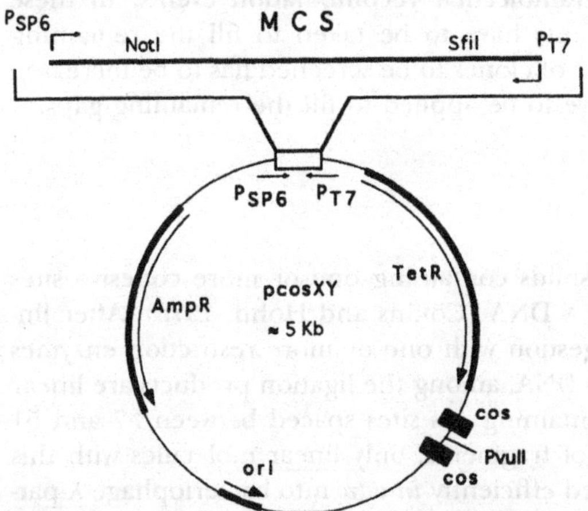

Figure 7.1 Map of a typical walking cosmid vector. This vector combines all the prerequisites needed for chromosome walking and restriction analysis of the insert DNA. *Not*I and *Sfi*I represent restriction sites which occur rarely in the genome to be cloned. They can be replaced by other sites if necessary.

ylation (methyltransferase corresponding to the restriction endonuclease) by a site-specific DNA protein complex consisting of an oligonucleotide providing the specificity and of RecA protein providing the stability. After removal of this complex, the cosmid can be linearized at the unmethylated restriction site by the appropriate enzyme.

For chromosome walking, labeled probes have to be synthesized from both ends of the cloned DNA which are then used in hybridization experiments against the library to identify overlapping clones. This can be achieved by the presence of phage SP6 or phage T7 promoters flanking the MCS. From these promoters, radioactively labeled RNA molecules are synthesized which are specific for both ends of the cloned fragment (Melton *et al.*, 1984). An example for cosmids carrying bacteriophage promoters is the Lorist vector family (Cross and Little, 1986). Alternatively, primers flanking the MCS can be used to synthesize DNA probes by primer extension, which are specific for the ends of the cloned fragment (Mizukami *et al.*, 1993).

Cosmid vectors containing two cos sites separated by a unique restriction site can be used to prepare two different vector arms (Bates and Swift, 1983). Ligation with these arms reduces the background of colonies arising from infective packaged vector concatamers lacking cloned DNA, and thereby increases the cloning efficiency. Presence of two antibiotic resistance markers (e.g., resistance against ampicillin and tetracycline) improves the stability of the cosmid clones during propagation (see below).

Numerous cosmid vectors are available, serving specific needs. For instance, a single-copy vector capable of cloning fragments of the same size range as the usual cosmids offers the ability to stably maintain sequences which are prone to delete or rearrange in multicopy cosmid vectors (Kim *et al.*, 1992); recently, a cosmid was developed for generating sets of nested deletions of cloned inserts and DNA sequencing templates *in vivo* (Wang *et al.*, 1993). It pays off in the long run to give much attention to the selection of the most suitable vector. Consulting monographs like "Molecular Cloning" (Sambrook *et al.*, 1989) and "Cloning Vectors" (Pouwels *et al.*, 1985) or catalogs from manufacturers of products for research in molecular biology might be a good start.

For establishment of an ordered cosmid bank of *M. pneumoniae* a specially designed cosmid, pcosRW2, was constructed which fulfils all the mentioned criteria (Wenzel and Herrmann, 1988). Restriction endonuclease *Eco*RI was chosen as cloning enzyme. This enzyme was shown to have an appropriate distribution of recognition sequences on the *M. pneumoniae* genome. The largest genomic *Eco*RI DNA fragments were smaller than 30 kbp, well below the maximal cloning capacity for the cosmid vector. This vector also contains resistance markers for both ampicillin and tetracycline. It proved essential to include both antibiotics in the growth medium, since deletions were more frequently observed using ampicillin or tetracycline alone (Wenzel and Herrmann, 1988).

IV. Construction of the Library

A. Vector Preparation

As mentioned, cosmids carrying two cos sites allow fast preparation of vector arms. In the first step, the cosmid is linearized by restriction with the endonuclease cutting between the two cos sites to give blunt ends, followed by calf intestine alkaline phosphatase (CIP) treatment. Restriction with the cloning enzyme yields two vector arms which then can be ligated to a partially digested genomic DNA.

1. Material

If not otherwise stated we always use deionized H_2O and store the solutions at room temperature.

Agarose gels: 0.5–1.5% standard agarose (melting temperature 87°C); 0.8–1% low-melting agarose (melting temperature 65°C) in 1× E buffer. Mix the agarose with 1× E buffer and melt it in a microwave oven. Cool it down to about 65°C before pouring the gel. Run low-percentage (0.5%) standard agarose gels or low-melting agarose gels in a cold room if available.

Alkaline phosphatase (CIP) buffer, 10×: 500 mM Tris-HCl, pH 9.0, 10 mM MgCl2, 1 mM ZnCl$_2$, 10 mM spermidine HCl. Divide in aliquots and store at −20°C.

E buffer, 20× (buffer for agarose gel electrophoresis): E buffer, 1× (working solution): 40 mM Tris base, 20 mM Na acetate, 2 mM ethylenediaminetetraacetic acid (EDTA). To prepare 1 liter stock solution (20×) dissolve 96.8 g Tris base, 54.4 g Na-acetate · 3 H_2O, and 12.8 g EDTA in 900 ml of H_2O, adjust the pH to 8.3 with glacial acetic acid, and add H_2O to 1000 ml.

Escherichia coli for plating and growth of cosmids: We used the following strains of *E. coli:* ABLE (Stratagene), for cloning of toxic products; DH1 (Hanahan, 1983); DH5 (Hanahan, 1983); DH5α (Hanahan, 1983); HB101 (Bolivar and Backman, 1979).

EDTA (ethylenediaminetetraacetic acid, disodium salt), 0.5 M, pH 8.0: Mix 93 g disodium EDTA ·2H_2O into 400 ml of water using a magnetic bar. Adjust the pH to 8.0 with 10 N NaOH. Add H_2O to 500 ml.

Enzymes: Restriction endonucleases are available from many sources. We purchase T4 DNA ligase from Boehringer Mannheim and New England Biolabs, and T4 kinase from Boehringer Mannheim.

LB medium (Luria–Bertrani medium): Dissolve 10 g bacto-tryptone, 5 g bacto-yeast extract, and 10 g NaCl in 800 ml H_2O by heating, adjust the pH to 7.4 with 1 N NaOH after the medium has cooled down, add H_2O to 1000 ml. Divide in smaller portions and autoclave to sterilize. For preparing plates add 15 g bacto-agar/1000 ml. Add antibiotics in solution after cooling to 60°C, just before pouring the plates.

λG buffer (for dilution of packaged cosmids): 50 mM Tris-HCl, pH 7.4, $MgSO_4$ 20 mM, NaCl 50 mM, 0.1% gelatin. To prepare 1 liter dissolve 1 g gelatin in 200 ml of H_2O by heating, mix with 50 ml 1 M Tris-HCl, pH 7.4, 20 ml 1 M $MgSO_4$, and 10 ml 5 M NaCl; adjust the volume to 1 liter with H_2O.

Ligation buffer, 10× (for T4 DNA ligase): 500 mM Tris-HCl, pH 8.0, 100 mM $MgCl_2$, 100 mM dithiothreitol (DTT), 10 mM ATP (for sticky-end ligation), 10 mM EDTA, pH 8.0; also prepare the same buffer without ATP. For blunt-end ligation mix both buffers 1:1 to lower final ATP concentration in the working solution to 0.5 mM. Divide in aliquots and store at −20°C.

Phage λ extracts for packaging cosmid cloned DNA: Stratagene.

Phenol buffered: Use redistilled phenol bottled under nitrogen, stored at −20°C, 100 g phenol; keep it in the original glass bottle, melt at 60°C, then add 0.12 g 8-hydroxychinoline (antioxidant), 25 ml of H_2O, and 6.25 ml of 1 M Tris base. Mix well and divide in aliquots; store at −20°C.

Qiagen, anion-exchange column for plasmid and cosmid purification (Qiagen, Germany): Follow the instructions of the manufacturer.

Restriction enzyme buffer, 10×: Use the buffers recommended and supplied by the manufacturers. Store at −20°C.

Sodium acetate, 3 M, pH 5.2: Dissolve 40.8 g of Na acetate · 3 H_2O in 80 ml of H_2O. Adjust pH to 5.2 with glacial acetic acid. Add H_2O to 100 ml; sterilize by autoclaving.

TE buffer (Tris/EDTA), pH 8: 10 mM Tris-HCl, pH 8.0, 0.1 mM EDTA, pH 8.0, prepared from a 100× TE stock solution of 1 M Tris-HCl, pH 8.0, 10 mM EDTA, pH 8.0; sterilize by autoclaving.

2. Procedure

(1) Linearize 20 µg of cosmid vector DNA in a final volume of 50–100 µl. In general, the DNA concentration for restriction enzyme digest should not exceed 200 µg/ml. The restriction enzyme should cut between

the cos sites and result in DNA with blunt ends (e.g., *Pvu*II). Verify digestion on an agarose gel.

(2) Extract the sample once with an equal volume of a 1:1 mixture of phenol/chloroform, followed by ethanol precipitation. If not otherwise noted, an ethanol precipitation is done as follows: Adjust the DNA solution to 0.3 M sodium acetate (final concentration) from a 3 M stock solution, pH 5.2, mix well, and add 2.5 vol of ethanol. Keep for 20 min on ice or at $-20°C$. Collect the DNA by centrifugation at $12,000g$ for 15 min at 4°C. Discard the supernatant, add 1 ml of 70% ethanol to the tube; mix well on a Vortex, centrifuge at $12,000g$ for 10 min. Repeat the washing steps. Remove the supernatant with a Pasteur pipette, traces of fluid may be removed by drying the tube at room temperature or 37°C. Dissolve the pellet in TE at a concentration between 100 and 200 μg/ml.

(3) Dephosphorylate the 5'-phosphorylated ends of the linearized vector (blunt ends) by treatment with CIP at a concentration of 0.5 U/ pmol DNA ends in 1× CIP buffer for 30 min at 37°C; 5 μg of a linearized 5-kbp-long cosmid vector corresponds to 1.5 pmol vector molecules or 3 pmol 5'-pholpharylated ends. Be aware that dephosphorylation of blunt ends and sticky ends with 3'-end overhang requires more enzymes than dephosphorylation of sticky ends with 5'-end overhang. Inactivate the enzyme by addition of EDTA (20 mM final concentration) and incubate at 68°C for 15 min. Extract twice with an equal volume of phenol/chloroform and precipitate the DNA with ethanol as explained above. Dissolve in TE buffer at a concentration of 200 μg/ml.

(4) Carry out a test ligation with an aliquot of the sample to check the effectiveness of the CIP treatment (see also Section IV.C). As a positive control set up a ligation including T4 polynucleotide kinase: mix 1 μg dephosphorylated, linearized cosmid vector and 2 μl 10× ligation buffer, adjust with TE buffer to 20 μl, and add 7 U T4 ligase; split into two 10-μl samples, add to one sample 0.5 U T4 polynucleotides kinase, and incubate both samples for 60 min at room temperature. Analyze the results on an 1% agarose gel. Include in the analysis as a control a sample with 0.05 μg untreated, linearized vector. Only the DNA from the sample with polynucleotide kinase included should be shifted up in the gel indicating ligation of the vector.

(5) Restrict the linearized vector DNA with a two- to threefold excess of the cloning enzyme (e.g., *Eco*RI), and verify digestion by agarose gel electrophoresis. Extract once with an equal volume of phenol/chloroform, precipitate with ethanol, and resuspend the DNA at a final concentration of 1 μg/μl in TE assuming a 90% yield of the original DNA. To verify, run a 1% agarose minigel with two samples of prepared vector, e.g.,

0.01 and 0.05 μg, and compare them to several probes of linearized vector with varying known (measured) DNA concentrations.

B. Preparation of Partially Restricted High-Molecular-Weight Genomic DNA

For construction of a cosmid library, high-molecular-weight bacterial DNA should be freshly prepared from growing cells. The DNA isolation is done by standard procedures taking precautions to avoid shearing the DNA to ensure that the DNA remains larger than 200 kbp (no pipetting with tips with narrow openings or vortexing of DNA, dialysis instead of alcohol precipitation, storage at 4°C, etc; see Chapter 1). To obtain maximal cloning efficiency, the size of the main fraction of partially digested DNA should be greater than the size of phage λ DNA (48.5 kbp). Optimal conditions for partial digestion have to be determined experimentally and the size of the products should be checked on low-percentage agarose gels (0.5%) or pulsed-field gels including phage λ DNA as a size marker added directly to the samples to be tested.

Numerous methods exist for performing partial digests (see Chapter 1); we have varied the time of digestion. When preparing a large-scale partial digest, aliquots of the reaction mixture should also be taken at different times around the optimal time point. These additional aliquots from the over- and underdigested samples are also pooled for size fractionation. This procedure ensures that fragments of a suitable size are also present from regions of the genome which have either more or less restriction sites than average.

During cosmid library construction, it is very important to avoid ligation of noncontiguous DNA fragments during cloning. There are two commonly used methods of avoiding these artefacts which prevents self-ligation. Those are treatment of the partially digested genomic DNA with alkaline phosphatase or size fractionation either by sucrose gradient centrifugation or by preparative agarose gel electrophoresis. Ligation of two properly size-selected DNA molecules prior to the addition of cosmid arms would yield DNA which is too large to be packaged by the λ packaging extracts. We recommend combining both methods. Only if the DNA to be cloned is not available in sufficient amounts would we treat the partially digested DNA with alkaline phosphatase and abandon the size fractionation.

1. Procedure: Preparation of Partially Digested Genomic DNA

(1) Do a pilot experiment with 25 μg DNA. Calculate the amount of enzyme needed to digest the DNA completely based on the theoretical

frequency at which the enzyme would cut. Take aliquots from the digestion covering several (5 to 10) time points between 0 and the calculated time for complete digestion. At the end of digestion add some extra enzyme to make sure you have a reference sample of completely digested DNA. Check the size distribution of the genomic DNA under these conditions. If you have samples which are satisfactory (that is, the main fraction of the DNA is between 40 and 60 kbp), use these condition for the scaled-up experiment. There are some difficulties, like the viscosity of the DNA solution, which make it impossible to measure the exact DNA concentration. Therefore, take several samples above and below the calculated optimal time points.

(2) Digest 250 µg of genomic DNA with the restriction enzyme to be used for cloning. Remove aliquots at different times (e.g., 1, 2, 4, 8, 16, and 32 min), and stop the digestion by adding EDTA to a final concentration of 20 mM.

(3) Check the average size of the different fractions by electrophoresis on a 0.5% agarose or pulsed-field gel. The main fraction of the genomic DNA should have a size range between 40 and 60 kbp. (λ DNA used as a size marker should be heated to 60°C for 10 min and cooled quickly in ice water just before application to the gel.)

(4) Pool the fractions of interest and extract with an equal volume of phenol/chloroform, do a standard ethanol precipitation and dissolve the DNA in TE buffer, adjust to 1× CIP buffer and dephosphorylate the 5'-ends by addition of CIP at a concentration of 0.03 U per microgram genomic DNA. Incubate at 37°C for 30 min. Inactivate the CIP as described above. The calculation of the CIP concentration is based on the assumption that the average size of the sample DNA is about 50 kbp and 1 µg of this DNA therefore has about 0.06 pmol 5'-terminal phosphorylated ends. The CIP concentration should be 0.05 U/pmol 5' ends. The DNA concentration can be estimated by comparing the UV-induced fluorescence of ethidium bromide-stained DNA of the sample with that of reference DNA by a spot test (Sambrook *et al.*, 1989).

(5) Prepare a 15–40% w/v sucrose gradient in a Beckman SW 28 (or equivalent) tube. Load onto each tube up to 250 µg genomic DNA in no more than 300 µl volume and centrifuge at 26,000 rpm for 18 hr at 15°C in a Beckman SW 28 rotor. As an alternative, run a preparative 1% agarose (low-melting) gel (see Chapter 1) and isolate the DNA in the range of 40–50 kbp. Use phage λ DNA or a mixture of linear homopolymers of a 10-kbp plasmid as size marker. Agarose gels offer greater discrimination in sizing, but sucrose gradients permit preparation of larger amounts of DNA.

(6) Collect 0.5-ml fractions from the bottom of the gradient and analyze aliquots of any second fraction on a 0.5% agarose or pulsed-field gel. Compare with phage λ DNA. Size markers must be in a similar concentration of sucrose and salt, and we therefore recommend addition of λ DNA as a size marker directly to these samples.

(7) Pool the fractions of interest and dialyze against TE at 4°C.

(8) Precipitate the DNA with ethanol and resuspend at a final concentration of 0.5 μg/μl in TE. Concentration could be calculated by comparative electrophoresis on 1% agarose minigels with samples of defined amounts of λ DNA as references. In this case, a 1% agarose gel is better than a 0.5% gel, since the DNA fragments of the samples varying in size will migrate as a single band, which facilitates the calculation of the DNA concentration.

C. Ligation

Set up the following reaction mixtures (the DNA concentrations should be between 200 and 300 μg/ml):

4 μg vector-DNA (≈1 pmol)
3 μg genomic DNA (≈0.1 pmol fragment assuming a size of 40 kbp)
2 μl 10× ligation buffer
7 U (Weiss) T4 DNA ligase[a]
Add TE to a final volume of 20 μl
Incubate at 16°C for at least 12 hr
Store at 4°C

Remove a sample (5%) before adding T4 ligase and again after the ligation period for comparison by electrophoresis. Successful ligation is indicated by shift-up of DNA bands. If no ligation is seen, extract with phenol/chloroform, precipitate the DNA, and repeat the ligation.

D. *In Vitro* Packaging

Since only small quantities of packaging extracts are required, we recommend use of commercial packaging extracts. They should have packaging efficiencies better than 10^8 plaque-forming units (pfu)/μg λ DNA.

Package an aliquot of the ligation mixture containing 0.5–1 μg DNA, as recommended by the manufacturer. Dilute packaged cosmids with 50 μl of a suitable buffer, e.g., λ G buffer. Store the packaged cosmids at 4°C. (They can be stored for several months.)

---

[a]Be sure to have the right T4 ligase concentration. Not all manufacturers use the Weiss unit for defining enzyme activity (Weiss *et al.*, 1968).

E. Infection

(1) Mix an aliquot ($\approx$10 μl $\approx$0.1–0.2 μg DNA) of the packaging reaction with 200 μl of a freshly prepared overnight culture of an appropriate *E. coli* strain (e.g., *E. coli* HB101, DH5, DH5α). The bacteria has to be grown in LB medium in the presence of 10 mM MgCl$_2$ and 0.2% maltose. Alternatively, the bacteria of the overnight culture were collected by centrifugation for 10 min at 3000g at 4°C and the pellet was resuspended in 1/2 vol of precooled 10 mM MgSO$_4$ solution. These "Mg cells" can be used for 1 week. Store at 4°C.

(2) Incubate for 15 min at 37°C. Add 1 ml growth medium (LB) and the appropriate antibiotic (e.g., 50 μg/ml ampicillin, 10 μg/ml tetracycline). Incubate 1 hr at 37°C with shaking. Centrifuge the suspension for 2 min in an Eppendorf centrifuge. Discard the supernatant. Resuspend the pellet in 200 μl LB medium by repeated, careful pipetting and plate 5, 20, and 50 μl and the remainder on LB agar plates with appropriate antibiotics. Incubate overnight at 37°C.

Two aspects have to be emphasized. First, the optimal molar ratio of genomic DNA fragments to cosmid arms is probably in the range of 1:3 or 1:4, but since it is essential to supply every fragment to be packaged with one cos site at each end, a molar ratio of 10:1 (vector arms to fragment) should be used. This compensates also for miscalculations of the fragment concentration. Even after sizing, the molar concentration of fragments can be determined only approximately. We recommend carrying out several pilot ligation reactions with varying amounts of genomic DNA at a constant vector concentration and checking the reaction by electrophoresis. An excess of vector arms is indicated by DNA bands which correspond to the three possible ligation products of these arms, i.e., left–left, left–right, and right–right. If the genomic fragments are in excess, then most of the vector arms should be shifted to higher-molecular-weight DNA. Cloning efficiency depends largely on the packaging efficiency of extracts from lysogenic *E. coli* strains. When these extracts are prepared in the laboratory, packaging efficiency should have a minimum of 10^8 pfu per microgram test DNA. Good extracts have efficiencies of 10^9 pfu. They are commercially available, though rather expensive. A yield of 10^5 recombinants from 1 μg of insert DNA is a satisfying result.

Occasionally, *E. coli* cells are doubly infected. DNA preparations from these clones contain different cosmids, as can be seen in agarose gels by the appearance of two bands of circular supercoiled DNA migrating closer together as the supercoiled and relaxed form of the same cosmid DNA. These two cosmids can be separated by isolating DNA and repackaging *in vitro*, if one is interested in pure preparations of each of these two cosmids.

F. Amplification and Storage

For long-term maintenance of the library, cosmid-infected *E. coli* cells should be stored at $-70°C$ in suspension as a mixture, as separated clones in multiwell dishes, or on specially treated replica filters (Wenzel and Herrmann, 1988; Sambrook *et al.*, 1989). It is also advisable to store (up to several months) the original library in the form of packaged cosmids at 4°C.

(1) Pick an appropriate number of single colonies (according to the formula of Clarke and Carbon) and propagate them in multiwell dishes containing growth medium with two antibiotics overnight at 37°C. These minicultures may then be stored at $-70°C$ after addition of glycerol to a final concentration of 20%. From this stock, individual cultures can be transferred into a growth medium containing two antibiotics and incubated overnight at 37°C.

(2) For storage of cosmid libraries as pools of bacteria, prepare plates with sufficient (200–500) colonies of transduced bacteria as described in step E2. Scrape the colonies together using a few milliliters of medium. Collect the cells by centrifugation, resuspend them in medium, add glycerol to 20% v/v, and store aliquots at $-70°C$. Alternatively, after infection as described in steps E1 and E2, add 20 ml growth medium (LB) with the appropriate antibiotics and incubate until the culture reaches the late log phase. Add glycerol to 20% and store as aliquots of 0.5 ml at $-70°C$.

(3) Isolation of cosmid DNA is done by the same procedures as plasmid isolation, for instance by the alkaline lysis method (Sambrook *et al.*, 1989). For restriction analysis crude DNA preparations (= DNA after ethanol precipitation and three washing steps with 70% ethanol) are usually good enough. If needed (e.g., DNase activities) include a phenol/chloroform extraction or purify the DNA by anion exchange chromatography (e.g., QIAGEN).

(4) Cosmids of doubly infected *E. coli* cells can be separated by repackaging about 1 µg of the isolated cosmid DNAs. Repackaging is also useful for counter selection against unstable cosmid clones which undergo deletion.

(5) Store ($-70°C$) an aliquot of cosmid DNA of any cosmid clone you want to keep. We recommend performing restriction analyses on all primary DNA isolates and comparison of all subsequent DNA preparations with the DNA patterns of the original isolates. Use a restriction enzyme

which gives a pattern of ca. 10–20 bands, so that individual bands can easily be recognized.

G. Preparation of Filters for Screening/Hybridization

(1) Transfer cells from the multiwell dishes using a replicating block onto nitrocellulose (or nylon) filters which are placed on agar plates containing the appropriate antibiotic. Incubate overnight at 37°C. The diameter of the colonies should not be larger than 2 mm. A replicating block can be made by fixing blunt needles (stainless steel) to a metal plate or other material which can be autoclaved.

(2) Place the filters for 3 min onto 3MM Whatman paper saturated with 10% SDS. Transfer to 3MM paper saturated with denaturing solution (0.5 M NaOH, 1.5 M NaCl). After 5 min transfer again to new 3MM Whatman paper saturated with 0.5 M Tris-HCl, pH 8, 1.5 M NaCl, and neutralize for 5 min. Dry at room temperature and fix the DNA to the filters by incubation for 1 hr at 80°C or by exposure to UV light for crosslinking.

(3) To prepare filters containing only cosmid DNA, spot 2 µl of a solution of cosmid DNA (10 µg/ml) corresponding to 4×10^8 copies of cosmid DNA onto a filter. Proceed as described in step 2, but without SDS treatment. If there are problems with sensitivity, increase the DNA concentration by a factor 10.

V. Sorting the Library

A. Comparison of Clones by Restriction Analysis

To establish a contiguous set of clones covering long stretches of DNA or even the complete bacterial genome, two basic methods can be applied, namely comparative restriction analysis and chromosome walking.

Comparative restriction analysis can be done in different ways, all of which exploit the same principle; two clones of a library are very likely to overlap, if they contain a limited number of restriction sites for one or more restriction endonucleases located at the same distance from each other. The most frequently used procedures are restriction fragment mapping (Smith and Birnstiel, 1976), fingerprinting (Coulson *et al.*, 1986; Knott *et al.*, 1989; Eiglmeier *et al.*, 1993) or the "landmarking" technique (Charlebois *et al.*, 1991). For fingerprinting, the cloned DNAs are digested by a restriction enzyme with a 6-bp specificity, and the ends of the resulting fragment are radioactively labeled. After digestion with a second 4-bp specific restriction endonuclease it is possible to separate and characterize the resulting small-labeled fragments very precisely by acrylamide gel electrophoresis and to compare the DNA patterns from different clones.

The landmarking technique can be done without any labeling and relies on standard agarose gel analysis. Clones are first digested preferably with a 6-bp specific endonuclease A, if possible with the original cloning enzyme. Clones are then digested with a second endonuclease B. This is chosen as cutting only rarely, for example on average only once in 30 kbp. Of the fragments generated by digestion of one cosmid with enzyme A, statistically only a small number (between zero and three fragments) will also be cut by enzyme B. The very specific pattern of those fragments which are produced by double digestion are easily recognized. Any of these fragments which have a restriction site for the rarely cutting endonuclease is called a "landmark." Generally one common landmark is sufficient for defining two overlapping clones.

Finally, Kohara *et al.* (1987) applied the Smith/Birnstiel technique to construct restriction maps from the *E. coli* genome for eight different enzymes and to align about 1400 phage λ clones to an ordered library comprising almost the entire genome. Since the Smith/Birnstiel technique is useful for aligning clones as well as for restriction site mapping we will describe this technique in detail and refer for the fingerprinting and landmarking technique to the literature.

Restriction analysis of a certain cosmid clone can be done according to a method originally described by Smith and Birnstiel (1976): First, the number and size of cloned restriction fragments to be mapped are determined by complete digestion and agarose gel electrophoresis. Then, the cosmid clone is linearized at a unique restriction site outside of the cloned DNA (see Section III). Aliquots of the linearized molecules are digested to different extents with the enzyme selected for mapping. These partially cut samples are separated on agarose gels, blotted, and hybridized to a labeled fragment of vector DNA. This probe is derived entirely from one side or the other of the unique site used to linearize the clone (Fig. 7.2).

The results show a ladder of DNA fragments which have the same unique end. By repeating these analyses in pairs with all the neighboring intermediate DNA fragments, the correct order of restriction fragments as well as the orientation of the cloned insert can be deduced. The order of restriction fragments produced by restriction enzymes other than the cloning enzyme can be determined similarly.

Fragment data from different enzymes are then combined by a computer program and compared with the alignments of other clones of the cosmid library (Kohara *et al.*, 1987). This method has the advantage of being fast and simple if single clones are considered. However, if the num-

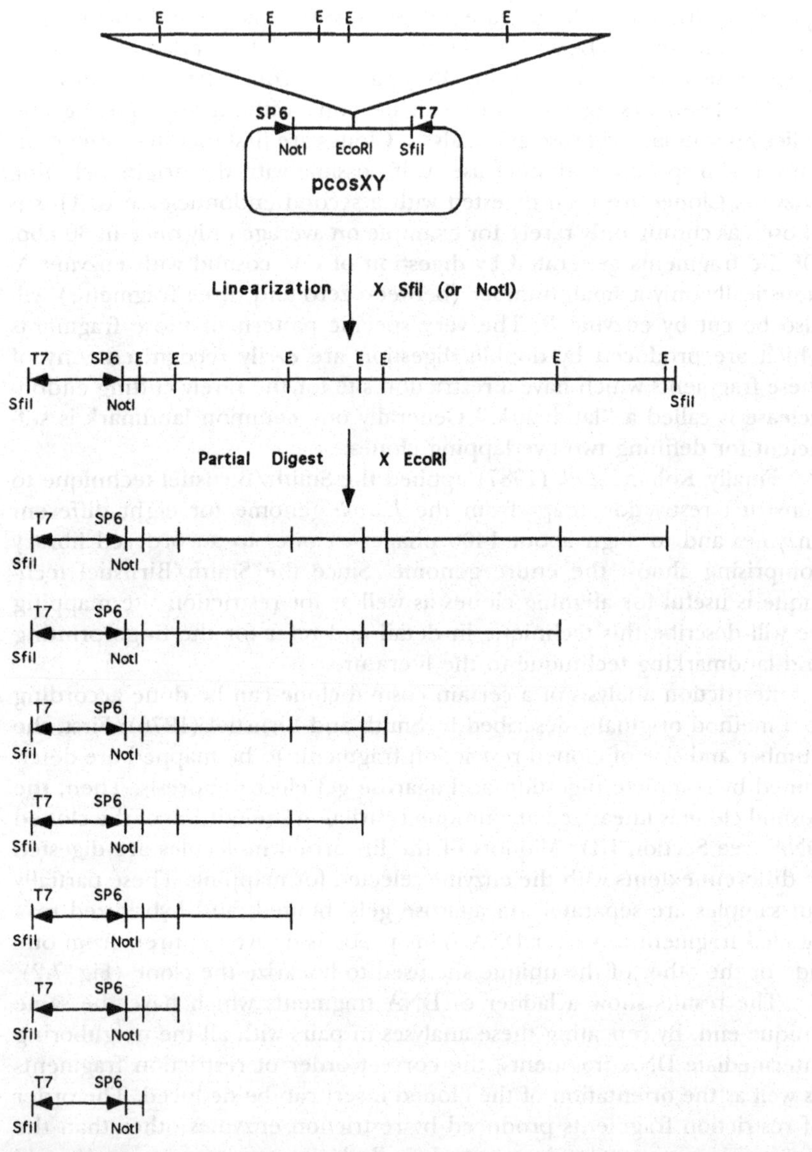

Figure 7.2 Schematic drawing showing the principle of restriction mapping according to Smith and Birnstiel. As an example, an *Eco*RI restriction mapping of a DNA region cloned in pcos XY is shown. The cosmid clone was linearized at the unique *Sfi*I site, digested partially with *Eco*RI, analyzed on a 0.6% agarose gel, blotted, and hybridized against labeled vector DNA. Only those intermediate fragments which carry the vector DNA on their ends show up. To confirm the order of *Eco*RI fragments repeat the analysis with DNA linearized at the unique *Not*I site.

ber of clones which have to be analyzed is very large, even by computer the alignment is not trivial.

For all three methods mentioned, it is necessary that the data on the DNA banding pattern be entered into a computer for evaluation by pattern matching analysis (Sulston *et al.*, 1988).

1. Material

If not otherwise stated we use deionized H_2O and store the solutions at room temperature.

Enzymes: Restriction endonucleases are available from many sources (e.g., Boehringer Mannheim, New England Biolabs and Stratagene). Calf intestinal alkaline phoshatase and phage SP6 and phage T7 DNA-dependent RNA polymerases were from Boehringer Mannheim.

Denaturing solution for hybridization experiments: 0.5 M NaOH, 1.5 M NaCl; prepare 1 M NaOH and 5 M NaCl from stock solutions.

Denhardt's reagent, 50×: 5 g of Ficoll 400, 5 g of polyvinylpyrrolidone, and 5 g of bovine serum albumin are dissolved in 500 ml of H_2O. The solution is filtered through a 0.2-μm pore size nitrocellulose filter. Store at $-20°C$ in aliquots.

Neutralizing solution for hybridization experiments: 0.5 M Tris-HCl, pH 8.0, 1.5 M NaCl; prepare 2 M Tris-HCl, pH 8.0, and 5 M NaCl from stock solutions.

Nitrocellulose 0.45-μm filter: Schleicher & Schüll.

Nylon membrane: Biodyne transfer membrane was from PALL.

Salmon sperm DNA: Prepare stock solution by dissolving salmon sperm DNA in TE buffer at concentration between 1 and 10 mg/ml. Stir on a magnetic stirrer for several hours at room temperature to dissolve the DNA. Sonicate the very viscous DNA solution until the DNA can be pipetted without problems. Extract the DNA twice with a 1:1 mixture of chloroform/phenol. Do a standard ethanol precipitation. Redissolve the DNA in TE. DNA used in hybridization experiments should be boiled for 15 min, divided in aliquots, and stored at $-20°C$. DNA used for standardization in Smith–Birnstiel experiments should be stored at $-20°C$ without boiling at a concentration of 1 mg/ml.

RNase inhibitor: RNA guard was from Pharmacia, and RNase inhibitor was from Boehringer Mannheim.

SDS (sodium dodecyl sulfate), 5%: Dissolve 5 g of SDS in 100 ml of deionized H_2O.

SSC buffer, 20×: 3 M NaCl, 0.3 M sodium citrate, pH 7.2. To prepare 1 liter dissolve 175.3 g NaCl and 88.2 g Na citrate · $2H_2O$ in 800 ml of H_2O, adjust to pH 7.2 with 1 N NaOH, and add H_2O to 1000 ml.

Sucrose gradient 15–40% (w/v): Buffer for sucrose gradient: 20 mM Tris-HCl, pH 8.0, 1 M NaCl, 5 mM EDTA. To prepare 500 ml mix 250 ml of H_2O with 75 g (200 g) sucrose, sterilize the sucrose solution, add to the cooled-down sucrose solution 10 ml of 1 M Tris-HCl, pH 8.0, 100 ml of 5 m NaCl, and 2.5 ml of 1 M EDTA, adjust with sterile H_2O to 500 ml, and mix well.

Transcription buffer, 10× (for T7 and SP6 RNA polymerases): 400 mM Tris-HCl, pH 7.5, 60 mM $MgCl_2$, 100 mM DTT (dithiotheritol) 40 mM sperimidine-HCl, 50 mM ATP, GTP, and CTP. UTP is added only as a ^{32}P-labeled compound. Mix from stock solutions. Store at $-20°C$ in aliquots.

2. Procedure

(a) Smith–Birnstiel Technique

(1) Linearize 2 μg of cosmid DNA (a crude DNA preparation as described in Section IV.F is sufficient) by cleavage at a unique restriction site (e.g., *Sfi*I in the multiple cloning site). Extract with an equal volume of phenol/chloroform, precipitate the DNA with ethanol, and resuspend in 20 μl TE.

(2) Do a partial digest with the desired enzyme. Mix 0.75 μg of linear cosmid DNA with a constant amount (5 or 10 μg) of sonicated calf thymus DNA or salmon sperm DNA (this guarantees an almost constant number of restriction sites in each digestion and helps to standardize the reaction conditions) and digest with the appropriate restriction endonuclease. Samples are taken at different time points, e.g., 0.5, 1, 2, 4, and 8 min, and loaded onto a 0.6% agarose gel), together with a size marker, preferably linear homopolymers of a sequenced DNA, e.g., a plasmid. The size of the plasmid depends on the size of the restriction fragments to be mapped. Since the estimation of the size of DNA intermediates is critical, run a sample of the size marker in every fourth slot of the gel. Sometimes it is also helpful to vary the agarose concentration.

(3) Transfer the gel to a nylon filter and hybridize against radioactively labeled vector DNA, to visualize all the intermediate DNA fragments which have the vector DNA at one end (Fig. 7.2). The labeled probe should also hybridize to the size marker. This is no problem since many cosmids and plasmids share common sequences like genes coding for resistance to antibiotics.

(4) Establish a calibration curve with the size marker and use it to calculate the sizes of all intermediate fragments (bands). The difference in size between two neighboring intermediates corresponds to one defined restriction fragment generated by a complete digest of the corresponding cosmid DNA with the same endonuclease. Measure the distance between all the neighboring intermediate fragments (bands), correlate it

to restriction fragments of the complete digest, and deduce the order of fragments (Fig. 7.2).

(5) Repeat steps 1–4 with the same DNA but linearize the cosmid DNA with a restriction enzyme which cuts the multiple cloning site on the other side of the site of insertion as the enzyme used in the first analysis (Fig. 7.2).

B. Chromosome Walking

Another widely applied approach is known as chromosome walking (Bukanov and Berg, 1994; Birkenbihl and Vielmetter, 1989; Wenzel and Herrmann, 1988). Starting from selected clones, labeled probes specific for the ends of the insert DNA are synthesized and used as probes in hybridizations against a representative library. Clones hybridizing with one of the probes are picked and the degree of overlap is determined by restriction analysis. These clones then serve as templates for the synthesis of new probes; by subsequent analysis, contigs are produced which in an ideal case can be linked together to form a circle, if the bacterial genome to be cloned exists in circular form. Linking endstanding cosmids to a circle is a strong indication that the entire genome is contained in the ordered cosmid library.

1. Procedure: Chromosome Walking Using RNA Probes

(1) Probe synthesis (*in vitro* transcription)

1.5 μl	10× transcription buffer
0.2–0.4 μg	Cosmid DNA (from one clone, crude preparation)
1 μl	RNase inhibitor (30 U/μl)
1 μl (10 μCi)	[α-^{32}P]UTP (3000 Ci/mmol)
1 μl	SP6 or T7-RNA polymerase (10–20 U/μl)
H$_2$O: add 15 μl	

(2) Mix the components at room temperature. Incubate for 1 hr at 40°C (SP6-RNA polymerase) or 37°C (T7-polymerase), respectively.

(3) Remove the nonincorporated nucleotides by gel filtration through a Sephadex G-100 column (pasteur pipette) or spin column.

(4) Measure the amount of incorporated radioactivity in a liquid scintillation counter; 10^5–10^7 Cerenkov counts (cpm) incorporated in RNA per microgram cosmid DNA is a suitable specific activity.

(5) Screening of the library: For hybridization of the dotted cosmid library (either DNA or colonies; see Section IV.G. 3) with the labeled RNA probes, prehybridize two identical cosmid filters (one for each probe) in a solution of 50% formamide, 5× SSC, 5x Denhard's solution, 100 μg/ml denatured sonicated salmon sperm DNA, and 0.05% SDS for 3–6 hr at

46°C. The hybridization is done in the same solution complemented with the appropriate ^{32}P-labeled RNA probe (10^5–10^6 cpm/10 ml solution). We normally hybridize overnight. The length of the RNA should be between 1000 and 2000 nucleotides.

(6) Wash the filters for 30 min at 68°C in 2× SSC, 0.1% SDS. Repeat wash in 1× SSC, 0.1% SDS, and finally in 0.5× SSC, 0.1% SDS. Dry the filters at room temperature. Autoradiographic signals are obtained by exposure of the filters to X-ray film.

(7) Identify positive clones and determine the degree of overlap to the cosmid clone from which the RNA probes were synthesized. This is done by restriction analysis with the same endonuclease used for cloning.

(8) The desired overlapping clone should have the following features: (i) a positive hybridization signal with only one of the two RNA probes; (ii) a relatively small degree of overlap, leading to a minimum number of clones necessary to cover the complete genome. The minimal overlap is one shared restriction fragment produced by the endonuclease used for cloning. This fragment should be larger than 300 bases, so that it can be identified easily on an ethidium bromide-stained agarose gel of the complete digest of the cosmid clones to be compared.

(9) Take positive clones and repeat steps 1–8.

(10) Instead of using the very specific RNA probes it is also possible to label the entire insert from individual cosmid clones and screen the cosmid library with them. This serves to walk in both directions simultaneously from the original clone. Newly identified clones must be analyzed to determine the degree of overlap by comparative restriction analysis (Birkenbihl and Vielmetter, 1989).

A high-resolution genomic restriction map can be constructed by aligning the restriction maps created for individual cosmids with those of overlapping clones. Furthermore, individual cosmids can be used as probes in Southern blotting experiments of genomic DNA which was restricted by enzymes cutting less frequently than the cloning enzyme. Data obtained from this macrorestriction analysis confirm the order of cosmid clones and help to eliminate clones containing noncontiguous fragments.

In the case of *M. pneumoniae* we decided to use the walking strategy with a library representing five times coverage of the genome. Two cosmids with different *Eco*RI restriction patterns were chosen as starting points for walking. A contiguous set of 34 cosmid clones covering virtually the complete genome was constructed (Wenzel and Herrmann, 1989). One gap remained which was analyzed first by methods described below and then closed by cloning a bridging fragment in a phage λ vector. A total of 143 genomic *Eco*RI fragments were identified and aligned. As a fidelity check, cosmid clones were aligned to a macrorestriction map of 25 genomic *XhoI* fragments

VI. Mapping Problems

The most severe problem for both the restriction mapping and walking strategies is the presence of highly repetitive DNA sequences in the target genome. Regarding restriction mapping, determination of a true overlap is rendered more difficult. The difficulty can be overcome by increasing the number of restriction enzymes. This leads to a more detailed map, but considerably more time is needed for the analysis. In chromosome walking, a considerable number of nonadjacent clones are identified when probes containing repetitive sequences are used. This often requires isolating specific fragments as probes which lack repetitive DNA. In other cases, it may be necessary to analyze the repeated elements in detail by subcloning, restriction analysis, or DNA sequencing. It also might be help-ful to prescreen the cosmid library with a repetitive sequence specific probe to identity in advance all the cosmids which carry this repetitive sequence. Though these procedures are time-consuming, as a "by-prod-uct" new information about the nature, number, and distribution of re-petitive DNA elements in a bacterial genome are obtained.

Additional problems are encountered when specific genomic regions either are underrepresented in the library or fail to be cloned in a par-ticular vector system (Kohara *et al.*, 1987; Knott *et al.*, 1988, 1989; Wenzel and Herrmann, 1989). Whereas the problem of underrepresentation can be solved by simply increasing the number of colonies to be screened, the second problem requires a more detailed analysis.

First, the nonclonable region has to be characterized in detail. Spe-cific probes from endstanding cosmids may help to identify a linking ge-nomic restriction fragment. For cloning this fragment a simple measure might be to use a host cell which reduces the copy number of the cosmids, and thereby the number of toxic products (e.g., *E. coli* ABLE, Stratagene). If this fails one should try to clone the fragment in question in a different vector system (for example, low-copy-number plasmid or cosmid or vector systems based on bacteriophages including M13 and phage λ, which re-quire different cell components for growth and might tolerate the DNA region in question). Another advantage of the phage system is that phages potentially can be propagated with low yields in a host bacterium, even if the phage carries a cloned fragment which is toxic to the bacterium. Usu-ally, the use of an alternative cloning system should help to overcome this problem. As an example in *M. pneumoniae*, a region of 16.4 kbp failed to be cloned into cosmids. By hybridization, it was possible to link the end-standing cosmid clones through identification of genomic *PstI* and *XbaI* fragments covering the gap region. This region was characterized further by genomic restriction analysis with additional restriction enzymes and was then subsequently cloned into plasmids and a phage λ-derived vector system (Short *et al.*, 1988) resulting in a plasmid clone and two phage λ clones closing the gap. The final analysis showed that a 1.3-kbp-long DNA fragment could only be cloned in a phage λ or phage M13/fd-derived cloning system but not in standard plasmids or cosmids.

If changing the cloning system does not help and if the nonclonable region is not too large (up to about 30 kbp), it may be possible to amplify these sequences by polymerase chain reaction using primers derived from sequences of the endstanding cosmids (Cheng *et al.*, 1994).

During chromosome walking, sometimes "gaps" (or pseudo-gaps) are observed. A pseudo-gap occurs when two clones are immediately adjacent, but do not overlap. It arises from large adjacent genomic restriction fragments exceeding the packaging capacity of phage λ. This is actually not a gap, but one has to prove that both fragments are adjacent. In addition, pseudogaps occur because the amount of sequence shared between overlapping clones is so small that the overlap is not recognized. In either case, these clones can be linked by their hybridization to a common genomic restriction fragment. To verify linkage, a single genomic restriction fragment must be identified which hybridizes with probes from the clones which flank the gap. After subcloning one or more of these linking fragments they have to be analyzed by further restriction analysis or DNA sequencing to verify unambiguously the linkage of the two adjacent cosmid clones.

Procedure for Linking Adjacent Cosmid

(1) Set up complete digests of 5 μg genomic DNA with different restriction enzymes at an analytical scale. It is useful to choose enzymes which cut 6-bp restriction sites in the target genome to produce fragments of greater size, increasing the chance of identifying linking fragments.

(2) Separate the genomic restriction fragments on a 1% agarose gel, blot, and hybridize separately with ^{32}P-labeled RNA probes from the endstanding cosmids (see above).

(3) If the clones which flank the gap are truly adjacent, at least one, and sometimes more, genomic restriction fragments will hybridize with probes from both clones

(4) These linking fragments which hybridize probes from both sides of a gap can be cloned into plasmid or phage λ-derived vector systems by

standard procedures and analyzed further. The procedure for linking adjacent fragments or bridging a true gap between two endstanding cosmids is very similar. The main difference is that closing a true gap is facilitated by a preceeding analysis on the size of the gap.

VII. Summary

For constructing an ordered cosmid library the following points should be considered:

(i) Verify whether the genome of your choice is linear or circular. So far, only a few linear bacterial chromosomes are known (see Fonstein and Haselkorn, 1995).

(ii) Carefully select a versatile vector and a suitable cloning enzyme. Endonucleases which recognize 6-bp restriction sites are more convenient for restriction mapping. Enzymes recognizing 4-bp restriction sites ensure a better random fragmentation. Do pilot experiments with several restriction enzymes. Be sure that the largest fragments are below 30 kbp.

(iii) Calculate the number of clones needed to cover statistically (99% probability) the complete genome. Take this number of cosmid clones as a working set and try to align all these clones before you go back to other clones of your collection. If you cannot close a circle with the ends of the linear contig, test a limited number of additional clones (as a rule of the thumb not more than 10 times that statistically called for). If this is not sufficient you should define the size of the gap(s) between the individual contigs and search for genomic restriction fragments which could close the gaps, trying to clone them using additional vector systems. If this also fails, the DNA bridging the gap might be amplified by PCR on genomic DNA (Cheng *et al.*, 1994).

(iv) At the beginning of the analysis one should start to prepare cosmid DNA minipreps (1.5 ml cell suspension) from 100 to 300 cosmids and do a comparative restriction analysis with an enzyme which cuts the cosmids 10–15 times (ideally it should be the cloning enzyme). Overlapping cosmids can be detected "by eye," but it is much better to perform a computer-assisted evaluation, since one can identify more precisely clones which overlap and clones without overlap.

(v) Establishing contigs can be accomplished through the use of both landmarking and walking procedures, carried out in parallel. "Landmarking" analysis should be performed with candidates believed to overlap among the clone selected for restriction analysis. Chromosome walking analysis should be initiated with 10–20 or more clones without apparent overlap. Prepare specific RNA probes and test your complete working set by dot blot analysis for positive clones. A useful positive clone reacts with the RNA probe from only one end of a cloned insert. Repeat then the analyses with the next set of 100–300 cosmids. Take only those cosmids from the working set for the "landmarking" technique which were not identified in the walking analysis. Repeat the restriction analysis. To continue the walk take clones identified from the walking hybridization and prepare new probes from these clones for a subsequent round of dot blot analysis. Continue until all clones of your walking set have been analyzed. If you do not succeed in linking the entire library, continue as recommended (ii).

Once established, an ordered cosmid library from a bacterial genome serves as a valuable resource which is easy to handle and facilitates a variety of detailed analyses. Most importantly, a genomic library permits construction of a high-resolution physical map which can be converted to a detailed genetic map by identifying and localizing genes of interest on the physical map. With recent improvements in sequencing strategies and techniques the method of choice will be DNA sequencing of subclones of cosmids, entire cosmids, or even the entire genome.

Acknowledgments

We thank I. Schmid for preparing the manuscript and the Bundesministerium für Forschung und Technologie (BCT 0381/5) for financial support.

References

Bates, B. F., and Swift, R. A. (1983). Double cos site vectors: Simplified cosmid cloning. *Gene* **26,** 137–146.

Birkenbihl, R. P., and Vielmetter, W. (1989). Cosmid-derived map of *E. coli* strain BHB2600 in comparison to the map of strain W3110. *Nucleic Acids Res.* **17,** 5057–5069.

Bolivar, F., and Backman, K. (1979). Plasmids of *Escherichia coli* as cloning vectors. *In* "Methods in Enzymology" (R. Wu, ed.), Vol. **68,** p. 245. Academic Press, New York.

Bukanov, N. O., and Berg, D. E. (1994). Ordered cosmid library and high-resolution physical-genetic map of *Helicobacter pylori* strain NCTC11638. *Mol. Microbiol.* **11,** 509–523.

Charlebois, R. L., Schalkwyk, L. C., Hofmann, J. D., and Doolittle, W. F. (1991). Detailed physical map and set of overlapping clones covering the genome of the archaebacterium *Haloferax volcanii* DS2. *J. Mol. Biol.* **222,** 509–524.

Cheng, S., Fockler, C., Wayne, M., and Higuchi, R. (1994). Effective amplification of long targets from cloned inserts and human genomic DNA. *Proc. Natl. Acad. Sci. U.S.A.* **91,** 5695–5699.

Clarke, L., and Carbon, J. (1976). A colony bank containing synthetic Col El hybrid plasmids. *Cell (Cambridge, Mass.)* **9**, 91–99.

Collins, J., and Hohn, B. (1978). Cosmids: A type of plasmid gene-cloning vector that is packageable in vitro in bacteriophage lambda heads. *Proc. Natl. Acad. Sci. U.S.A.* **75**, 4242–4246

Coulson, A., Sulston, J., Brenner, S., and Karn, J. (1986). Toward a physical map of the genome of the nematide *Caenorhabditis elegans. Proc. Natl. Acad. Sci. U.S.A.* **83**, 7821–7825.

Cross, S. H., and Little, P. F. R. (1986). A cosmid vector for systematic chromosome walking. *Gene* **49**, 9–22.

Eiglmeier, K., Honoré, N., Woods, S. A., Cuafron, B., and Cole, S. T. (1993). Use of an ordered cosmid library to deduce the genomic organization of *Mycobacterium leprae. Mol. Microbiol.* **7**(2), 197–206.

Feiss, M., and Siegele, D. A. (1979). Lambda replacement vectors carrying polylinker sequences. *Virology* **92**, 190–200.

Ferrin, L. J., and Camerini-Otero, R. D. (1991). Selective cleavage of human DNA: RecA-assisted restriction endonuclease (RARE) cleavage. *Science* **254**, 1494–1497.

Fonstein, M., and Haselkorn, R. (1995). Physical mapping of bacterial genomes. *J. Bacteriol.* **177**, 3361–3369.

Hanahan, D. (1983). Studies on transformation of *Escherichia coli* with plasmids. *J. Mol. Biol.* **166**, 557–580.

Kim, U. J., Shizuya, H., de Jong, P. J., Birren, B., and Simon, M. I. (1992). Stable propagation of cosmid sized human DNA inserts in an F factor based vector. *Nucleic Acids Res.* **20**, 1083–1085

Knott, M., Rees, D. J. G., Cheng, Z., and Brownlee, G. G. (1988). Randomly picked cosmid clones overlap the pyrB and oriC gap in the physical map of the *E. coli* chromosome. *Nucleic Acids Res.* **16**, 2601–2612.

Knott, V., Blake, D. J., and Brownlee, G. G. (1989). Completion of the detailed restriction map of the *E. coli* genome by the isolation of overlapping cosmid clones. *Nucleic Acids Res.* **17**, 5901–5912.

Kohara, Y., Akiyama, K., and Isono, K. (1987). The physical map of the whole *E. coli* chromosome: Application of a new strategy for rapid analysis and sorting of a large genomic library. *Cell (Cambridge, Mass.)* **50**, 495–508.

Koob, M., and Szybalski, W. (1990). Cleaving yeast and *Escherichia coli* at a single site. *Science* **250**, 271–273.

Koob, M., Burkiewicz, A., Kur, J., and Szybalski, W. (1992). RecA-AC: Single-site cleavage of plasmids and chromosomes at any predetermined restriction site. *Nucleic Acids Res.* **20**, 5831–5836.

Melton, D. A., Krieg, P. A., Rebagliati, M. R., Maniatis, T., Zinn, K., and Green, M. R. (1984). Efficient in vitro synthesis of biologically active RNA and RNA hybridization probes from plasmids containing a bacteriophage SP6 promoter. *Nucleic Acids Res.* **12**, 7035–7056.

Mizukami, T., Chang, W. I., Garkavtsev, I., Kaplan, N., Lombardi, D., Matsumoto, T., Niwa, O., Kounosu, A., Yanagida, M., Marr, T. G. *et al.* (1993). A 13 kb resolution cosmid map of the 14 Mb fission yeast genome by nonrandom sequence-tagged site mapping. *Cell (Cambridge, Mass.)* **73**, 121–132.

Morowitz, H. J. (1984). The completeness of molecular biology. *Isr. J. Med. Sci.* **20**, 750–753.

Pouwels, P. H., Enger-Valk, B. E., and Brammar, W. J. (1985). "Cloning Vectors: A Laboratory Manual." Elsevier, Amsterdam and New York.

Quiang, B. Q., and Schildkraut, I. (1984). A type II restriction endonuclease with an eight nucleotide specificity from *Streptococcus fimbriatus. Nucleic Acids Res.* **12**, 4507–4516.

Sambrook, J., Maniatis, T., and Fritsch, E. F. (1989). "Molecular Cloning: A Laboratory Manual," 2nd ed. Cold Spring Harbor Lab. Press, Cold Spring Harbor, NY.

Seed, B., Parker, R. C., and Davidson, N. (1982). Representation of DNA sequences in recombinant DNA libraries prepared by restriction enzyme partial digestion. *Gene* **19,** 201–209.

Short, J. M., Fernandes, J. M., Sorge, J. A., and Huse, W. D. (1988). Lambda ZAP: A bacteriophage lambda expression vector with in vivo excision properties. *Nucleic Acids Res.* **16,** 7583–7600.

Smith, H. O., and Birnstiel, M. L. (1976). A simple method for restriction site mapping. *Nucleic Acids Res.* **3,** 2387–2398.

Sulston, J., Mallett, F., Staden, R., Durbin, R., Horsnell, T., and Coulson, A. (1988). Software for genome mapping by fingerprinting techniques. *CABIOS* **4,** 125–132.

Wang, G., Blakesley, R. W., Berg, D. E., and Berg, C. M. (1993). pDUAL: A transposon-based cosmid cloning vector for generating nested deletions and DNA sequencing templates in vivo. *Proc. Natl. Acad. Sci. U.S.A.* **90,** 7874–7878.

Wang, Y., and Wu, R. (1993). A new method for specific cleavage of magabase-size chromosmal DNa by lambda terminase. *Nucleic Acids Res.* **21,** 2143–2147.

Weiss, B., Jacquemin-Sablon, A., Live, T. R., Fareed, G. C., and Richardson, C. C. (1968). Enzymatic breakage and joining of deoxyribonucleic acid. VI. Further purification and properties of polynucleotide ligase from *Escherichia coli* infected with bacteriophage T4. *J. Biol. Chem.* **243,** 4543.

Wenzel, R., and Herrmann, R. (1988). Physical mapping of the *Mycoplasma pneumoniae* genome. *Nucleic Acids Res.* **16,** 8323–8336.

Wenzel, R., and Herrmann, R. (1989). Cloning of the complete *Mycoplasma pneumoniae* genome. *Nucleic Acids Res.* **17,** 7029–7043.

Wenzel, R., Pirkl, E., and Herrmann, R. (1992). Construction of an restriction map of *Mycoplasma pneumoniae* and localization of selected genes. *J. Bacteriol.* **174,** 7289–7296.

8

Construction of P1 Artificial Chromosome (PAC) Libraries from Lower Vertebrates

Chris T. Amemiya, Tatsuya Ota, and Gary W. Litman

I. Introduction

Large-fragment plasmid cloning systems are increasingly valuable for genomic mapping and analysis. Both the P1 artificial chromosome (PAC) and F-factor-based bacterial artificial chromosome (BAC) systems allow cloning and propagation of relatively large DNA fragments (up to 300 kb) as bacterial plasmids in *Escherichia coli* (Shizuya *et al.*, 1992; Ioannou *et al.*, 1994). Neither PACs nor BACs can accommodate the megabase-sized inserts cloned with yeast artificial chromosomes (YACs); however, the higher cloning efficiency, improved fidelity, and greater ease of handling of PACs and BACs are attractive for physical mapping projects that do not require spans of many megabases. This chapter will focus on construction and analysis of PAC genomic libraries from lower vertebrate species, where this system has proven particularly advantageous for resolving issues involving segmental arrangement of antigen receptor genes which exhibit novel chromosomal organization.

The PAC vector pCYPAC-2 (Fig. 8.1) contains several features of the original P1 bacteriophage-derived pAD10-*Sac*BII plasmid described by Pierce *et al.* (1992a), including its single-copy origin of replication and the positive selection of insert-containing transformants using the *Sac*BII gene. The single-copy origin of replication assures stable and faithful propagation of large plasmids, with comparatively low frequencies of deletions or other rearrangements (Shizuya *et al.*, 1992; Ioannou *et al.*,

NONMAMMALIAN GENOMIC ANALYSIS: A PRACTICAL GUIDE

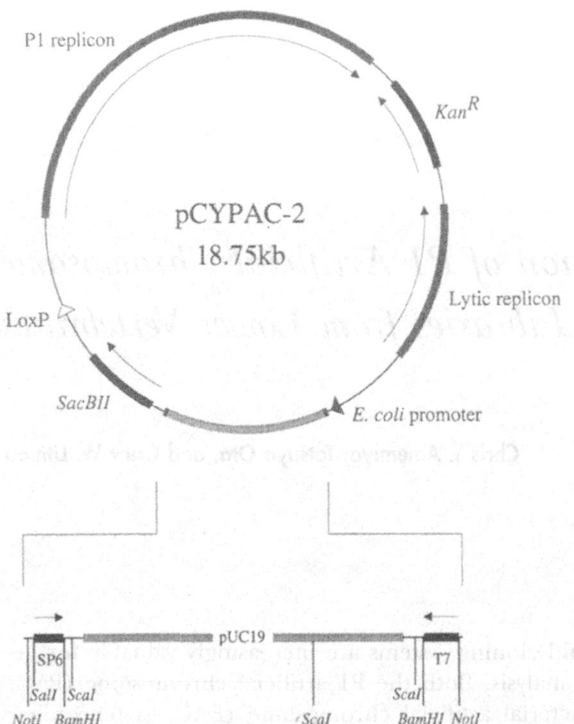

Figure 8.1 Diagram of the pCYPAC-2 cloning vector. Notable features of the vector are summarized in Ioannou *et al.* (1994) and are described here briefly. The "P1 replicon" is the origin of replication used for single-copy propagation. The "lytic replicon" provides multi-copy replication when cultures are grown in the presence of 1 m*M* IPTG (Sternberg and Cohen, 1989; Sternberg, 1990). The enlarged area shows the region between the two *Not*I sites, including the pUC19 insert (Ioannou *et al.*, 1994), which confers ampicillin resistance and facilitates production of large quantities of vector DNA by virtue of its high-copy replicon. For PAC cloning, *Mbo*I (or *Sau*3A) partial digests are cloned into the *Bam*HI sites, which are flanked by SP6 and T7 RNA promotor sequences. When lacking an insert, however, the *E. coli* promotor will drive transcription of the *Sac*BII gene, whose gene product is toxic to the cells in the presence of sucrose (positive selection). *Kan*^R is the kanamycin phosphotransferase gene (kanamycin antibiotic resistance); LoxP is a short nucleotide sequence used for LoxP-Cre homologous recombination (Sternberg, 1990). Arrows indicate direction of replication or transcription. The complete nucleotide sequence of pCYPAC-2 (deduced and melded from multiple sources) is available from the authors (C.T.A.) upon request.

1994). Unlike conventional P1 cloning (Sternberg, 1990) in which *in vitro* packaging strictly limits the length of DNA segments that may be cloned, PAC clones are generated by electroporation. Transformation of large DNA is relatively efficient and permits recovery of a wide range of clone sizes. Thus, while P1 clones may contain inserts averaging 80 kb and ranging up to 95 kb, PACs can average 150 kb and range over 300 kb (Ioannou *et al.*, 1994). pCYPAC-2 also contains SP6 and T7 RNA promoter sites for ease in restriction mapping (see below), for producing RNA probes by *in vitro* transcription, and for direct sequencing of insert ends using conventional primers. The vector also contains *Not*I restriction enzyme sites flanking the SP6 and T7 promoters, facilitating restriction mapping as well as assessment of the sizes of inserts.[1] The vector also contains an intact pUC19 plasmid for generating large quantities of pCYPAC-2 vector DNA (using the pUC origin of replication). Subsequently, the pUC portion is destroyed via *Sca*I cleavage during preparation of the vector for cloning, thereby eliminating its involvement in the actual ligation of large inserts.[2]

Several intriguing genomics problems are unique to lower vertebrates (referring to fishes, amphibians, and reptiles), including the observed differences in genome structure (Hudson *et al.*, 1980; Medrano *et al.*, 1988; Cross *et al.*, 1991), fundamental differences in chromosomal architecture (Schempp and Schmid, 1981; Amemiya and Gold, 1986, 1987; Medrano *et al.*, 1988; Schmid and de Almeida, 1988; Gold and Li, 1991), and conservation of duplicate gene loci (Morizot, 1983, 1990; Lundin, 1993). Further, phylogenetic comparisons of genomic structure and organization of selected genes (or gene families) in lower vertebrates can reveal patterns of molecular evolution (Fjose *et al.*, 1988; Shamblott and Chen, 1992; Amemiya *et al.*, 1993; Anderson *et al.*, 1994) and uncover conserved sequence motifs that potentially are involved in gene expression and regulation (Aparicio *et al.*, 1995). In addition, geneticists are increasingly using certain fishes (pufferfish, zebrafish) as model systems for positional cloning of human genes (Brenner *et al.*, 1993; Little, 1993; Baxendale *et al.*, 1995), for identifying DNA stretches that may be important for gene regulation (Aparicio *et al.*, 1995), and for examining genes and developmental processes involved in early embryogenesis (Molven *et al.*, 1991; Krauss *et al.*, 1993; Kahn, 1994). Our interest in lower vertebrates is spurred by our studies on the early origins and diversification of rearranging genes of the immune system, i.e., immunoglobulins and T-cell receptors (Amemiya and Litman, 1990, 1991; Litman *et al.*, 1993; Amemiya *et al.*, 1993; Ota and Nei, 1994; Rast and Litman, 1994). We have shown previously that the basic organization of these genes differs between major evolutionary lineages (Litman *et al.*, 1993; Rast and Litman, 1994); thus,

1. Note: We have also employed the *Not*I sites for cloning size-separated *Not*I fragments (C. T. Amemiya, unpublished).

2. It is not absolutely imperative to use *Sca*I to destroy the pUC19 region of pCYPAC-2. In lieu of *Sca*I cleavage we have used *Bam*HI digestion and preparative gel electrophoresis to eliminate the pUC19 portion, and subsequently used this vector preparation to produce PAC libraries using both sea urchin and pufferfish DNA. This alternative procedure for vector preparation is available from the authors (C.T.A., T.O.) upon request.

the PAC cloning system is particularly advantageous for characterizing long-range linkage relationships within and among components of these multigene families.

In this chapter, we present procedures for generating PAC libraries from lower vertebrates. These include protocols for vector preparation, embedding of high-molecular-weight (HMW) DNA, generation of *Mbo*I partial digests, library construction, routine isolation of PAC DNA from minicultures, and characterization (restriction mapping) of PAC clones.

II. Materials

A. Supply Checklist for PAC Cloning[3]

- Electroporator for bacterial transformation
- Electroporation cuvettes
- Electrocompetent cells—*E. coli* DH10B strain (Gibco/BRL)
- Pulsed-field gel unit (CHEF; Bio-Rad); field-inversion gel unit is useful for analytical sizing gels
- Millipore filters for drop dialysis (type VS, 0.025 μm)
- GELASE enzyme (Epicentre Technologies)
- T4 DNA ligase (New England Biolabs (NEB), 400 U/μl)
- SeaPlaque GTG low-melting-point agarose (FMC)
- Incert low-melting-point agarose (FMC)
- Lithium dodecyl sulfate (USB)
- Kanamycin
- Ampicillin
- 96-well (or 384-well) microtiter dishes, sterile, flat- or round-bottom
- Sterile toothpicks
- Disposable petri dishes (15 × 100 mm; 15 × 150 mm)
- Multichannel pipetter, 10 to 200 μl
- SOC medium (Gibco/BRL)
- 14-ml snap cap culture tubes, polystyrene or polypropylene
- LB medium
- Glycerin (Ultrapure, USB)
- LB-agar with 5% sucrose
- Sarkosyl
- Proteinase K

3. This list does not include common laboratory equipment or supplies. Manufacturers are listed for certain products only. These are merely recommendations and reflect the authors' personal preferences based on experience; products from other manufacturers should be interchangeable in many instances.

- No. 1 glass coverslips, 22 × 22 mm, 22 × 50 mm
- Calf intestinal phosphatase (Boehringer-Mannheim)
- *Not*I enzyme (NEB)
- *Sca*I enzyme (NEB)
- *Bam*HI enzyme (NEB or Gibco/BRL)
- *Mbo*I enzyme (NEB)
- Plug molds (Bio-Rad)
- Sterile TE and 0.5× TE
- Cut (large-orifice) tips for P200 pipetman, P1000 pipetman
- 96-pin (or 384-pin) transfer tool

B. Stock Solutions

1. Plasmid Isolation Solution 1 (P1)

15 mM Tris, pH 8
10 mM EDTA, pH 8
100 μg/ml RNase A
For 100 ml:
   750 μl 2 M Tris, pH 8
   2.0 ml 0.5 M EDTA, pH 8
   H_2O to 100 ml
   Autoclave; cool to room temperature
   Add 1 ml of 10 mg/ml stock RNase A
   Store at 4°C

2. Plasmid Solution 2 (P2)

0.2 N NaOH
1% SDS
For 500 ml:
   465 ml H_2O
   10 ml 10 N NaOH
   25 ml 20% SDS
   Filter sterilize; store at room temperature

3. Plasmid Solution 3 (P3)

3 M potassium acetate, pH 5.5
Autoclave; store at room temperature

4. Calf Intestinal Phosphatase (CIP) Buffer, 10×

500 mM Tris, pH 9
10 mM $MgCl_2$
1 mM $ZnCl_2$
10 mM spermidine
For 1 ml:
   500 μl 1 M Tris, pH 9
   10 μl 1 M $MgCl_2$

1 μl 1 *M* $ZnCl_2$
10 μl 1 *M* spermidine
479 μl H_2O
Filter sterilize; store at 4 or $-20°C$

5. Phosphate-Buffered Saline (PBS)

137 m*M* NaCl
2.7 m*M* KCl
4.3 m*M* $Na_2HPO_4\cdot 7H_2O$
1.4 m*M* KH_2PO_4
Make a 10× PBS solution as follows and dilute to 1× (or as necessary).
For 1000 ml:
  80.0 g NaCl
  2.0 g KCl
  11.5 g $Na_2HPO_4\cdot 7H_2O$
  2.0 g KH_2PO_4
  H_2O to 1000 ml
  Adjust pH to 7.3 and autoclave; store at room temperature

6. Cell Lysis Solution

1% lithium dodecyl sulfate
10 m*M* Tris (pH 8)
100 m*M* EDTA (pH 8)
For 500 ml:
  5 g lithium dodecyl sulfate
  2.5 ml 2 *M* Tris, pH 8
  100 ml 0.5 *M* EDTA, pH 8
  H_2O to 500 ml
  Filter sterilize; store at room temperature

7. Agarose Block Storage Solution (20% NDS)

0.2% *N*-laurylsarcosine
2 m*M* Tris, pH 9
0.14 *M* EDTA, pH 9

8. NDS

Mix 350 ml dH_2O, 93 g EDTA, and 0.6 g Tris base; adjust pH to >8 with ~100–200 pellets of NaOH. Dissolve 5 g of *N*-laurylsarcosine with 50 ml H_2O then add to the EDTA/Tris solution. Adjust the pH to 9 with NaOH and bring volume to 500 ml with H_2O. Filter sterilize and store at 4°C. For 20% NDS, dilute 1/5 with sterile H_2O. Store at 4°C.

9. MboI Equilibration Buffer

100 m*M* NaCl
50 m*M* Tris, pH 7.9
1 m*M* DTT

For 500 ml:
  10 ml 5 M NaCl
  12.5 ml 2 M Tris, pH 7.9
  0.5 ml 1 M DTT
  H_2O to 500 ml
  Autoclave; store at 4°C

10. Oligonucleotide Hybridization Solution

6× SSC
5× Denhardt's solution
0.05% sodium pyrophosphate
0.5% SDS
For 50 ml:
  15 ml 20× SSC
  5 ml 50× Denhardt's solution
  0.5 ml 5% sodium pyrophosphate
  1.25 ml 20% SDS
  H_2O to 50 ml

11. Oligonucleotide Wash Solution

6× SSC
0.05% sodium pyrophosphate
0.1% SDS
For 500 ml:
  150 ml 20× SSC
  5 ml 5% sodium pyrophosphate
  2.5 ml 20% SDS
  H_2O to 500 ml

III. Protocols

A. General Precautions

As has been documented for YAC cloning (Burke *et al.*, 1987; Nelson and
Brownstein, 1994), multiple factors influence the efficiency of PAC cloning. Of paramount importance is the quality of both insert and vector
DNAs. Also, as a multistep process, succesful PAC cloning requires success
at each of the sequential procedures; PAC cloning can be difficult since
failure at any one step can compromise the entire cloning effort. Hence,
extreme care must be taken in all steps of the cloning procedure. However, simple quality control tests exist for each of the steps, making success
likely and diagnosis of any trouble rapid. Although PAC cloning remains
largely empirical in that some amount of optimization is required for any
new cloning project, the following guidelines maximize the chances of
efficiently producing large-insert PAC clones:

(1) All reagents for PAC cloning, including restriction enzymes, agarase, T4 DNA ligase, H₂O, and buffers, should be used only for PAC cloning and set aside specifically for this purpose.

(2) When handling insert DNAs—during agarose embedding steps, partial restriction digestion, agarase treatment, or the ligation/drop dialysis/transformation steps—it is essential to wear gloves and treat these samples as if they were RNA to avoid any potential for DNase digestion. Likewise, because of the very large sizes of insert DNAs, these samples should not be frozen at $-20°C$ or mechanically sheared by overpipetting or vortexing. It is critical that large-bore pipet tips[4] are used exclusively for every step in which high-molecular-weight (HMW) insert and vector-insert DNAs are being processed.

(3) With regard to the cloning vector, it is advantageous to aliquot quality-tested, prepared vector (i.e., after restriction digestion and dephosphorylation as described below) into smaller quantities and to store the aliquots frozen in TE (or as ethanol precipitates) rather than to repetitively thaw and refreeze the master stock.

B. Preparation of PAC Vector for Cloning

The following procedure is used for preparing pCYPAC-2 vector that is suitable for PAC cloning. In this procedure, the vector DNA is cleaved with *Sca*I (to destroy the pUC19 portion) and with *Bam*HI (to generate the cloning site). Subsequently, the vector DNA is treated with CIP and purified via spin-dialysis. Several quality control experiments are provided so as to troubleshoot the critical steps and to assure the production of the best possible vector preparation.

(1) Grow an overnight culture of bacteria containing the pCYPAC-2 plasmid using LB medium supplemented with kanamycin (25 μg/ml) and ampicillin (50 μg/ml). Isolate plasmid DNA using standard CsCl ultracentrifugation (Sambrook *et al.*, 1989) or a commercially available plasmid purification column (e.g., Qiagen). Resuspend the DNA to a final concentration of 1 μg/ml.

(2) Since both *Sca*I and *Bam*HI enzymes can produce "star" activity, it is advisable to titrate the specific batches of enzymes on the same batch of vector DNA to be used for cloning. Digest 1-μg aliquots of pCYPAC-2 with varying amounts of the respective enzymes for 1 hr, then electrophorese on an analytical agarose gel (0.8%). *Bam*HI yields two fragments

---

4. These are P200 and P1000 wide-orifice tips in which the ends have been cut off for specifically pipetting HMW DNAs. We recommend purchasing these from commercial sources rather than cutting the ends off manually.

(16,013 and 2738 bp), whereas *Sca*I yields three fragments (16,049, 1768, and 934 bp). The least amount of enzyme (per microgram) required for complete digestion in 1 hr should be used for preparing the vector for cloning. If it appears that there is star activity (trace bands or smearing), use a different batch of enzyme and repeat the titration.

(3) Digest 20–50 μg of pCYPAC-2 with *Sca*I in a 100- to 200-μl reaction for 1 hr using the amount of *Sca*I selected from the above titration experiment. After digestion, run a small aliquot (1 μl) on an analytical agarose gel. If the digest appears to be complete and is devoid of smearing (i.e., no star activity), proceed to step 4 below. If the digest appears to be incomplete, transfer contents to a new tube, add additional *Sca*I (half the quantity as before), incubate for an additional 1 hr, and check for completeness of the digest again by running an analytical agarose gel. If the digest does not appear to be complete, allow the digest to proceed for a longer period (1 hr or more) and check for completeness of digest by gel electrophoresis as above. If smearing is evident (star activity), discard contents and set up another digestion using fresh DNA.

(4) Extract the DNA using a phenol series[5] (Sambrook *et al.*, 1989), add sodium acetate to a final concentration of 0.3 *M*, and precipitate for ≥30 min at −20°C using 2 vol of absolute ethanol. Microfuge for 10 min and wash the DNA with two 1-ml changes of cold (−20°C) 70% ETOH before air-dying. Resuspend the pellet with 50–100 μl H_2O.

(5) Digest the (*Sca*I-digested) DNA for 1 hr with *Bam*HI in a 100- to 200-μl reaction using the amount of enzyme selected from the above titration experiment. Extract the DNA using a phenol series, ethanol precipitate, and wash the pellet twice (as above) before air-drying.

(6) Resuspend the double-digested DNA with 200 μl H_2O and aliquot into five or six aliquots for dephosphorylation.[6] Reactions are performed in a volume of 100 μl in which varying amounts of CIP (Boehringer-Mannheim) are used. Previous experience has shown that employing a CIP concentration between 0.01 and 0.5 unit/μg will generally yield a good vector prep for cloning. Each 100-μl reaction is composed of 4–5 μg of cut vector DNA, 10 μl of 10× CIP buffer (see Stock Solutions) and varying quantities (e.g., 0.01, 0.05, 0.1, 0.2, and 0.5 unit/μg) of CIP enzyme (diluted in 1× CIP buffer). Incubate at 37°C for 1 hr.

---

5. Sequential extraction with tris-saturated phenol, phenol–chloroform–isoamyl alcohol (25:24:1), and chloroform–isoamyl alcohol (24:1).

6. Because the effectiveness of the phosphatase reaction is highly variable with each batch of enzyme and vector, optimal conditions must be established empirically. This is accomplished by performing multiple phosphatase reactions using different amounts of phosphatase for each, and testing each condition by ligation of the vector to a test insert.

(7) Individual CIP reactions are terminated by adding 5 μl 20% SDS, 2 μl 0.5 *M* EDTA (pH 8), 2 μl proteinase K (10 mg/ml stock), and 91 μl H₂O, followed by incubation at 56°C for 30 min.[7]

(8) Samples are extracted using a phenol series as above. Individual samples are then brought to 2 ml with TE and spin-dialyzed through Centricon-30 or Centricon-100 (Amicon) units following the manufacturer's recommendations.[8] This step is essential to remove any of the small *Sca*I-*Bam*HI fragments (generated in the double digestion step above) that would greatly reduce PAC cloning efficiency.

(9) After spin-dialysis the volumes of the respective retentates are measured (usually around 40 μl) and a 1-μl aliquot of each is electrophoresed on an analytical 0.8% agarose minigel to quantify the yield. Quantitation is made by comparison of the fluorescence intensity of the sample (after staining with ethidium bromide) to that of the individual bands of known amounts (100–500 ng) of λ-*Hin*dIII standards that are run in parallel. Volumes are adjusted to assure that DNA concentrations are equivalent between samples (approximately 25 ng/μl).

(10) To determine which of the different dephosphorylated vector preparations should be used for PAC cloning experiments, ligation reactions are set up for each of the respective vector preps, including a test insert DNA[9] that should clone very easily, thus providing a critical assessment of the cloning efficiency of the preps. Ligation reactions [consisting of 50 ng of each dephosphorylated vector prep and either 50 ng of test insert DNA or no insert DNA (vector-only control)] are done in a volume of 50 μl. These reactions use 200 units (0.5 μl) of T4 ligase (NEB) and the 10× ligation buffer supplied with the product. Reactions are incubated at 15°C overnight or in a bath of room temperature water (4 liters) placed in a refrigerator overnight to allow gradual cooling to 4°C.

(11) Ligation reactions are terminated by heating at 65°C for 5 min and chilled briefly on ice. Samples are then pulse microfuged for 5–10 sec and drop-dialyzed[10] against a large volume (≥50 ml) of 0.5× TE for 1.5–2 hr at room temperature. Samples are recovered from the drop-dialysis membranes and placed in sterile 1.5-ml Eppendorf tubes. Sample loss (2–10 μl) is inevitable at this stage.

7. Alternatively, CIP reactions may be terminated by addition of 1/10th vol of 200 m*M* EGTA (or EDTA), followed by heating at 65°C for 10 min.

8. Microcon-50 (Amicon) units also have been found to be effective for this step.

9. The test insert DNA is prepared by digesting genomic DNA (e.g., HELA) with *Bam*HI, extracting with a phenol series, ethanol-precipitating, then resuspending with TE to 50 ng/μl.

10. This is done by transferring the sample using a wide-bore pipet tip to the shiny side of a Millipore type VS membrane disc (0.025 μm, 13 mm) floated on 50+ ml of 0.5× TE in a large (15 × 150 mm) petri dish.

(12) One microliter of each dialyzed ligation is electroporated using 20 μl of electrocompetent DH10B cells[11] (Gibco/BRL) following the manufacturer's recommendations for electrotransformation of *E. coli*.[12] Transformed cells are transferred immediately to 12-ml snap cap tubes containing 1 ml of SOC medium, and incubated at 37°C, with shaking at 250–300 rpm, for 1 hr.

(13) Ten- and 100-μl aliquots ($\sim10^{-5}$ and 10^{-4} μg, respectively) are plated (in duplicate) onto each of three selective plates: LB-agar containing ampicillin (50 μg/ml) and kanamycin (35 μg/ml), LB-agar containing kanamycin (35 μg/ml),[13] and LB-agar containing kanamycin (35 μg/ml) and sucrose (5%). Plates are incubated for $\sim$12 hr at 37°C.

(14) Colonies are counted. In a good vector preparation, there should be virtually no colonies on LB-amp-kan plates regardless of whether ligations contain inserts. If colonies are present, this suggests that the initial restriction digests were incomplete such that the pUC19 region was not excised and destroyed. The comparison of the number of colonies between LB-kan and LB-kan-sucrose for those ligations containing insert DNA provides an indication as to the efficiency of dephosphorylation; a nearly equivalent number is highly desired and would suggest that dephosphorylation was very effective.[14] Cloning efficiencies of greater than 2×10^7 colonies per microgram of vector are achieved easily for ligations of test DNA ($\geq$200 and $\geq$2000 colonies, respectively, from the 10- and 100-μl kan-sucrose platings). The no-insert (self-ligation) control serves as another measure of the efficiency of dephosphorylation, and thus one should see a wide range in number of colonies on the LB-kan plates for the different vector preps but very few colonies on the LB-kan-sucrose plates (regardless of dephosphorylation).

11. The DH10B strain of *E. coli* has a cell wall mutation that renders it highly permeable to electroporated molecules and has been found to be far superior to any other *E. coli* strains for transforming large DNA (Shen *et al.*, 1995; C. T. Amemiya, unpublished data).

12. We employ a Gibco/BRL *E. coli* Cell Porator, 1.5-mm gap cuvettes, and a "medium" setting (field strength = 16.66 kV/cm). Other electroporators (Bio-Rad, BTX) have been used with comparable success.

13. Note that this kanamycin concentration is higher than the 25 μg/ml used for liquid cultures. We have often observed slowly growing (false positive) colonies when the lower kanamycin concentration is used on LB-agar plates.

14. The sucrose in the medium provides positive selection for insert-containing clones (Pierce *et al.*, 1992a). Nonrecombinant clones have a functional *SacBII* gene that converts sucrose to levan, which is toxic to *E. coli*. In recombinants, however, the insert interrupts transcription of the *SacBII* gene and thus allows growth of the clones. Therefore, a vector prep with as much as 50% more transformants on the LB-kan plates could still be used for PAC cloning (albeit with lower cloning efficiencies) as long as sucrose selection is maintained.

In summary, the appropriate batch of dephosphorylated vector to use for PAC cloning is that which shows: (1) virtually no background clones from contamination of uncut vector; (2) very low numbers of vector self-ligation transformants on LB-kan plates; (3) a nearly equivalent number of recombinants on LB-kan and LB-kan-sucrose plates; and (4) a cloning efficiency using *Bam*HI-digested HELA DNA (and the ligation conditions outlined above) upwards of 2×10^7 colonies per microgram of vector.

Finally, a cautionary note should be made regarding *Bam*HI star activity and the generation of PAC vector artifacts. The *Sac*BII-based vectors (pAd10-*Sac*BII, pCYPAC-2) are very susceptible to *Bam*HI star digestion (Pierce and Sternberg, 1992; P. J. de Jong, unpublished) and use of such vector preps for large-insert cloning will result in a substantial number of clones that do not contain inserts yet still grow on sucrose plates. These artifacts tend to be smaller in size than intact vector molecules and often have lost their *Not*I sites. Unfortunately, the proportion of these molecules relative to insert-containing clones increases as the insert sizes increase. This effect is due to the much higher transformation efficiency of smaller molecules relative to large DNA produced by electroporation, and makes detection of these vector artifacts very difficult solely from the above experiments. Thus, precautionary measures should be taken to minimize overdigestion of the PAC vector including the use of high-quality (previously tested) enzyme stocks and by limiting the restriction enzyme units and digestion time.

C. Embedding of Erythrocytes in Agarose Blocks for High-Molecular-Weight DNA

All vertebrates (with the exception of mammals) have nucleated erythrocytes, which are excellent sources of HMW DNA. The following method (adapted from Southern *et al.*, 1987) uses the detergent *lithium dodecyl sulfate* instead of SDS/proteinase K, for lysis of cells and disruption of proteins. The resultant DNA is highly amenable for restriction digestion and for genomic library construction. The method described here employs Bio-Rad plug molds (75 µl per plug); however, other embedding formats are effective.[15] It is assumed that the quantity of DNA to be embedded per plug will be ~10 µg (although it may be advantageous to embed larger quantities in certain instances).

15. We also have prepared HMW DNAs in agarose "beads" via a standard paraffin oil embedding method (Imai and Olson, 1990). Beads treatment was similar to that of the method described here for agarose blocks and the subsequent DNAs successfully were used for generating PAC clones from a shark and a pufferfish.

(1) Blood to be embedded (>1 ml) should be collected with a heparinized syringe, and the cells should be washed once or twice in PBS.[16] Appropriate serial dilutions should be made and the number of cells determined using a hemacytometer (most vertebrates have >10^8 cells per milliliter of blood). Knowing the (approximate) genome size,[17] one can estimate the number of cells needed for 10 μg of DNA (the amount to be embedded per plug[18]). Calculate how many blocks can be made for the cell suspension on hand.

(2) If necessary, pellet the cells at 1000 rpm in a table top centrifuge, and resuspend with PBS to one-half the final volume of blocks (37.5 μl × total number of blocks to be made). Equilibrate the cell suspension at 37°C for 5 min.

(3) Prepare enough molten 2% Incert agarose (in PBS) for the number of blocks to be made. Equilibrate the agarose at 45°C.

(4) Working quickly (and keeping the cells at 37°C), add an equal volume of the equilibrated agarose to the cell suspension. Mix briefly and distribute aliquots into the Bio-Rad plug molds[19] using a P-1000 pipetter. Do not distribute individual 75-μl aliquots repetitively, but rather, fill a P-1000 tip with suspended cells and quickly dispense contents to the top of each well. Place the plug mold apparatus at 4°C for 5–10 min in order to allow the agarose to solidify.

(5) Remove the tape from the bottom of the plug molds and use the plunger (supplied with molds) to release individual plugs into cell lysis solution (2–3 ml per plug). Plugs can be treated *en masse* in a 50-ml conical centrifuge tube or individually in wells of 24-well microtiter dishes.

(6) Incubate at 37°C for 1 hr with occasional swirling.

(7) Replace the cell lysis solution and incubate overnight at 37°C.

(8) Supplant the cell lysis solution with 20% NDS (2–3 ml per plug) and place on a gentle shaker for 2 hr at room temperature.

16. Note that the PBS concentration may need to be adjusted for the species under study. This is best done empirically using a series of PBS concentrations and by microscopically observing cells, i.e., good morphology and absence of hemolysis should be observed. For example, the optimal PBS concentration for the zebrafish (a freshwater fish) is around 80–90% that of mammalian 1× PBS, whereas that for marine fishes is much higher.

17. Genome sizes for selected vertebrates may be found in "Molecular Biology LABFAX" (Brown, 1991).

18. It is important to note that the state of the cells also will come into play if dealing with a cell population that can undergo replication. Whereas nucleated erythrocytes are largely in the 2C stage (DNA not replicated), whole embryos or a mammalian leukocyte suspension will likely contain a substantial fraction of cells that are at the 4C stage (i.e., replicated—two chromatids per chromosome). For this reason, we suggest using 3C as the quantity of DNA per cell when dealing with a replicating cell population.

19. These molds are supplied with the bottoms sealed with plastic tape. However, the molds can be reused several times, employing autoclave tape to seal the bottoms each time.

(9) Transfer plugs to a large volume (20 blocks/50 ml) of fresh 20% NDS. Plugs can be stored at 4°C for at least 1 year.

D. Partial Restriction Digests and Size-Selection of DNA

The cohesive ends generated from partial digests of the four-base cutters (*Mbo*I, *Sau*3A, and *Dpn*II) are compatible with those of the *Bam*HI ends of the PAC vector. We have successfully used partial digests with both *Sau*3A and *Mbo*I to generate PAC clones from lower vertebrate species. The following protocol uses *Mbo*I, which appears to be easier to control with respect to partial digestion kinetics. In this method, large quantities of agarose-embedded HMW DNA are digested for varying times using a tiny amount of *Mbo*I enzyme. After terminating the reactions, aliquots are analyzed by pulsed-field gel electrophoresis (PFGE) to identify the reactions containing the most suitable size distribution of partial digestion products. DNAs from these selected reactions are then pooled and separated in a preparative pulsed-field gel. Successive fractions (slices) containing DNA of different sizes are removed from the preparative gel. To confirm the size of the DNA in each preparative gel fraction, a small portion is rerun on an analytical pulsed-field-gel. Confirmation of sizes is necessary because the actual length of DNA molecules in any region of the preparative gel may vary considerably from the apparent size relative to the size standard (see Chapter 1). Gel fractions containing DNA of the desired size range are then treated with agarase (e.g., GELASE) to digest the agarose prior to ligation.

(1) Transfer several blocks of embedded DNA (from 20% NDS storage buffer) to 50 ml of sterile 1× TE buffer.[20] If possible, you should use at least 4 blocks (~40 μg) per time point. Therefore, for five time points, you will need at least 20 blocks.

(2) Allow blocks to equilibrate for 30 min at 4°C on a platform rocker or gyrotory shaker (gentle agitation). Replace with 50 ml of fresh 1× TE and rock for 30 min at 4°C.

(3) Replace 1× TE with 50 ml of *Mbo*I equilibration buffer (see Stock Solutions) and rock for 30 min at 4°C. Replace with 50 ml of fresh *Mbo*I equilibration buffer and rock for 30 min at 4°C.

(4) Transfer four blocks to each 1.5- or 2-ml Eppendorf tube (one tube per time point). Bring the volume in each tube to 1 ml with *Mbo*I equilibration buffer using graduations on the side of the Eppendorf tubes. Place tubes on ice.

---

20. Teflon spoonulas and disposable bioloops bent at a 90° angle, 1 cm from the end, are good utensils for scooping and handling the agarose blocks.

(5) Make a dilution of *Mbo*I enzyme with *Mbo*I equilibration buffer to a concentration of 0.04 unit/μl (for 5 unit/μl *Mbo*I stock, this is a 1/125 dilution). Add 10 μl of this dilution to each of the tubes containing DNA to be digested. Tap tubes gently to mix and allow to equilibrate on ice for 30 min. The final enzyme concentration of approximately 0.01 unit/μg was empirically determined using two different lots of *Mbo*I from NEB.

(6) Add 10 μl of 1 *M* $MgCl_2$ to each tube. Tap tubes gently to mix and allow to equilibrate on ice for 30 min.

(7) Place tubes in a 37°C water bath. Remove individual tubes at 20 or 30 min intervals, and transfer contents to 10 ml LDS solution (cell lysis solution) to terminate digests. Allow blocks to sit in LDS solution at room temperature for at least 15 min before transferring to 20% NDS for long-term storage at 4°C.[21]

(8) For each time point, take a sample slice (around one-fifth of a block) for analytical PFGE analysis. Likewise, take a comparable-sized aliquot from an undigested block for use as a no-digest control. Equilibrate the gel slices in two 30-min changes of 10 ml electrophoresis buffer (1× TAE or 0.5× TBE) before loading the pulsed-field gel. Include appropriate standards such as lambda concatemers (50-kb ladder) and low-range PFG markers (NEB). It is not necessary to load yeast chromosomes as these are, for the most part, larger than the sizes one is interested in for PAC cloning. Suggested gel conditions[22] are: 1% agarose (0.5× TBE or 1× TAE), 120° orientation angle, 6 V/cm, 14°C buffer temperature, 20 hr run time with the switch time linearly ramped from 10 to 60 sec. After electrophoresis, stain the gel in ethidium bromide, destain appropriately, and photograph. A representative gel is shown in Fig. 8.2. As expected, the increase in the digestion time shifts (lowers) the mass distribution of molecules. From the analytical gel one empirically determines and selects which digests to pool for the preparative gel.[23] It is suggested that blocks from two or three time points be pooled for the preparative gel. For example, digestions 3–5 were chosen from the experiment shown in Fig. 8.2.

(9) Equilibrate (gentle shaking at 4°C) the pooled blocks for the preparative gel in two 30-min changes of 50 ml sterile 1× TAE buffer.

21. Alternatively, partial digests can be terminated and blocks stored in 0.5 *M* EDTA (pH 8). Blocks can be stored in either solution for at least 1 month.

22. We use a CHEF DRIII (Bio-Rad) unit; therefore, the suggested parameters here and elsewhere in the chapter may require adjustments if another PFGE unit is used.

23. Note that if it appears as if all fractions are overdigested, the enzyme concentration should be lowered. Conversely, if all fractions show insufficient digestion products in the appropriate size range, the enzyme concentration should be increased. In both instances, go back to step 1 and use a different enzyme concentration.

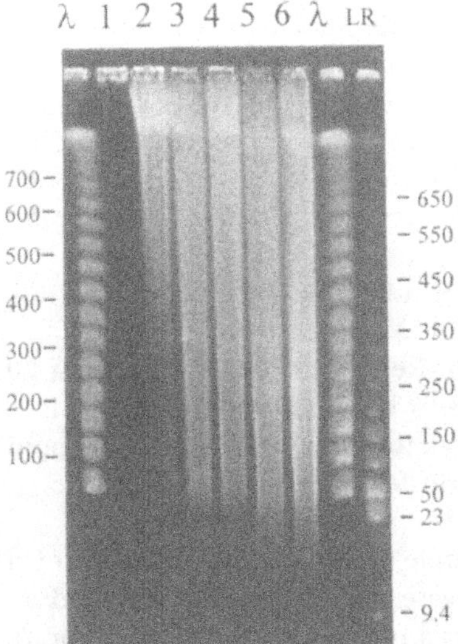

Figure 8.2 Analytical pulsed-field gel of *Danio rerio* (zebrafish) *Mbo*I partial digests. Forty microgram aliquots of agarose-embedded zebrafish DNA were digested in a volume of 1 ml with 0.01 unit of *Mbo*I per microgram for varying time periods (20 to 100 min). Reactions were terminated by transferring blocks to 1% LDS (cell lysis solution). Small slices were taken from digests at each time point, equilibrated briefly in 1× TAE buffer, and electrophoresed in a 1% agarose pulsed-field gel. The gel was electrophoresed at 14°C in a CHEF DRIII unit (Bio-Rad) with a reorientation angle of 120° using 1× TAE at 6 V/cm, and a linear switch time ramp from 20 to 60 sec for 20 hr. After electrophoresis, the gel was stained with ethidium bromide. Molecular weight markers were lambda concatemers (λ) and low-range PFG standards (LR) (NEB). Track 1 represents a zero-time control. Tracks 2–6 represent digestion times of 20, 40, 60, 80, and 100 min, respectively. Sizes are given in kb.

(10) Cast the preparative gel by overlaying 1% SeaPlaque low-melting point agarose (1× TAE) over a thin base (2 mm) of hardened 2% LE agarose (1× TAE). The 2% agarose allows easier handling of the preparative gel and serves as a base for resting the bottom of the gel comb. After the low-melting-point agarose has hardened, remove the gel comb and use a coverslip to carefully remove the agarose separating adjacent wells in the middle of the gel such that a wide preparative well is formed, the width of which is adjusted to accommodate the total volume of pooled blocks of the preparative digests. The volume of the preparative well is determined by the thickness of the gel (typically 6–8 mm), and the length (typically 2 mm) and width of the preparative well. Thus, a preparative well with the dimensions 6 × 2 × 50 mm will have a volume of 600 μl.

(11) While the preparative gel is solidifying, rinse the gel chamber and all tubing extensively with distilled water prior to addition of electrophoresis buffer. The electrophoresis buffer for the preparative gel (1× TAE) should be autoclaved to prevent degradation of the DNA by contaminating nucleases.

(12) Transfer the pooled blocks to a sterile petri dish and remove as much liquid as possible. Use a coverslip to cut the agarose into small (<1 mm³) blocks (Wang and Schwartz, 1993). Transfer the agarose pieces to a 1.5- or 2-ml Eppendorf tube, and place at 65°C for 5–10 min to melt the gel. Using a wide-bore pipet tip and a P1000, quicky transfer the viscous mass to the makeshift preparative well. Load appropriate standards (lambda concatemers or low-range PFG markers; NEB) on each side of the preparative well and seal the wells with molten SeaPlaque agarose (in 1× TAE). Suggested gel conditions are: 1% agarose (1× TAE), 120° orientation angle, 5.3 V/cm, 14°C buffer temp, 30 hr run time with the switch time linearly ramped from 10 to 60 sec.

(13) After electrophoresis, use a gel scoop/slicer to cut along the outer edges of the preparative track so that each outer gel portion contains a small amount of preparative digest as well as the flanking standards. Stain these outer portions with ethidium bromide but leave the center portion unstained. On a UV transilluminator, place the two outer portions of the gel together and photograph along with an adjacent fluorescent ruler. An example of such a picture (without the ruler) is shown in Fig. 8.3 for a preparative gel of partially digested *Danio rerio* (zebrafish) DNA.

(14) From the photograph, use the molecular size markers and the fluorescent ruler as guides for determining the area of the preparative gel that contains the DNA in the desired size range for PAC cloning (~80 to ~400 kb).[24]

(15) Use a 22 × 50-mm coverslip to excise the desired fractions (3–5 mm) from the preparative gel. Note that it is also advisable to take a fraction of ~20 kb average size for use as a positive control in the cloning steps below. Cut off a small aliquot (~60 μl equivalent) from each fraction for an analytical gel and store the remainder in 15-ml screw cap centrifuge tubes containing 10–15 ml of 20% NDS or 0.5 M EDTA at 4°C.

(16) Run an analytical PFG on the aliquots of the fractions using electrophoresis conditions as in step 8 above. An example of such a gel is shown in Fig. 8.4.

---

24. To aid in better visualization of the region for excision, the fluorescent molecular weight standards in the outer gel portions can be marked by placing a half-toothpick vertically into each band (or by stabbing with a small amount of India ink).

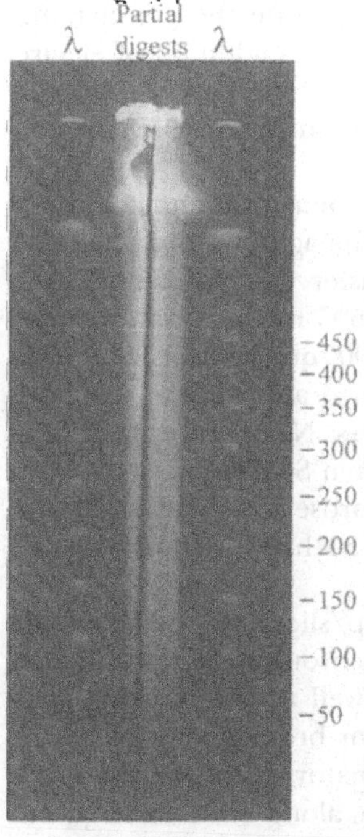

Figure 8.3 Stained portion of a pulsed-field preparative gel of *Danio rerio* (zebrafish) *Mbo*I partial digests. Blocks containing partially digested DNA of the desired size range were equilibrated in 1× TAE buffer and cut into small pieces using a coverslip. After melting at 65°C for 10 min, the combined samples were loaded into a preparative well in a 1% SeaPlaque low-melting-point agarose gel with a 2% LE agarose base (in 1× TAE buffer). Lambda concatemers (NEB) were loaded in the tracks flanking the pooled partial digests. The gel was run at 14°C using a reorientation angle of 120° in 1× TAE buffer at 5.3 V/cm for 30 hr using 10–60 sec switch time. After electrophoresis, vertical cuts were made on both sides of the region containing the size-separated partial digests, and the outer portions were stained with ethidium bromide (the middle section was neither stained nor subjected to UV light). The photograph shows both sides of the gel placed next to one-another. The broad distribution of the ethidium bromide-stained material in the middle confirms the expected distribution of partial digests in the range of 50 to >400 kb.

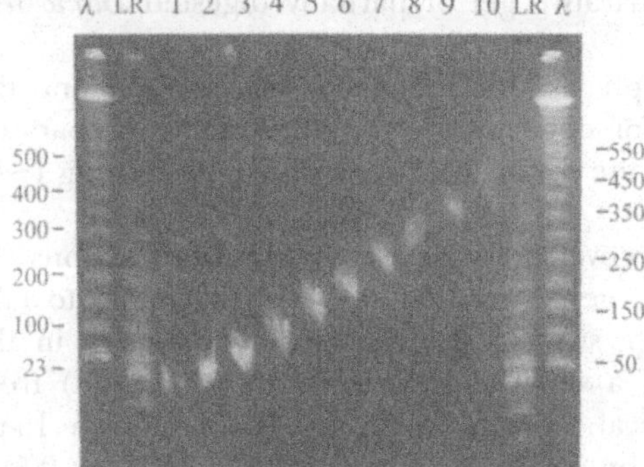

Figure 8.4 Analytical pulsed-field gel of *Spheroides nephelus* (pufferfish) *Mbo*I partial digest fractions. Samples (ca. 6 × 6 × 1.5 mm) were sliced from individual fractions of a preparative gel. These agarose pieces were equilibrated for 30 min in 0.5× TBE buffer and loaded directly (without melting) in a 1% agarose gel. The gel was electrophoresed at 6 V/cm at 14°C, in 0.5× TBE using 10–70 sec switch time, for 20 hr. Size markers were lambda concatemers (λ) and low-range PFG markers (LR) (NEB); sizes are given in kb. Sample tracks 1–10 are aliquots of fractions 1–10 from the preparative gel. Fraction 1 (~20 kb average fragment size) was used as a ligation control for PAC cloning; fractions 4, 5, and 6 were used in constructing a PAC library.

(17) Remove 1/4 to 1/3 of the desired size fraction and transfer to a 50-ml conical centrifuge tube filled with 1× TE buffer containing 50 mM NaCl.[25] Allow to equilibrate for 1 hr at 4°C by placing the tube on a platform rocker or gyrotory shaker. Replace the buffer with 50 ml of fresh 1× TE/50 mM NaCl, and continue to equilibrate for another 1 hr at 4°C.

(18) Transfer the samples to sterile petri dishes and remove as much liquid as possible. Use coverslips to chop the respective agarose samples into small pieces ($<$ 1 mm^3).

(19) Transfer the agarose pieces to 1.5- or 2-ml Eppendorf tubes, incubate at 70°C for 15 min, then place tubes at 45°C for 15 min. Add 1 unit of GELASE per 200 μl of molten gel, mix gently using the pipet tip, and incubate at 45°C for 1–2 hr.

(20) Ensure that all the agarose has been digested by placing tubes on ice and visually observing if any of the material resolidifies. If undigested agarose remains, repeat step 18.

(21) Run a 5- to 10-μl aliquot of each sample on an analytical 0.8% agarose minigel along with 25, 50, and 100 ng of λ *Hin*dIII markers in order to quantify the DNAs. For this quantitation, PFGE is not necessary and is not recommended, since the goal is to maintain all the DNA in a single sharp band. The concentration of DNAs should be at least 1 ng/μl.

The preparation of high-quality partial digest fractions is of paramount importance for successful PAC cloning. If the concentration of DNA (step 20) is too low ($<$1 ng/μl), it is advisable to repeat the entire procedure using a larger number of pooled blocks.[26] If, on the other, the concentration of resultant DNAs is very high ($>$5 ng/μl) this could also create problems. After analytical gel electrophoresis of respective fractions (step 15) one should not see material migrating in front of the presumed lower size limit of the respective fraction. If such smearing is evident, the prior preparative gel may have been overloaded, resulting in aberrations during the run. If, after subsequent PAC cloning (see **Generation of PAC Clones,** below) numerous smaller-sized clones ($<$40 kb) are generated from fractions thought to contain a higher average fragment size, a second size-fractionation should be done. If such is the case, pool two or three of the most appropriate fractions and run a second size-fractionation preparative gel starting from step 9 (above).

25. NaCl (50 mM) has been found to be effective in reducing degradation of HMW DNAs during melting of agarose/GELASE treatment, and does not have detectable effects on subsequent ligation to the PAC vector.

26. Pierce and Sternberg (1992) have reported using *n*-butanol extraction for concentrating sucrose gradient-separated *Sau*3A partial digests; we have not been as successful using this method for concentrating dilute DNAs from agarase-treated fractions.

E. Generation of PAC Clones

The efficiency with which PAC clones are generated is determined, in part, by the frequency of large, intramolecular ligations (circles); however, the actual kinetics of the reaction are difficult to model. Theoretical predictions suggest that the optimal concentration that will favor intramolecular ligations is ≤1 µg/ml for molecules over 100 kb (Collins and Weissman, 1984; Collins, 1988). In practice, a number of factors may cause deviation from the theoretical predictions due to: (1) the inevitable formation of insert-only circles, insert–insert doublets, and inserts with two vector molecules; and (2) the extreme bias of electrotransformation toward smaller transformants.

The following ligation conditions have successfully produced PAC clones from a variety of lower vertebrate species. However, owing to the above considerations, these serve only as a starting point; investigators should vary these conditions with respect to vector-insert concentrations in order to further optimize cloning efficiency, if possible.

(1) Set up 50-µl ligation reactions as follows using three or four different fractions of partial digests.

5.0 µl 10× T4 ligase buffer (supplied with NEB T4 ligase)
2.0 µl (50 ng) *Bam*HI-CIP'd pCYPAC-2 vector[27]
× µl (~50 ng) *Mbo*I partial digests[28]
0.5 µl (200 units) T4 ligase (NEB)
H₂O (if needed) to 50 µl

In addition, set up one reaction using a smaller-sized fraction (20 kb, if available) or *Bam*HI-digested DNA (see **Preparation of PAC Vector for Cloning**) as a ligation control. Ligations are incubated at 15°C overnight or in a room-temperature bath (4 liters) placed in a refrigerator overnight to allow gradual cooling to 4°C.

(2) Reactions are terminated by heating at 65°C for 5 min and chilled briefly on ice. Samples are then microfuged for 5–10 sec and drop-dialyzed (as above, see **Preparation of PAC Vector for Cloning**) against a large volume (≥50 ml) of 0.5× TE for 1.5–2 hr at room temperature.[29] Samples are recovered from the drop-dialysis membranes and placed in sterile 1.5-ml Eppendorf tubes.

(3) One microliter of each ligation is electroporated using 20 µl of electrocompetent DH10B cells (Gibco/BRL) following the manufacturer's recommendations for electrotransformation of *E. coli*. Transformed cells

27. This is an approximately 10:1 molar excess of vector to insert for fragments ~130 kb. This 10:1 vector to insert ratio has been shown empirically (Shizuya *et al.*, 1992; Ioannou *et al.*, 1994) to yield good numbers of human BAC and PAC clones, respectively.

28. Note that this may comprise the bulk of the ligation if the DNA concentration is very low.

29. It is essential to use wide-bore pipets during all transfer steps.

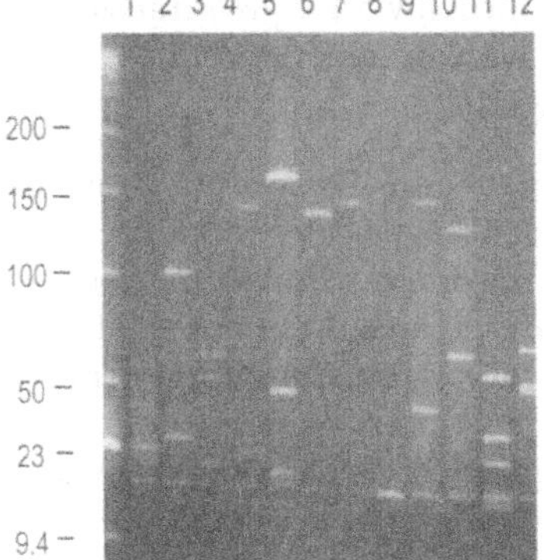

Figure 8.5 Analytical pulsed-field gel of representative PAC clones of *Spheroides nephelus* (pufferfish) digested with *Not*I. Selected PAC clones were chosen from a ligation which used fraction 6 (see Fig. 8.4). Transformants were grown in 2-ml minicultures as per the protocol in the text and 4 μl of respective DNA was digested with *Not*I. Digests were electrophoresed at 14°C in a 1% agarose gel in 0.5× TBE buffer, at 6 V/cm and 1–15 sec switch time, for 20 hr. Sizes of the low-range PFG marker (NEB) are given in kb. Tracks 1–7 and 9–12 are *Spheroides nephelus* clones; track 8 is a vector-only control. The ∼16-kb band apparent in all tracks is attributable to the vector. In this example the size of cloned inserts ranges from ∼140 kb (track 6) to ∼240 kb (track 5).

are immediately transferred to 12-ml snap cap tubes containing 1 ml of SOC medium,[30] and incubated at 37°C, 250–300 rpm, for 1 hr.

(4) Aliquots (100–200 μl) are plated (in duplicate) onto K/S plates, i.e., LB-agar containing kanamycin (35 μg/ml) and sucrose (5%). Plates are incubated for 12–15 hr at 37°C.

(5) Count colonies and estimate the cloning efficiencies (relative to vector).

(6) Pick several clones from transformations of the various ligations (size fractions) and initiate plasmid minicultures (see below, **Rapid Isolation of PAC Clones**).

(7) Evaluate the average insert sizes of the transformants by *Not*I digestion (see below), and subsequent analytical PFGE. A representative gel is shown in Fig. 8.5.

(8) Vary the ligation conditions (vector and insert concentration) for those fractions which give the desired insert sizes (100+ kb) in order to increase the cloning efficiency.[31] In addition, it may be advantageous to alter (lower) the electroporation field strength in order to increase the cloning efficiency. It has been determined empirically that a slightly lower

30. Do not substitute LB for SOC medium; the latter is far superior for generating PAC recombinants.

31. Alter the vector and insert concentrations by a factor of 2 or 3.

field strength (10–12 kV/cm) than used normally for *E. coli* transformation (i.e., 16–17 kV/cm) yields a higher average insert size but a lower number of transformants (Sheng *et al.*, 1995).[32]

Finally, determine the best size fractions and ligation and transformation conditions for generation of a PAC library of desired representation. This decision will likely involve a compromise between the number of recombinants generated, the size of inserts, and the number of electroporations required. That is, while one always attempts to generate a library of clones containing inserts as large as possible, the cloning efficiency for generating the largest PACs is poor. Thus the amount of time and effort required to produce a library of adequate coverage with very large inserts may be prohibitive. Consider the following hypothetical situation. If a certain fraction of size-selected DNA yields 100 clones per tranformation with an average insert size of 150–200 kb using 10 kV/cm field strength, 100 transformations would be required to generate 10^4 clones. If, by increasing the field strength to 14–16 kV/cm, the average insert size decreases to 130–150 kb but the number of clones increases to 500/electroporation, only 20–23 transformations would be required for the equivalent genomic coverage (as indicated above). Finally, any mapping project benefits most from a library with deep coverage, permitting many clones to be isolated for any region and minimizing the occurrence of missing clones, or gaps.

F. Library Construction and Screening

Once it has been determined how many clones are required for the library, the requisite number of electroporations are done *en masse*, and after growth of the transformants in SOC they are either: (1) plated out onto selective media (S/K); or (2) archived by addition of sterile glycerin to 10%, transferring contents to a cryotube, flash-freezing in a dry ice–ethanol bath, and storing at −70°C. The latter is useful in that once a good set of ligations is generated, electroporations should be done im-

32. In our own experience, a transformation that normally gives ~100-kb inserts with 16.6 kV/cm field strength yields inserts at least 20–30 kb larger using 10 kV/cm field strength, but with a fivefold (or greater) decrease in transformation efficiency. Also, regardless of field strength, our experience as well as that of others (Sheng *et al.*, 1995; B. W. Birren, unpublished) suggests an upper threshold in readily electroporated clone sizes and an apparent exponential diminution in transformation efficiency with incrementally larger clones; i.e., it is very easy to generate clones with average insert sizes around 80–120 kb, but it is more difficult to produce clones averaging 130–150 kb without noticeable (nonlinear) reduction in transformation efficiency. This observation is probably due to an electroporation phenomenon.

mediately so as to avoid possible nuclease degradation. Moreover, the investigator may not be ready to pick or handle recombinants, and thus, storage and recovery of the SOC-expressed material is very convenient. Recovery of the −70°C frozen material is done by slowly thawing each tube on ice for 10–15 min and subsequently plating onto selective (S/K) media.[33]

The desired genomic coverage of the library largely dictates the manner in which the library is stored. In cases where the genome size is reasonably small (≤5 pg per haploid genome) or if complete coverage of the genome is not necessary, it is recommended that the library be arrayed, i.e., each clone possesses a unique address in the library (Shepherd *et al.*, 1994). This is done by growing individual clones in wells of 96- or 384-well microtiter dishes. Thus, if 20,000 clones are required for ~1× coverage of a genome, a ~5× coverage library will require ~1042 96-well plates or ~260 384-well plates. Manual picking of individual clones is done easily by stabbing colonies grown on selective (S/K) plates with the end of a sterile toothpick and inoculating wells containing growth medium. However, this can be very labor-intensive and alternatively the work can be contracted to commercial companies which utilize colony picking robots. Clones are grown in LB medium containing kanamycin (25 µg/ml) and glycerin (7.5–10%), working replicas are made using 96- or 384-pin replicator tools, and plates are subsequently stored at −70°C.

Alternatively, if the genome size of the organism is very large, if the investigator does not have the resources for storing a large number of microtiter plates at −70°C, or if it is not anticipated that the library will be used repeatedly, clones can be stored as pools, each primary pool containing up to hundreds of clones (Pierce *et al.*, 1992b). In order to generate pools, SOC-expressed material is plated onto selective (S/K) agar plates at a desired density and all clones on the plate are scraped into media containing glycerin (Pierce *et al.*, 1992b). Another way of generating pools (when the number of clones per pool is small, i.e., ≤20) is to directly inoculate SOC-expressed material into microtiter wells containing S/K liquid media. This assumes that the titer of recombinants in the SOC-expressed material is known. The biggest disadvantage of archiving PAC libraries as pools is that clones often have unequal growth rates such that eventual recovery of a single clone is troublesome and requires several rounds of screening. Moreover, it is somewhat difficult to make a faithful replica of the library without losing clones during the replication process.

33. We have observed from 0 to 20% loss of recombinants during cryopreservation. The quality of glycerin appears to greatly affect recoveries from cryopreserved material (C. T. Amemiya, unpublished data); thus, only the highest quality glycerin should be utilized for this purpose.

Screening of PAC libraries is done using PCR and/or hybridization. For PCR screening, the library is subdivided into pools, and PCR is performed on DNAs isolated from the pools. Each pool can encompass several thousand clones and various hierachical and multidimensional pooling schemes have been published for arrayed libraries (Green and Olson, 1990; Amemiya *et al.*, 1990, 1992; Bentley, 1992). Such an approach is generally two- or three-tiered, the last round of screening usually involving PCR on rows and columns of microtiter dishes (Heard *et al.*, 1989) or direct colony hybridization (Nizetic *et al.*, 1991; Pierce and Sternberg, 1992). For hybridization screening (arrayed libraries only), high-density colony filters are made using robotics workstations (Bentley *et al.*, 1992; Olsen *et al.*, 1993; see Chapter 11). The advantage of this method over PCR- and pooling-based methods of screening is that it is a one-tiered approach, i.e., clones are immediately identified after hybridization and there is no need for subsequent rounds of screening. The chief disadvantages involve the high cost of producing the filters and the limitations on their reuse for repeated probings.

G. Rapid DNA Isolation from PAC Clones

This is a rapid alkaline lysis miniprep method (Birboim and Doly, 1979) for isolating DNA from large PAC clones and is a modification of a commercial (Qiagen tip[34]) method that uses no organic extractions or columns. The method works very well for analytical restriction digests of PAC clones and can be scaled up. With slight alterations this procedure also has been used for routine analyses of M13 RF, plasmid, and cosmid DNAs.

(1) Using a sterile toothpick, inoculate a single isolated bacterial colony into 2 ml LB media containing 25 μg/ml kanamycin. If starting from archived plates, use 1–5 μl of bacterial suspension to start the culture. Use a 12–15 ml snap-cap polypropylene tube. Grow overnight (up to 16 hr) shaking at 250–300 rpm at 37°C.

(2) Transfer 1.5 ml of culture to a 2-ml Eppendorf tube and centrifuge for 1 min.

(3) Discard supernatant. Resuspend (vortex) each pellet in 0.3 ml P1 solution. Add 0.3 ml of P2 solution and gently shake tube to mix the contents. The appearance of the suspension should change from very turbid to almost translucent.

(4) Add 0.3 ml P3 solution slowly to each tube and shake gently during addition. A thick white precipitate consisting of *E. coli* DNA and protein will form. Invert the tube several times to mix the solution. If the

34. It should be noted that, if necessary, highly purified PAC DNA can be produced using Qiagen tips; however, large losses are incurred.

precipitate suspension is not uniform (i.e., is clumpy) keep mixing the contents. Place tube on ice for 5 min.

(5) Spin in a microcentrifuge for 10 min at room temperature to pellet the white precipitate.

(6) Remove tubes from centrifuge and using a P1000 transfer ~0.7–0.8 ml of the supernatant to a 2-ml Eppendorf tube. (Note—If white precipitate is transferred over, respin for 5 min and transfer supernatant to a new tube.) Add 0.8 ml ice-cold isopropanol. Mix by inverting tube a few times.

(7) Spin in microfuge (room temperature) for 15 min at maximum speed (14,000 rpm).

(8) Remove supernatant and add 0.5 ml of ice-cold 70% ethanol to each tube. Invert tubes several times to wash the DNA pellets. Spin in microfuge for 5 min. Optional—repeat step 8.

(9) Remove as much of the supernatant as possible. Occasionally, pellets will become dislodged from tube, thus it is better to *carefully* aspirate off the supernatant rather than pour it off.

(10) Air-dry pellets at room temperature (do not vacuum-dry in a SpeedVac because resuspension of DNA becomes difficult). When the DNA pellets turn from white to translucent in appearance, i.e., when most of the ethanol has evaporated, resuspend each in 40 μl TE. Do not use a narrow-bore pipet tip to mechanically resuspend the DNA. Allow the solution to sit in the tube with occasional tapping of the bottom of the tube (or by gentle vortexing) to suspend DNA.

(11) Use 3–5 μl for digestion with *Not*I or other rare-cutter enzymes. There are *Not*I sites flanking the SP6 and T7 promotor regions of the pCYPAC-2 vector; therefore, this is a very useful enzyme for analysis of insert size and for partial digest restriction mapping. Use 7–10 μl for digestion with a more frequent cutter such as *Bam*HI or *Eco*RI.[35]

H. Restriction Mapping of PAC Clones

The large insert size of a PAC clone, versus that of a λ-phage or a cosmid, makes restriction analysis by conventional double-digestion methods less feasible. The method that we have found to be most useful employs partial digestion to determine physical distances of restriction sites relative to the ends of the insert DNA, and can be employed when internal *Not*I sites do not occur in the PAC insert to be mapped.[36] An outline of the method is

35. The greater amount required relative to that for rare cutters is due to the lower quantity of DNA per band when visualizing on an analytical gel stained with ethidium bromide.

36. The method works optimally if there are no internal *Not*I sites; however, the method still can be used when there is a single *Not*I site within the insert. It should be noted that the larger cloned inserts have a greater chance of possessing internal *Not*I sites. Moreover, species with higher GC base compositions are more likely to have a higher frequency of *Not*I sites.

given in Fig. 8.6 in which mapping is subdivided into six steps: (1) excision of insert DNA by *Not*I digestion, (2) partial restriction enzyme digestion of the insert DNA, (3) pulsed-field gel electrophoresis, (4) Southern blotting, (5) hybridization with end probes, and (6) construction of restriction map. *Not*I cleaves external to the SP6 and T7 RNA promoter regions, thereby allowing detection of end-containing fragments by Southern hybridization with radioactive SP6 and T7 probes (Fig. 8.6). It must be emphasized that every clone constitutes a separate mapping project and that experimental conditions need to be designed specifically for each effort. Thus, some degree of improvisation is necessary for every project.

The selection of restriction enzymes is of particular importance in constructing restriction maps. It is desirable to use restriction enzymes that cleave at a moderate (5–15) number of sites per clone; restriction enzymes that generate numerous (>15) fragments should be avoided

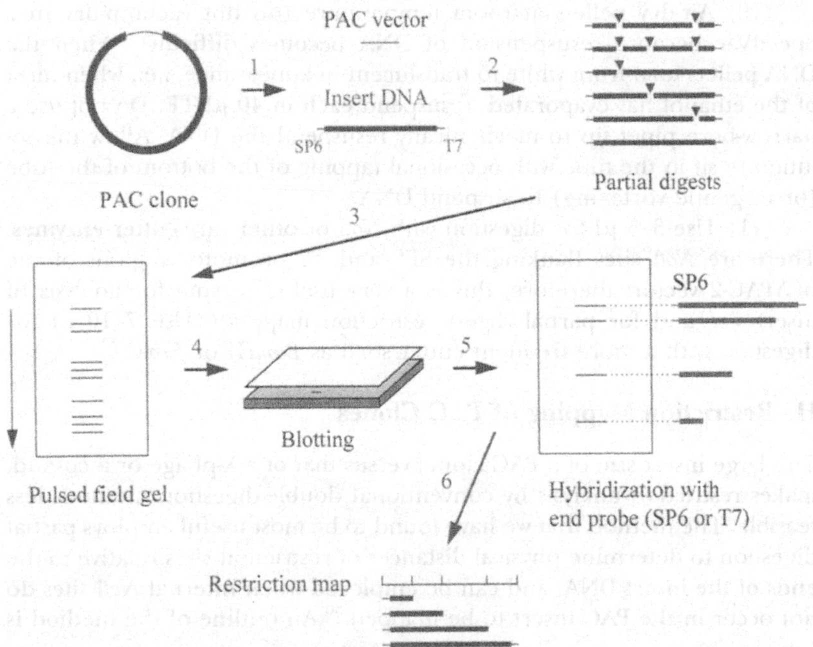

Figure 8.6 Schematic view of a partial digestion restriction mapping strategy using oligonucleotide hybridization with SP6 and T7 primers. (1) Excision of insert DNA by *Not*I digestion. (2) Partial digestion of insert DNA by restriction enzymes of choice. (3) Pulsed-field gel electrophoresis. (4) Southern blotting. (5) Southern hybridization with SP6 or T7 primers. (6) Construction of restriction map. See text for details.

as these fragments are very difficult to orient. To identify the exact number and approximate size of restriction fragments, complete restriction digests of PAC clones should be assayed first. Detection of restriction fragments is facilitated by using a simple, nonradioactive, end-labeling method as discussed in Ota and Amemiya (1996) and below.

(1) An analytical *Not*I digest should be performed first on each PAC clone to be restriction mapped. This digest is necessary to determine the approximate size of the cloned insert and verify that there is, at most, one internal *Not*I site (preferably none). Set up 20-μl digests for each respective PAC clone consisting of:

3–5 μl of PAC miniprep DNA (see above)
2 μl of 10× reaction buffer 3 (NEB)
2 μl of 10× BSA (NEB)
1 μl (10 units) of *Not*I (NEB)
H_2O to 20 μl

Digest for 1 hr at 37°C and run in an analytical pulsed-field gel using electrophoresis conditions as outlined above (**Generation of PAC Clones**) or in Fig. 8.5.[37]

(2) It is best to initiate mapping studies having sufficient DNA available from the outset. "Scale up" the rapid miniprep method proportionately (above) to 100 ml (use a Sorvall RC5B or equivalent high-speed centrifuge). Resuspend the final pellet in 500 μl of TE, extract the DNA with a phenol series, ethanol precipitate, and wash the pellet with 70% ethanol. After air-drying, resuspend the pellet in 100–200 μl of TE and determine the DNA concentration by UV spectrophotometry or by gel electrophoresis and visual comparison to known quantities of DNA. If necessary, ethanol precipitate the DNA again, wash with 70% ethanol, and adjust the final concentration to 200–500 ng/ml with TE. A yield of >200 ng/ml of starting culture should be expected for a 100- to 150-kb clone.

(3) Digest 2–10 μg of PAC DNA with *Not*I in a 20- to 100-μl reaction (using 10 units of *Not*I per microgram) for ≥1 hr (overnight). Electrophorese a small aliquot (1 μl) on an analytical 0.8% agarose minigel in order to verify complete digestion (the ~16-kb vector band should appear distinct). Extract the DNA using a phenol series, ethanol precipitate, wash the pellet with 70% ethanol, and air dry. Resuspend the DNA in 20–100 μl of TE or H_2O.

37. Note that the clones in Fig. 8.5 were selected due to their larger sizes (>170-kb average insert size) and most have internal *Not*I sites. This species (*Spheroides nephelus,* pufferfish) has a greatly reduced genome size and an apparent higher density of *Not*I sites relative to mammals (C. T. Amemiya, unpublished observations).

(4) Establish appropriate conditions for partial digestion of PAC DNA by digesting ~0.1-μg aliquots of *Not*I digests (above) with chosen[38] restriction enzymes by varying digestion times or serial dilution of the enzyme.[39] Terminate reactions according to the manufacturer's recommendations for the particular restriction enzyme and electrophorese these partial digests in an analytical 0.8% agarose minigel; select a partial digest condition that generates a uniform distribution of fragments.

(5) For analytical mapping purposes, digest 0.5 μg of *Not*I digests with the chosen partial digest condition(s). After, digestion, label the fragments using a fill-in or replacement reaction with biotinylated nucleotides as outlined in Ota and Amemiya (1996).[40] Run a pulsed-field gel on the partial digests along with appropriate standards such as lambda concatemers or a 5-kb ladder (Gibco/BRL). For resolving 5- to 150-kb restriction fragments, we recommend 1% agarose (0.5× TBE), 14°C buffer temperature, 1.0–6.0 sec switch time, 6 V/cm, for 17 hr (however, the exact PFGE conditions employed are dependent on the sizes and distribution of fragments for the respective enzyme and may need to be modified to suit the experiment). Complete digests of insert DNA with or without prior *Not*I digestion (and end-labeled with biotin) should be included on the gel in order to estimate the sizes of the fragments and to identify the end fragments (Fig. 8.7).

(6) After electrophoresis, stain the gel with ethidium bromide and photograph. Southern blot the gel to an uncharged nylon membrane (e.g., Hybond-N, Amersham) using the manufacturer's recommendations. After transfer, covalently affix the DNA to the nylon membrane by UV crosslinking.[41]

38. These are restriction endonucleases that do not cut too frequently in genomic DNA and which cleave the vector sequence only once or none at all. These include *Bam*HI, *Xba*I, *Sal*I, *Xho*I, *Eag*I, and *Sac*I. It is recommended that clones be assayed for frequency of sites prior to setting up partial digests.

39. Our starting point for partial digestion (irrespective of restriction enzyme) is 0.1–0.5 units of restriction enzyme per 100 ng of DNA for 1–5 min.

40. A single biotinylated nucleotide (biotin-14-dATP, biotin-14-dCTP; Gibco/BRL) is incorporated at the ends of restriction fragments by fill-in reactions (5′ ends) or by replacement reactions (3′ or blunt ends) using Klenow or T4 polymerase, respectively. These reactions are performed in the same tube as the restriction digests, without replacing the buffer. Fill-in reactions are done at room temperature for 5–10 min and use a final dNTP concentration of 4 μ*M*; replacement reactions are done at 37°C for 10–15 min and use a final dNTP concentration of 10 μ*M*. After Southern blotting, the biotin-labeled fragments are detected using chemiluminescence, which allows visualization of much smaller quantities of DNA with greater uniformity of restriction fragments (relative to ethidium bromide fluorescence).

41. If this is done using a UV transilluminator, it is recommended that the efficacy of crosslinking be titrated.

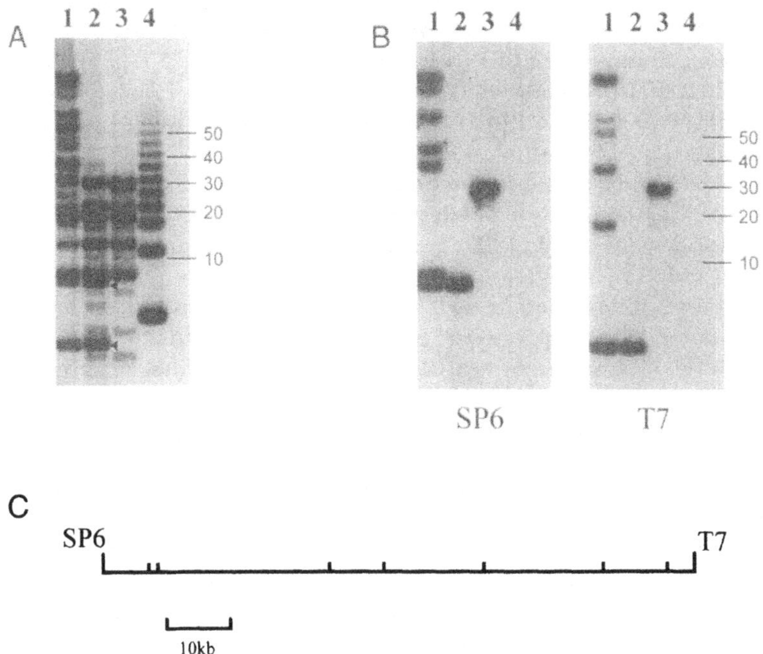

Figure 8.7 Restriction mapping of the *Xba*I sites of a *Heterodontus francisci* (horned shark) immunoglobulin heavy-chain-containing PAC clone (33A9; ~95 kb). (A) Visualization of *Xba*I restriction digests separated in a pulsed-field gel. Track 1: partial *Xba*I digestion of *Not*I-digested insert DNA; 0.5 µg of *Not*I-digested PAC DNA was partially digested with 1.5 units of *Xba*I for 3 min. Track 2: complete *Xba*I/*Not*I digestion of PAC DNA (~0.1 µg). Track 3: complete *Xba*I digestion of PAC DNA (~0.1 µg). Track 4: 5-kb ladder (Gibco/BRL). All samples (including the 5-kb ladder) were end-labeled according to Ota and Amemiya (1996) using 5′ overhang fill-in reactions with biotin-14-dATP and/or biotin-14-dCTP (Gibco/BRL) and the Klenow fragment of DNA polymerase. The pulsed-field gel (1% agarose) was electrophoresed in 0.5× TBE at 14°C, 6 V/cm, 0.7–4.2 sec switch time, for 13.5 hr. After electrophoresis, the DNA was transferred to an uncharged nylon membrane (Hybond-N, Amersham) using a conventional Southern blot method (Sambrook *et al.*, 1989); chemiluminescent detection (Phototope; NEB) was employed to identify the (biotinylated) restriction fragments. The two end-fragments of insert DNA were identified by comparing the *Xba*I restriction fragments with or without *Not*I digestion (tracks 2 and 3, respectively) and are marked with arrowheads. An additional *Xba*I fragment (~0.75 kb) was not identified in the pulsed-field gel but could be visualized by conventional gel electrophoresis (not shown). Sizes are given in kb. (B) Hybridization of the Southern blot in (A) with SP6 and T7 radioactive oligonucleotide probes, as indicated (see text for details). There are eight partial digest fragments that hybridize with the SP6 or T7 probes (tracks 1; the larger partial digests are not well-resolved). The end fragments denoted by arrowheads in (A) are clearly seen in tracks 2 for both the SP6 and T7 hybridizations. (C) *Xba*I restriction map generated for the 33A9 *Heterodontus francisci* PAC clone. Small bars denote the seven internal *Xba*I sites.

(7) Restriction fragments are visualized using a chemiluminescence system (e.g., Phototope, NEB) based on Ota and Amemiya (1996).

(8) In a 50-μl reaction, end-label 500 ng of SP6 or T7 oligonucleotide[42] with 50 μCi of $[\gamma\text{-}^{32}P]$ATP (3000–6000 Ci/mmol) using polynucleotide kinase (Sambrook *et al.*, 1989).

(9) Prehybridize the membrane in oligonucleotide hybridization solution (see Solutions, under **Materials**) for ≥4 hr at 42°C. Replace the hybridization solution, add the respective ^{32}P-labeled oligonucleotide, and hybridize at 42°C overnight.

(10) Wash the membrane with oligonucleotide wash solution (see Solutions, under **Materials**) once at room temperature and three or four times at 42°C (15 min each wash) and expose to X-ray film. After developing the film, strip the membrane by washing in a solution of $0.1\times$ SSC/ 0.1% SDS at 65°C, and hybridize with the other oligonucleotide probe (step 9). If the clone contains a known gene or marker, the membrane also can be hybridized to this probe in order to orient the locus within the clone.

(11) Estimate the sizes of bands and infer the relative order of the restriction fragments. Again, it is essential to identify and define the sizes of all restriction fragments prior to partial digestion, i.e., the sum of all the fragments must equal the total size of the insert.[43]

(12) It is useful both to construct maps for several restriction enzymes and to conduct double-digests to corroborate the inferred order of restriction sites.

IV. Summary

The PAC cloning system is a powerful approach for genome mapping. The protocols described here have been found to be effective for generating PAC libraries from lower (as well as higher vertebrate) species and for rapidly characterizing PAC clones (restriction mapping). We are using the PAC system for examining extended loci of the rearranging genes (immunoglobulins and T-cell receptors) in several species, including both teleosts (bony fishes) and chondrichthyians (cartilaginous fishes). The system is very tractable for addressing issues that are not feasible using λ or

42. SP6: 5′-GATTTAGGTGACACTATAG-3′; T7: 5′-TAATACGACTCACTATAGGG-3′.

43. In our experience, when the sum of fragment sizes does not equal the insert size, the problem has generally been the presence of doublets (or triplets) in the analytical gels of complete digests. The use of the biotin-chemiluminscence system to detect restriction fragments as well as employing different PFGE conditions for band separation can reduce this problem.

cosmid clones. The PAC cloning system also should prove to be useful for addressing a range of genomics problems that are unique to lower vertebrates and to other nonmammalian species.

Acknowledgments

We thank Ms. Barbara Pryor for editorial assistance and Dr. Pieter de Jong, in whose laboratory the first PAC cloning experiments were performed. Research was supported by the Center for Human Genetics, Boston University School of Medicine, and by NIH Grant R37-AI23338 (to G.W.L.).

NOTE ADDED IN PROOF. With regard to preparation of partially digested DNA for cloning, we have empirically determined that elution of DNAs from preparative fractions via "electroelution" into dialysis bags yields equivalent PAC cloning efficiencies as with the agarose digestion method described previously.

References

Amemiya, C. T., and Gold, J. R. (1986). Chromomycin A3 stains nucleolus organizer regions of fish chromosomes. *Copeia*, 226–231.

Amemiya, C. T., and Gold, J. R. (1987). Chromomycin staining of vertebrate chromosomes: Enhancement of banding patterns by NaOH. *Cytobios* **49**, 147–152.

Amemiya, C. T., and Litman, G. W. (1990). Complete nucleotide sequence of an immunoglobulin heavy-chain gene and analysis of immunoglobulin gene organization in a primitive teleost fish. *Proc. Natl. Acad. Sci. U.S.A.* **87**, 811–815.

Amemiya, C. T., and Litman, G. W. (1991) Early evolution of immunoglobulin genes. *Am. Zool.* **31**, 558–569.

Amemiya, C. T., Aslanidis, C., Alleman, J. A., Chen, C., and de Jong, P. J. (1990). Use of a multi-dimensional pooling scheme and ALU-PCR for cosmid contig-mapping in the myotonic dystrophy region (19q13.2–3). *Am. J. Hum. Genet.* **47**, 958.

Amemiya, C. T., Alegria-Hartman, M. J., Aslanidis, C., Chen, C., Nikolic, J., Gingrich, J. C., and de Jong, P. J. (1992). A two-dimensional YAC pooling strategy for library screening via STS and *Alu*-PCR methods. *Nucleic Acids Res.* **20**, 2559–2563.

Amemiya, C. T., Ohta, Y., Litman, R. T., Rast, J. P., Haire, R. N., and Litman, G. W. (1993). V$_H$ gene organization in a relict species, the coelacanth *Latimeria chalumnae*: Evolutionary implications. *Proc. Natl. Acad. Sci. U.S.A.* **90**, 6661–6665.

Anderson, M., Amemiya, C., Luer, C., Litman, R., Rast, J., Niimura, Y., and Litman, G. W. (1994). Complete genomic sequence and patterns of transcription of a member of an unusual family of closely related, chromosomally dispersed Ig gene clusters in *Raja*. *Int. Immunol.* **6**, 1661–1670.

Aparicio, S., Morrison, A., Gould, A., Gilthorpe, J., Chaudhauri, C., Rigby, P., Krumlauf, R., and Brenner, S. (1995). Detecting conserved regulatory elements with the model genome of the Japanese puffer fish, *Fugu rubripes*. *Proc. Natl. Acad. Sci. U.S.A.* **92**, 1684–1688.

Baxendale, S., Abdulla, S., Elgar, G., Buck, D., Berks, M., Micklem, G., Durbin, R., Bates, G., Brenner, S., Beck, S., and Lehrach, H. (1995). Comparative sequence analysis of the human and pufferfish Huntington's disease genes. *Nature Genetics* **10**, 67–76.

Bentley, D. R. (1992). The analysis of YAC clones. *In* "Techniques for the Analysis of Complex Genomes" (R. Anand, ed.), pp. 113–135. Academic Press, San Diego, CA.

Bentley, D. R., Todd, C., Collins, J., Holland, J., Dunham, I., Hassock, S., Bankier, A., and Gianneli, F. (1992). The development and application of automated gridding for efficient screening of yeast and bacterial ordered libraries. *Genomics* **12**, 534–541.

Birnboim, H. C., and Doly, J. (1979). A rapid alkaline extraction procedure for screening recombinant plasmid DNA. *Nucleic Acids Res.* **7**, 1513–1523.

Brenner, S., Elgar, G., Sandford, R., Macrae, A., Venkatesh, B., and Aparicio, S. (1993). Characterization of the pufferfish *(Fugu)* genome as a compact model vertebrate genome. *Nature (London)* **366**, 265–268.

Brown, T. A., ed. (1991). "Molecular Biology LABFAX." Bios Scientific Publishers, Oxford.

Burke, D. T., Carle, G. F., and Olson, M. V. (1987). Cloning of large segments of exogenous DNA into yeast by means of artificial chromosome vectors. *Science* **236**, 806–812.

Collins, F. S. (1988). Chromosome jumping. *In* "Genome Analysis: A Practical Approach" (K. E. Davies, ed.), pp. 73–94. IRL Press, Washington, DC.

Collins, F. S., and Weissman, S. M. (1984). Directional cloning of DNA fragments at a large distance from an initial probe: A circularization method. *Proc. Natl. Acad. Sci. U.S.A.* **81**, 6812–6816.

Cross, S., Kovarik, P., Schmidtke, J., and Bird, A. (1991). Non-methylated islands in fish genomes are GC-poor. *Nucleic Acids Res.* **19**, 1469–1474.

Fjose, A., Molven, A., and Eiken, H. G. (1988). Molecular cloning and characterization of homeobox-containing genes from Atlantic salmon. *Gene* **62**, 141–152.

Gold, J. R., and Li, Y. C. (1991). Trypsin G-banding of North American cyprinid chromosomes: Phylogenetic considerations, implications for fish chromosome structure, and chromosomal polymorphism. *Cytologia* **56**, 199–208.

Green, E. D., and Olson, M. V. (1990). Systematic screening of yeast artificial chromosome libraries by use of the polymerase chain reaction. *Proc. Natl. Acad. Sci. U.S.A.* **87**, 1213–1217.

Heard, E., Davies, B., Feo, S., and Fried, M. (1989). An improved method for the screening of YAC libraries. *Nucleic Acids Res.* **17**, 5861.

Hudson, A. P., Cuny, G., Cortadas, J., Haschemeyer, E. V., and Bernardi, G. (1980). An analysis of fish genomes by density gradient centrifugation. *Eur. J. Biochem.* **112**, 203–210.

Imai, T., and Olson, M. V. (1990). Second-generation apporach to the construction of yeast artificial-chromosome libraries. *Genomics* **8**, 297–303.

Ioannou, P. A., Amemiya, C. T., Garnes, J., Kroisel, P. M., Shizuya, H., Chen, C., Batzer, M. A., and de Jong, P. J. (1994). A new bacteriophage P1-derived vector for the propagation of large human DNA fragments. *Nat. Genet.* **6**, 84–89.

Kahn, P. (1994). Zebrafish hit the big time. *Science* **264**, 904–905.

Krauss, S., Concordet, J.-P., and Ingham, P. W. (1993). A functionally conserved homolog of the *Drosophila* segment polarity gene *hh* is expressed in tissues with polarizing activity in zebrafish embryos. *Cell (Cambridge, Mass.)* **75**, 1431–1444.

Litman, G. W., Rast, J. P., Shamblott, M. J., Haire, R. N., Hulst, M., Roess, W., Litman, R. T., Hinds-Frey, K. R., Zilch, A., and Amemiya, C. T. (1993). Phylogenetic diversification of immunoglobulin genes and the antibody repertoire. *Mol. Biol. Evol.* **10**, 60–72.

Little, P. (1993). Small and perfectly formed. *Nature (London)* **366**, 204–205.

Lundin, L. G. (1993). Evolution of the vertebrate genome as reflected in paralogous chromosomal regions in man and the house mouse. *Genomics* **16**, 1–19.

Medrano, L., Bernardi, G., Couturier, J., Dutrillaux, B., and Bernardi, G. (1988). Chromosome banding and genome compartmentalization in fishes. *Chromosoma* **96,** 178–183.

Molven, A., Njolstad, P. R., and Fjose, A. (1991). Genomic structure and restricted nerual expression of the zebrafish wnt-1 (int-1) gene. *EMBO J.* **10,** 799–807.

Morizot, D. C. (1983). Tracing linkage groups from fishes to mammals. *J. Hered.* **74,** 413–416.

Morizot, D. C. (1990). Use of fish gene maps to predict ancestral veretebrate genome organization. *In* "Isozymes: Structure, Function, and Use in Biology and Medicine" (Z.-I. Ogita and C. L. Markert, eds.), pp. 207–234. Wiley-Liss, New York.

Nelson, D. L., and Brownstein, B. H., eds. (1994). "YAC Libraries: A User's Guide." Freeman, New York.

Nizetic, D., Drmanac, R., and Lehrach, H. (1991). An improved bacterial colony lysis procedure enables direct DNA hybridisation using short (10, 11 bases) oligonucleotides to cosmids. *Nucleic Acids Res.* **19,** 182.

Olsen, A. S., Combs, J., Garcia, E., Elliot, J., Amemiya, C., de Jong, P. J., and Threadgill, G. (1993). Automated production of high density cosmid and YAC colony filters using a robotic workstation. *BioTechniques* **14,** 116–123.

Ota, T., and Amemiya, C. T. (1996). A nonradioactive method for improved restriction analysis and fingerprinting of large P1 artificial chromosome clones. *Genet. Anal.* (in press).

Ota, T., and Nei, M. (1994). Divergent evolution and evolution by the birth-and-death process in the immunoglobulin V_H gene family. *Mol. Biol. Evol.* **11,** 469–482.

Pierce, J. C., and Sternberg, N. (1992). The bacteriophage P1 cloning system. *In* "Techniques for the Analysis of Complex Genomes" (R. Anand, ed.), pp. 39–58. Academic Press, San Diego, CA.

Pierce, J. C., Sauer, B., and Sternberg, N. (1992a). A positive selection vector for cloning high molecular weight DNA by the bacteriophage P1 system: Improved cloning efficacy. *Proc. Natl. Acad. Sci. U.S.A.* **89,** 2056–2060.

Pierce, J. C., Sternberg, N., and Sauer, B. (1992b). A mouse genomic library in the bacteriophage P1 cloning system: Organization and characterization. *Mamm. Genome* **3,** 550–558.

Rast, J. P., and Litman, G. W. (1994). T-cell receptor gene homologs are present in the most primitive jawed vertebrates. *Proc. Natl. Acad. Sci. U.S.A.* **91,** 9248–9252.

Sambrook, J., Fritsch, E. F., and Maniatis, T. (1989). "Molecular Cloning: A Laboratory Manual," 2nd ed. Cold Spring Harbor Lab. Press, Cold Spring Harbor, NY.

Schempp, W., and Schmid, M. (1981). Chromosome banding in Amphibia: VI. BrdU-replication patterns in anura and demonstration of XX/XY sex chromosomes in *Rana esculenta. Chromosoma* **83,** 697–710.

Schmid, M., and de Almeida, C. G. (1988). Chromosome banding in Amphibia: XII. Restriction endonucease banding. *Chromosoma* **96,** 283–290.

Shamblott, M. J., and Chen, T. T. (1992). Identification of a second insulin-like growth factor in a fish species. *Proc. Natl. Acad. Sci. U.S.A.* **89,** 8913–8917.

Sheng, Y., Mancino, V., and Birren, B. W. (1995). Transformation of *E. coli* with large DNA molecules by electroporation. *Nucleic Acids Res.* **23,** 1990–1996.

Shepherd, N. S., Pfrogner, B. D., Coulby, J. N., Ackerman, S. L., Vaidyanathan, G., Sauer, R. H., Balkenhol, T. C., and Sternberg, N. (1994). Preparation and screening of an arrayed human genomic library generated with the P1 cloning system. *Proc. Natl. Acad. Sci. U.S.A.* **91,** 2629–2633.

Shizuya, H., Birren, B. W., Kim, U.-J., Mancino, V., Slepak, T., Tachiiri, Y., and Simon, M. (1992). Cloning and stable maintenance of 300-kilobase-pair fragments of human DNA in *Escherichia-coli* using an F-factor-based vector. *Proc. Natl. Acad. Sci. U.S.A.* **89,** 8794–8797.

Southern, E. M., Anand, R., Brown, W. R. A., and Fletcher, D. S. (1987). A model for the separation of large DNA molecules by crossed field gel electrophoresis. *Nucleic Acids Res.* **15**, 5925–5943.

Sternberg, N. (1990). Bacteriophage P1 cloning system for the isolation, amplification, and recovery of DNA fragments as large as 100 kilobase pairs. *Proc. Natl. Acad. Sci. U.S.A.* **87**, 103–107.

Sternberg, N., and Cohen, G. (1989). Genetic analysis of the lytic replicon of bacteriophage P1. II. Organization of replicon elements. *J. Mol. Biol.* **207**, 111–134.

Wang, Y. K., and Schwartz, D. C. (1993). Chopped inserts: A convenient alternative to agarose/DNA inserts or beads. *Nucleic Acids Res.* **21**, 2528.

The Selection of Chromosome-Specific DNA Clones from African Trypanosome Genomic Libraries

Sara E. Melville, Nancy S. Shepherd, Caroline S. Gerrard,

and Richard W. F. Le Page

I. Introduction

The creation of chromosome-specific libraries of large-insert clones of genomic DNA has proven valuable in a variety of approaches to genome mapping and analysis. In some cases this strategy has been employed because one chromosome is of particular interest, while in other cases the aim has been to rapidly select chromosome-specific markers and probes (Fuscoe *et al.*, 1989; Choo *et al.*, 1990; Budarf *et al.*, 1991; Saito *et al.*, 1991). For certain mammalian genomes this has been achieved by the use of somatic cell hybrids or by flow-sorting of chromosomes before cloning (Van Dilla and Deaven, 1990; Nižetić *et al.*, 1991; Miller *et al.*, 1992; Milan *et al.*, 1993). However, in some eukaryotes, such as the kinetoplastids, the chromosomes do not undergo the condensation necessary for either microscopic analysis or flow-sorting. This precludes many of the published procedures for chromosome purification and analysis. Nevertheless, the smaller eukaryotic genomes do have a big advantage: the chromosomes are fully resolvable by pulsed-field gel electrophoresis, thus giving us access to considerable amounts of DNA from any one chromosome. This DNA may be recovered from the gel and cloned to create a chromosome-specific library. This procedure works very well when cloning into vectors such as plasmids or bacteriophage lambda, but presents greater difficulty if larger inserts are required. However, the smaller size of the chromosomes permits efficient identification of such clones from a genomic library with the inclusion of a low proportion of false positives; we describe here the use of complex chromosomal probes to select clones which derive from an individual chromosome band, producing a smaller, chromosome-specific subgenomic library.

II. Materials

A. Section III.A

(1) SDM (semidefined medium)-79 (Brun and Schonenberger, 1979): Add, per liter, 7 g F-14 powder (Gibco BRL, Gaithersburg, MD), 2 g Medium 199 TC 45 powder (Wellcome, Kent, UK), 8 ml MEM amino acids (50×) without L-glutamine (Gibco BRL), 6 ml MEM nonessential amino acids (100×) (Gibco BRL), 1 g glucose, 8 g *N*-2-hydroxyethylpiperazine-*N'*-2-ethanesulfonic acid (HEPES), 5 g 3-[*N*-morpholino]propanesulfonic acid (MOPS), 2 g $NaHCO_3$, 100 mg Na pyruvate, 200 mg L-alanine, 100 mg L-arginine, 300 mg L-glutamine, 70 mg L-methionine, 80 mg L-phenylalanine, 600 mg L-proline, 60 mg L-serine, 160 mg L-taurine, 350 mg L-threonine, 100 mg L-tyrosine, 10 mg adenosine, 10 mg guanosine, 50 mg glucosamine-HCl, 4 mg folic acid, 2 mg *p*-aminobenzoic acid, 0.2 mg biotin, pH adjusted to 7.3 with 2 *N* NaOH and filter-sterilized with a 0.22-μm filter (Sartorius, Göttingen, Germany). Supplement with heat-inactivated (15 min at 56°C) 10% v/v fetal calf serum (FCS) (Gibco BRL, Bio Cult).

(2) Trypanosome dilution buffer (TDB): 118 m*M* NaCl, 1.2 m*M* KH_2PO4, 30 m*M* TES, 16 m*M* Na_2HPO4, 5 m*M* $NaHCO_3$, 5 m*M* KCl, pH 8.0 (for some organisms, e.g., *Leishmania* spp., phosphate-buffered saline (PBS) is often used).

(3) PBS: 137 m*M* NaCl, 2.7 m*M* KCl, 8 m*M* Na_2HPO_4, 1.75 m*M* KH_2PO_4, adjusted to pH 7.4.

(4) STE buffer: 100 m*M* NaCl, 50 m*M* EDTA, 100 m*M* Tris-HCl, pH 7.8.

(5) SDS (sodium dodecyl sulfate) 10% stock solution: dissolve 100 g solid in 900 ml distilled H_2O, heat to 65°C, make up to 1 liter, and adjust to pH 7.2. Filter-sterilize using a 0.2-μm filter.

(6) RNase stock solution: dissolve 10 mg solid pancreatic RNase A, DNA grade (Sigma Chemical Co., St. Louis, MO), in 1 ml sterile distilled H_2O. Heat to 95–100°C for 10 min. Cool, and store aliquots at −20°C.

(7) Proteinase K stock solution: dissolve 20 mg solid proteinase K isolated from *Tritirachium album* (Sigma) in 1 ml sterile distilled H_2O and store aliquots at −20°C.

(8) TE buffer: 10 mM Tris-HCl, pH 8, 1 mM EDTA.

(9) Partial digestion buffer, 10×: 100 mM Tris-HCl, 1 M NaCl, 10 mM dithiothreitol (DTT), pH 7.

(10) *Sau*3AI restriction endonuclease (New England Biolabs (NEB), Beverley, MA)

(11) Dephosphorylation buffer, 10×: 100 mM Tris-HCl, 500 mM NaCl, 100 mM MgCl$_2$, 10 mM DTT, pH 7.9.

(12) Ligation buffer, 10×: 500 mM Tris-HCl, pH 7.8, 100 mM MgCl$_2$, 100 mM DTT, 10 mM ATP, 250 μg BSA.

(13) T4 DNA ligase (NEB)

B. Section III.A.2

Sucrose gradients: A 10–40% sucrose gradient was prepared in 20 mM Tris·HCl, pH 8, 0.8 M NaCl, 10 mM EDTA, pH 8.

C. Section III.B.1

(1) 5× TBE electrophoresis buffer: 54 g Tris base, 27.5 g boric acid, and 20 ml 0.5 M EDTA, pH 8, in 1 liter distilled water.

(2) 5× TB(0.1)E electrophoresis buffer: 54 g Tris base, 27.5 g boric acid, and 2 ml 0.5 M EDTA, pH 8, in 1 liter distilled water.

D. Section III.B.2

(1) Many makes of low-melting-temperature agarose are available and we have used two with success: Seaplaque GTG (FMC, Rockland, ME) and Bio-Rad Low Melt preparative grade (Bio-Rad, Richmond, CA). It is safer to select an agarose which has been tested for *in agaro* enzyme activity.

(2) β-Agarase I (New England Biolabs).

(3) Agarase buffer: TE, pH 6.5.

E. Section III.C

(1) L-broth: 10 g bactotryptone, 5 g bacto-yeast extract (Difco Laboratories, Detroit, MI), 10 g NaCl in distilled H$_2$O, total volume 1 liter, adjusted to pH 7. Sterilize by autoclaving 20 min at 15 lb/in.[2]

(2) For plates, add 11 g bactoagar (Difco). For freezing plates also add 15% glycerol. [For long-term storage of bacteria on plates, viability is enhanced by use of the richer medium given in Hanahan and Meselson (1980), referred to as Terrific Broth in Sambrook *et al.* (1989): add 12 g bactotryptone, 24 g yeast extract, 11 g bactoagar, and 4 ml glycerol to 800

ml distilled H₂O. Sterilize as above. Once cool, add 100 ml 0.17 *M* KH₂PO₄, 0.72 *M* K₂HPO₄(filter-sterilized) and make up to 1 liter with sterile distilled H₂O.]

F. Section III.C.3

(1) 5× random priming reaction buffer: 400 m*M* Tris-HCl, pH 8, 40 m*M* MgCl₂, 80 m*M* β-mercaptoethanol, 1.6 *M* HEPES, pH 7, 0.1 m*M* each of dATP, dGTP, dTTP, store at −20°C.

(2) Random 9mer oligonucleotide primers in TE, store at −20°C.

(3) [α−³²P]dCTP (3000 Ci mmol⁻¹).

(4) Exo(-) Klenow DNA polymerase (Stratagene, La Jolla, CA): We prepare the chromosome probe using a Prime-It II kit from Stratagene, but other standard methods or kits may be suitable.

(5) The radioactively labeled DNA may be separated from unincorporated radionucleotides using a Sephadex G-50 column: two commercial systems we have used with success are NICK columns with Sephadex G-50 DNA grade (Pharmacia Biotech AB, Uppsala, Sweden) and NUCTRAP push columns (Stratagene).

(6) Prehybridization and hybridization solution: 0.5 *M* sodium phosphate, pH 7.4, 7% SDS. Prepare 1 *M* solutions of Na₂HPO₄ and NaH₂PO₄. To make 200 ml mix 77.4 ml 1 *M* Na₂HPO₄ and 22.6 ml 1 *M* NaH₂PO₄ (this will result in pH 7.4) and add 100 ml 14% SDS (filtered). Hybridization solutions based on Denhardt's solution (approximately equal Na⁺ concentration) work equally well.

G. Section III.E.1

(1) Specific vector amplification primers:

T3 (Supercos): 5′ GAAATTAACCCTCACTAAAGGG 3′
T7 (Supercos/p*Ad*10*sac*BII): 5′ GTAATACGACTCACTATAGGGC 3′
SP6 (p*Ad*10*sac*BII): 5′ CGACATTTAGGTGACACTATAG 3′

(2) Random amplification primer:

NS-2: 5′ GTCAGTCAGTCAGANNNNGAG 3′

This is the primer used by Wesley *et al.* (1994). Although we experimented with other sequences, this primer consistently gave the best results.

(3) 10× polymerase chain reaction (PCR) buffer: 400 m*M* Tris-HCl, pH 8.9, 15 m*M* MgCl₂, 100 m*M* ammonium sulfate.

    (4) 1.25 mM dATP, dCTP, dGTP, dTTP.
    (5) Taq polymerase: Amplitaq (Perkin Elmer).

III. Procedures

A. Creating Genomic Libraries of Trypanosome DNA

Very gentle lysis of the trypanosomes facilitates the extraction of minimally disrupted DNA. Cultures of procyclic trypanosomes were grown to a concentration of 2–5 × 10^7 live organisms per milliliter in SDM-79 with 10% (v/v) FCS. The cells were centrifuged and washed in an equal volume of trypanosome dilution buffer (TDB). Final resuspension was in 20 ml STE. SDS was added to 0.5% (w/v) and RNase A to 200 µg ml^{-1}, 2 hr at 37°C. Proteinase K was added to 2 mg ml^{-1}, 4 hr at 50°C. Sterile 4 M NaCl solution was added to make the DNA solution 1 M and left at 4°C overnight. The DNA solution was dialyzed at 4°C against 1 liter of TE with eight changes over 4 days, then transferred with minimum disruption to sterile tubes and stored at 4°C. This DNA solution was used for partial digestion, and also sheared for prehybridization of filters (see Section III.C.3).

    At all stages during the preparation of genomic libraries prior to manipulation of packaged DNA, wide-bore pipette tips should be used with genomic DNA fragments (e.g., Cell Saver tips from Alpha Laboratories, Eastleigh, UK, or cut 0.5 cm off the ends of standard tips). The DNA was partially digested with Sau3AI to produce fragments of a size suitable for cloning into cosmids (35–50 kb) or bacteriophage P1 (60–95 kb): 65 µl 10× buffer (without Mg^{2+}), 2.5 µl bovine serum albumin (BSA; 20 mg ml^{-1} stock), 22 µl sterile distilled H_2O, and 2.5 units Sau3AI restriction endonuclease were added to 500 µl trypanosome DNA (50 µg) in each of five sterile 1.5-ml microfuge tubes and stirred very slowly with the pipette tip. A sixth tube contained everything except the enzyme to act as a control for nonspecific degradation. The solution was allowed to stand at 4°C for 4 hr to allow diffusion of the enzyme through the high-molecular-weight DNA. The tubes were warmed to 30°C for 15 min; 58 µl 100 mM MgAc was added to each tube in turn and the reaction was allowed to proceed for 0.5, 1, 2, 3, or 4 min at 30°C. The control tube was incubated for 4 min also. At the end of each incubation, 60 µl 0.5 M EDTA, pH 8, was added and the tube was transferred immediately to a 70°C waterbath for 15 min, then placed on ice; 40 µl of each reaction was electrophoresed alongside 2.8 µg undigested genomic DNA and lambda ladder markers (Bio-Rad) through a 1% NA agarose gel in 0.5× TBE in a CHEF-DRII pulsed-field gel apparatus (Bio-Rad) at 200 V for 10 hr with a linearly ramped pulse time of 3 to 8 sec. The gel was stained with ethidium bromide. By comparison to the markers, the reaction giving the greatest density of DNA fragments in the desired size range (see above) was identified and the adjacent reaction containing slightly larger fragments was selected to prepare the cosmid and bacteriophage P1 libraries. The aliquot from the control tube should have the same appearance on electrophoresis as the undigested aliquot. If DNA is limiting, smaller reactions should be set up first to determine the approximate amount of enzyme required to achieve the desired partial digestion before setting up one or two 50 µg reactions; also if the frequency of cutting sites for the cloning enzyme is unknown, more small reactions over a greater range of incubation times or enzyme concentration may be required.

1. Cosmids

The preparation of cosmid libraries is described in Chapter 7 by Wenzel and Hermann. The MC7 cosmid library used in these experiments was prepared using genomic DNA from the procyclic culture form of a hybrid progeny clone rescued from a laboratory cross between two *Trypanosoma brucei* brucei field isolates. The full designation of this clone is F532/72 mcl 7 as described in Turner *et al.*, (1990) and Sternberg *et al.*, (1989).

The selected aliquot of partially digested genomic DNA was extracted with 670 µl buffered phenol and precipitated with 1500 µl ethanol. The dried pellet was resuspended in 240 µl sterile distilled H_2O with 60 µl 5× dephosphorylation buffer and 5 units calf intestinal phosphatase (CIP), and incubated at 37°C for 30 min. The solution was extracted with 300 µl buffered phenol, then 150 µl phenol and 150 µl chloroform, precipitated with 30 µl sodium acetate (3 *M*) and 750 µl 100% ethanol, then washed with cold (0°C) 70% ethanol. The dried pellet was resuspended in 50 µl sterile distilled H_2O.

The cosmid vector Supercos (Evans *et al.*, 1989; Stratagene) was prepared according to the manufacturer's instructions and the cloning ends were not dephosphorylated; 2.5 µg genomic DNA fragments in 10 µl was mixed with 2 µl 10× ligation buffer, 1 µg prepared vector arms, 2 µl 10 m*M* ATP, and 2 Weiss units of T4 DNA ligase in a total volume of 20 µl, and incubated at 16°C overnight. The ligated DNA was packaged using Gigapack XL (Stratagene) and introduced into *Escherichia coli* NM554 host cells according to the manufacturer's instructions. The cloning efficiency was 6.4×10^5 transformants μg^{-1} genomic DNA.

2. Bacteriophage P1

The use of bacteriophage P1 for the preparation of libraries of genomic DNA has been described (Sternberg, 1990; Pierce and Sternberg,

1993; Shepherd *et al.*, 1994). The trypanosome P1 library was prepared from DNA extracted from the procyclic culture form of a *T.b.brucei* field isolate, TREU927/4 (Turner *et al.*, 1990), as described in Section III.A. The DNA was partially digested with *Sau*3AI as described above and the selected partially digested DNA reaction (670 μl) was loaded onto an 18-ml sucrose gradient and centrifuged in an ultracentrifuge (Beckman, L8-70M) in an SW40 rotor at 18,000 rpm for 18 hr at 20°C; 0.5-ml fractions were collected and 40 μl of each was spot-dialyzed on filters (Millipore VSWP 02500) floating on 200 mls TE for 2 hrs. The aliquots were loaded into the wells of a pulsed-field gel and electrophoresed as described in Section III.A.1 in order to select a suitable size fraction (see above). The selected fraction was concentrated 10-fold by butanol extraction (Sambrook *et al.*, 1989) and spot dialyzed against TE as described above (ensure that no butanol remains before loading onto the filter or it will dissolve the filter). The ligation reaction was carried out in 15 μl with 200 ng genomic DNA fragments, 100 ng p*Ad10sac*BII (Pierce, *et al.*, 1992), 1.5 μl 10× ligation buffer, and 2 units T4 DNA ligase overnight at 14°C and the ligated DNA was packaged into *E.coli* strain NS 3529 according to published procedures (Pierce and Sternberg, 1993).

From one sucrose fraction, 1988 clones were obtained. The average insert size of 164 clones is 65 kb (median:68 kb). No mini-chromosome DNA was cloned (S. E. Melville and C. S. Gerrard, unpublished). This represents an estimated 3.7× haploid genome equivalent of chromosomes greater than 350 kb in size. (Gottesdiener *et al.*, 1990; S. E. Melville and V. Leech, unpublished results).

B. Probe Preparation

1. Pulsed-Field Gel Electrophoresis

The preparation of plugs of genomic DNA from protozoan parasites is described elsewhere in this volume by Morzaria. The plugs of *T.b.brucei* DNA were prepared with 2×10^8 trypanosomes from procyclic culture in SDM-79 medium with 10% FCS as described previously (Van der Ploeg *et al.*, 1984) except that the cultures were shaken at approximately 100 rpm at 28°C as this results in superior pulsed-field gel electrophoresis (PFGE) separations, possibly by preventing the formation of rosettes. Cultures should be grown only to late log phase: once the medium has changed from red to orange, monitor the organisms in culture twice daily under the microscope and harvest as soon as movement decreases or death (more than 1 organism in 10^3) is evident.

Figure 9.1 shows the separation of *T.b.brucei* chromosomes (F532/ 72 mcl 7) and indicates the 1.5-Mb chromosome which was isolated from the gel and used to prepare the probe. This separation was achieved in a 1.2% NA agarose gel in $1\times$ TB(0.1)E buffer in a CHEF DRII (Bio-Rad), using a linearly ramped pulse time of 1100 to 700 sec at 2.5 V cm^{-1} over 144 hrs. Lowering the EDTA concentration increases the mobility of the chromosomal DNA in PFGE. Its usefulness in the separation of trypanosome chromosomes was discovered empirically. The chromosome band chosen to illustrate the procedure is well separated from others and the level of ethidium bromide fluorescence suggests that it is likely to consist of a single haploid chromosome species. Many markers specific to this chromosome have been isolated (S. E. Melville and C. S. Gerrard, unpublished results) and the apparent homolog (to which all these markers also hybridize) is known to comigrate with other chromosomes in the band at approximately 1.1 Mb. The full extent of homology and the content of the extra 400 kb is not known.

2. Elution of the Chromosome and Preparation of a Complex Probe

We favor the use of agarase to separate DNA from its agarose substrate, although, since it is not necessary to prevent shearing of the DNA, several other methods may prove suitable. Agarase requires the use of a low-melting-point agarose, and since the use of such agaroses both alters the chromosome separation and renders DNA less visible after ethidium bromide staining, we prefer to cut the chromosome band from the pulsed-field gel in NA agarose, reset it into a gel of low-melting-point (LMP) agarose and move the chromosome into the LMP gel. We set the 1.5-Mb chromosome band into a gel of 1% low-melting-point agarose in $0.5\times$ TBE and electrophorese with a pulse time of 5–12 sec at 6 V/cm for 8 to 10 hr in a CHEF DRII (Bio-Rad).

Protocol 1: Preparation of the Probe

(i) Lay the gel on the clean surface of a UV transilluminator and cut a block of agarose containing the chromosome band from the gel using a clean razor blade. This may be done with UV illumination of the entire chromosome band (312 nm) but minimize exposure (no more than 2 min). Cut the block into small pieces, 0.2 × 0.4 cm, and dialyze against sterile 30–50 ml TE, pH 6.5, for an hour.

(ii) Remove the buffer and melt the agarose at 65°C.

(iii) Transfer the tube to 37°C, add 1 unit of agarase for every 200 μl volume, and incubate for at least 1 hr. Verify that digestion is complete by placing the tube on ice: if solid pieces of agarose are observed, melt again at 65°C, cool to 37°C, and add an additional 0.25 units enzyme/200 μl volume. Incubate for an hour at 37°C. Higher concentrations of agarose will require correspondingly more units of enzyme.

(iv) When fully digested, centrifuge at top speed (13,000 rpm) in a microfuge for 15 min and transfer the solution to a new tube. Some agarose lumps may be found at the bottom of the centrifuged tube and should be discarded.

(v) Add 1/10 vol 3 M sodium acetate, pH 5.2, then at least 2 vol of cold 100% ethanol (store at 4°C) and cool on ice for 15 min. Centrifuge at 13,000 rpm in a microfuge for 30 min, discard ethanol, taking care not to disturb the pellet.

(vi) Add 1 vol of cold 70% ethanol (stored at 4°C). Cool on ice for 15 min, recentrifuge at 13,000 rpm for 10 min, and discard ethanol.

(vii) Dry DNA pellet and resuspend in TE (see below).

At no point during this procedure should wide-bore tips be used, as we do not wish to retain the DNA at full size. The optimal volume in which to resuspend can only really be determined empirically; therefore it is best in the first instance to resuspend in a small volume (50 μl for the band shown in Fig. 9.1, see below). We have found it difficult to accurately determine the DNA content of these solutions by spectrophotometry or fluorometry, due to the presence of the oligosaccharides from the digestion of agarose.

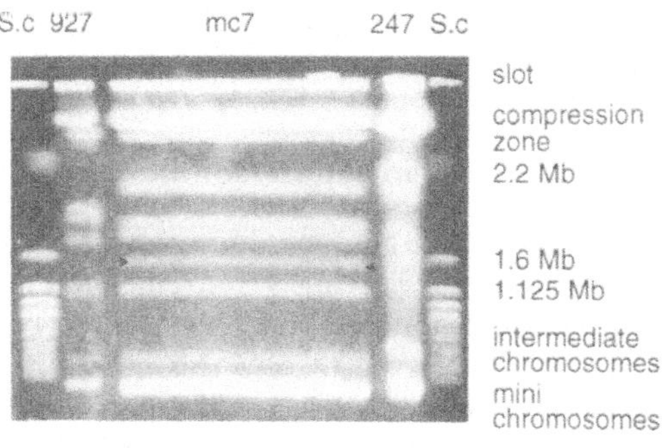

Figure 9.1 The chromosomes of trypanosome isolates 927 and 247 and of the hybrid clone MC7 were resolved by PFGE and stained with ethidium bromide. *S. cerevisiae* chromosomes were also separated to serve as size standards. The MC7 chromosome identified by the arrow was isolated from the gel and used to prepare a radioactive probe for the selection of DNA clones deriving from this chromosome and its homolog.

Spectrophotometric estimation may be improved by extracting the DNA solution with phenol and chloroform after agarase treatment, but this also results in loss of DNA. However, it is also difficult to decide theoretically how much DNA is required to make a very complex probe representing over 1 Mb of unique sequence DNA with an unknown level of repetitive sequence. It has proven more reliable to estimate the amount of DNA in the chromosome band excised from the gel and to carry out test probings of colony lifts with different amounts of labeled DNA in the range 50 to 250 ng (estimated from the amount loaded on the gel, see below) to determine the minimum required to obtain reliable replicate autoradiographs. Test probings in our laboratory may serve as a guide, although genomes of different sequence compositions may vary. The plugs were prepared with 2×10^8 trypanosomes in 100 μl. Therefore we estimate that the trypanosome plugs contained 19.4 μg of DNA (0.097 pg DNA per nucleus, Borst *et al.*, 1982) and $16 \times 1/2$ plugs were loaded on the gel, i.e., 155.2 μg. This was about the maximum amount it was possible to load without smearing on electrophoresis. The eluted chromosome band is thought to contain a single chromosome of approximately 1.5 Mb and therefore represents 1.875% of the estimated 80-Mb diploid nuclear genome (Gottesdiener *et al.*, 1990). The isolated band therefore contained an estimated 2.91 μg of DNA. This was resuspended in 50 μl TE after ethanol precipitation and 1.5 μl was added to each labeling reaction to produce the radioactive probe. Each probe preparation was sufficient to hybridize to two replica colony lifts of the cosmid or four of the P1 libraries (see Section III.C) in 10 ml hybridization solution.

The DNA may be denatured by boiling and a radioactive probe prepared by standard random priming methods (Feinberg and Vogelstein, 1983; see Sections II and III.C.3).

C. Preparation of Colony Lifts and Hybridization Conditions

There are different ways of storing and using genomic libraries, depending on the equipment available and the experimental requirements. The packaged phage used to generate cosmid libraries may be stored at 4°C with some loss in titre over a year or more (the titre of this library decreased by 27% over 2 years; some researchers have reported a greater loss than this). Alternatively, primary or amplified cosmid libraries in *E. coli* may be maintained as glycerol stocks at -70°C or in liquid nitrogen for 5 years or more. If the culture is not defrosted when taking samples, there is little loss of viability. Master filters of randomly plated clones may be created by plating the bacteria directly onto nylon filters and storing the filters on LB-agar plates at 4°C for up to 1 month or on LB-agar plates with 15% glycerol at −70°C for a year or more. If replica filters are prepared from these master filters, then the replicas can be used in hybridization experiments and the selected clones retrieved later from the master filter. Finally, colonies may be picked directly into 96-well plates containing 200 µl L-broth with 10% glycerol, either individually or in pools of clones, and maintained at −70°C.

P1 libraries may also be stored as packaged phage, as glycerol stocks at −70°C or in liquid nitrogen, or as colonies on filters at −70°C. However, these libraries are not simple to produce. It is therefore prudent to consider carefully the means of storage and replication of a P1 library. Since the number of clones required for good coverage of small genomes is small enough to allow the storage of clones individually in microtitre plates, this is more convenient for most purposes: the production of replica libraries and replica filters is considerably simplified by the use of replicating tools; the dangers of the overrepresentation of some clones and loss of others, whether through the amplification of clones in culture or the pooling of clones in 96-well plates, are avoided, and the later retrieval of selected clones from the library is greatly facilitated. The P1 library consisting of a 3.5× haploid genome content of the trypanosome requires only 21 96-well plates, or 6 384-well plates, for which manual replicating tools are available (for example, from Sigma Chemical Co., St. Louis, MO, or Genetix, Wimborne, UK). The trypanosome libraries used here to illustrate the techniques were treated in two separate ways: a portion of the cosmid library was plated randomly, replica filters were created from the one plating, and the chromosome-specific clones of interest were retrieved from the first plate and stored as a sublibrary in 96-well plates. The rest of the library was stored as an unamplified glycerol stock at −70°C. The P1 library was arrayed by picking individual colonies into separate wells of 96-well plates, such that each clone has a unique coordinate, and high-density filters were created by robot.

1. Preparation of Replica Filters by Random Plating

An aliquot of the total library is plated and several copies are made from the master copy. The master copy is then placed onto LB-agar with 15% glycerol and stored at −70°C.

Protocol 2: Creation of Replica Filters

(i) Place a filter onto a dried LB-agar plate containing the appropriate antibiotic (50 µg/ml ampicillin for Supercos) and allow to moisten. Most makes of membrane may be used (we use Hybond-N from Amersham), but all are somewhat toxic to bacteria. This may be lessened by soaking the membrane in distilled water and sterilizing by autoclaving, but the toxicity cannot be removed entirely.

(ii) Spread bacteria directly onto the filter, leaving a 0.5-cm gap at the edge of the filter. It is best to titre the library first by spreading a range of dilutions onto filters. The titre will be reduced by at least 20%, possibly up to 50%, compared to plating directly onto LB-agar. The desired density of plated bacteria will depend on the hybridization experiments. For chromosomal probings we plate a 5× haploid genome equivalent (5000 clones) onto a filter of 20 × 20 cm on 22 × 22-cm plates. This is a very low density, but it facilitates the retrieval of the correct clones and simplifies later rescreening procedures (see Section III.D.1). Incubate right side up for 1 hr at 37°C, to allow full absorption of the liquid, then turn upside down. Incubate overnight at 37°C. At this low density overnight growth can be accommodated; if plating at a higher density (2–3×), allow the colonies to grow until they are 0.5 cm in diameter (~12 hr).

(iii) To prepare a replica filter from this original filter, lay a second 20 × 20 cm onto an LB-agar plate with ampicillin and allow to prewet. Lay the master filter colony side up onto a piece of 3MM paper on a glass plate, lay the second filter on top (do not adjust its position if you have failed to superimpose it perfectly), mark the orientation of both filters with respect to one another by piercing them in an asymmetrical pattern with a sterile needle (the resulting pinpricks may be marked with India ink later to make them more visible), cover with more 3MM paper and another glass plate, and press down hard and uniformly with your body weight to ensure even transfer of colonies from the lower filter to the upper.

(iv) Carefully peel the filters apart with a single, firm motion (help from a colleague may be necessary with 20 × 20-cm filters) and place them on LB-agar plates with 50 μg/ml ampicillin. Incubate both plates upside down at 37°C for several (5–7) hrs, until the colonies have grown to the size of those on the master plate originally. Two replica filters are required for each probing with the chromosomal probe. The replicating procedure may be carried out several times from the original filter to produce several replica filters if the colonies on the master filter are allowed a period of regrowth between each transfer (we routinely produce four copies; more may be possible but the efficiency of transfer reduces and colonies are lost). The original filter may then be kept at 4°C until the hybridizations are completed, and the selected clones retrieved from the filter before discarding the plate, or it may be frozen at −70°C on medium containing 15% glycerol until required.

(v) The simplest and cheapest method for the treatment of filters in preparation for hybridization is also one of the most effective. Prepare four trays containing pads of three or four layers 3MM paper slightly larger than the filters on which the bacteria are growing. Saturate one pad with 3% SDS, another with denaturation buffer, and two with neutralization buffer. Pour away excess liquid. Pour 300–500 ml 2× SSC into a separate tray.

(vi) Lay the filters with the colonies upward on the SDS-soaked pad for 3 min, then transfer onto the denaturation buffer for 7 min, then onto each of the neutralization buffer pads for 3 min.

(vii) Finally, immerse the filter in the 2× SSC and ensure that the bacterial debris is removed from the filter. This may be done by removing the filter from the liquid, wiping off the bacterial debris using a tissue with short, sharp movements, and washing again in fresh 2× SSC. Alternatively, bacterial colonies may be retained at this stage and washed off in prehybridization (Section III.C.3).

(viii) Allow the nylon filter to air-dry completely at room temperature or in an incubator at 65°C or less.

(ix) Fix the DNA to the filter with 0.12 J cm^{-2} 312-nm UV light or as recommended by the manufacturers.

2. Preparation of an Arrayed P1 Library and Replica High-Density Filters

Arrayed libraries are created in multiwell plates. Clones prepared from organisms with very large genomes may be pooled in order to reduce the number of plates required. However, this necessitates further screening steps when selecting clones from the library and differential growth in the wells may lead to the loss of some clones. If possible, it is simplest and most useful to organize a library with one clone per well. (Pierce and Sternberg, 1993; Shepherd *et al.*, 1994).

Protocol 3: Creation of an Arrayed P1 Library

(i) Fill each well of a set of multiwell plates with 200 µl L-broth containing 25 µg/ml kanamycin and 10% glycerol. We pour the medium into a sterile petri dish and use an eight-channel pipette (Anachem) with sterile, disposable tips to fill the wells. (Robotics are available from Beckman.)

(ii) Using sterile toothpicks, pick each individual P1 clone from an agar plate directly into the medium in a well. Leave occasional blank wells to act as controls for medium sterility and cross-contamination, and to create an irregular pattern on the filters. Cover with a sterile multiwell plate cover. Incubate overnight at 37°C.

(iii) Create replica libraries using manual replicating tools or a robot to transfer clones to new multiwell plates containing the same medium. We replicate small libraries manually as 25 plates can be replicated in half an hour. The metal 96-pin replicating tool (Sigma) is sterilized by immersion of the pins in 100% ethanol, then flamed with a bunsen burner.

Plastic 96- (or 384-) pin tools (Genetix) are sterilized by immersion of the pins in 100% ethanol. Excess ethanol is shaken off manually and the small amount remaining does not adversely affect the bacterial stocks. Robots replicate at approximately the same rate: if there are 100 or more plates they require less operator time. Incubate overnight at 37°C.

(iv) Seal the edges of the plates with lids using plastic film (e.g., Saranwrap) and store all plates in racks or on a level surface at −70°C.

Filters (20 × 20 cm) representing four 96-well plates can be prepared by hand using the 96-pin replicating tool. Generation of colony filters containing a higher density of clones requires use of robotic devices. High-density filters of the trypanosome library were produced using a Beckman robot. This allowed the transfer of P1 clones from 96-well plates (1988 separate clones) onto one 12 × 8-cm filter. The colonies were allowed to grow at 37°C until they were ~0.75 mm in diameter, but not touching their nearest neighbors. We incubate P1 filters for 16 hr. Some of the filters were treated as described above except that the bacterial debris was not wiped away, and others were treated by the method of Olsen *et al.* (1993) (see Section IV).

3. Hybridization of Gel-Purified DNA to Cosmid and P1 Filters

The radioactive probe and hybridization solutions may be prepared by one of several standard procedures (Sambrook *et al.*, 1989). We use a random priming reaction to incorporate radioactive dCTP into the chromosomal DNA.

Protocol 4: Radioactive Labeling by Random Priming

(i) Add 0.25 OD units random 9mer oligonucleotide primers to DNA aliquot (for amount see Section III.B.2) and make up to 25 µl with sterile H_2O.

(ii) Heat to 95–100°C for 5 min.

(iii) Centrifuge briefly and add 10 µl 5× reaction buffer, 5 µl $[\alpha-{}^{32}P]dCTP$ (3000 Ci/mmol) and 5 units Exo(-) Klenow DNA polymerase.

(iv) Incubate at 37°C for 10 min.

(v) Separate labeled DNA from unincorporated radionucleotides. We use NICK columns (see Section II).

(vi) Hybridization may be carried out in bottles or bags as preferred, with constant agitation of the hybridization fluid.

There are several parameters which require careful consideration.

(1) The time allowed for hybridization to take place. Since the probe is a very complex one, consisting of over 1 Mb unique sequence DNA,

the common hybridization time of 16 hr may be insufficient to allow complete hybridization of single-copy sequences. We have found that an incubation time of 30–40 hr to allow hybridization of the labeled 1.5-Mb probe to a 5× library of the 40-Mb haploid trypanosome genome results in consistent, detectable signals on replica filters.

(2) The final washing of the filters should be at high stringency. All filters shown here have been washed in 2× SSC for 15 min at room temperature. Preheat the following solutions to 65°C. Wash filters in 2× SSC, 0.1% SDS for 15 min at 65°C, 1× SSC, 0.1% SDS for 15 min at 65°C, with a final wash in 0.1× SSC, 0.1% SDS for 15 min at 65°C. Although the exposure time is lengthened (at least 2×) compared to that required after lower stringency washing (final wash at 1–2× SSC), there will be fewer false positives.

(3) The amount of dispersed repetitive sequence in the genome. Although the technique described here is recommended for use with genomes of low repeated DNA content, many genomes contain repeated sequences which will increase the number of false positives. In the case of trypanosomes, highly dispersed repeated sequences were thought to be few, with the exception of two elements, RIME and INGI, the frequency of which had been estimated at 200 copies each per genome, i.e., every 200 kb (Hasan *et al.*, 1984; Kimmel *et al.*, 1987). Also, variable surface glycoprotein genes are found in several sites in the genome: these exist as gene families with some cross-homology, and with associated repeated sequence (Beals and Boothroyd, 1992).

If the precise nature of the repeated sequence fraction is known, unlabeled denatured copies of these sequences may be added to the hybridization solution to preassociate with the repeated fraction of the probe and to compete for hybridization sites on the filters (see discussion in Section III.H). If the amount or nature of the repeated sequence in the genome is uncharacterized, sheared, unlabeled, denatured genomic DNA may be used in the prehybridization solution or in pretreatment of the probe in order to reduce hybridization of the probe to moderate or highly repetitive DNA on the filter by competition or preassociation (Sealey, *et al.*, 1985; Pinkel *et al.*, 1988). The filters are incubated with hybridization solution containing boiled, sheared genomic DNA for a defined period before introducing the radioactive probe. Any DNA sequence which is present in the genome in high copy number will locate and hybridize to its complement either in solution or on the filter in a shorter time than will unique sequence. Alternatively, denatured genomic DNA may be allowed to preassociate with the denatured probe in a volume of ca. 1 ml for a specific period before the probe is added to the hybridization solution, or in the hybridization solution before applying it to the filters (Sealey *et al.*, 1985; Pinkel *et al.*, 1988).

If the sequence complexity and the nature of the repeated complement are unknown, it is difficult to calculate the amount of genomic DNA which should be added and pilot experiments may be necessary. Several published papers discuss the factors affecting DNA reassociation and methods for estimating the approximate amount of genomic DNA required to maximize the contrast ratio (hybridization to single copy sequences to hybridization to dispersed repeated sequence) (e.g., Britten *et al.*, 1974; Sealey *et al.*, 1985, Pinkel *et al.*, 1988). For reproducible rates of reassociation the parameters which must be controlled are the concentrations of salt and DNA, the temperature, and the fragment size. Such experiments have proven successful using mammalian DNA, where the level of repeated sequence is very high (e.g., 500,000 copies Alu/human genome, or 5%). Sealey *et al.* reported success in preassociation of human probes using Cot (the product of DNA concentration and time) values of 10 to 100, in this case 10 mg ml^{-1} preassociation for 1 or 10 min, but loss of signal at Cot 1000. Pinkel *et al.* describe successful competitive inhibition of hybridization to human repeat sequence and give a semiquantitative discussion of the parameters they used to calculate the level required. The following calculation is performed as described in that paper.

The concentration of dispersed sequences is increased by a larger factor than that of chromosome-specific sequences by the addition of genomic DNA to the hybridization fluid. However, use of too much will decrease the intensity of specific hybridization to an unacceptable level. They showed that most benefit was observed at levels of $Q < 5$, where Q is the ratio of unlabeled to labeled copies of chromosome-specific sequences. Where m_p is the mass of probe and m_b that of genomic DNA, and f_i is the fraction of the genomic DNA represented by the ith chromosome, the ratio of unlabeled to (uniformly) labeled copies of the chromosome-specific sequences (whether single or multicopy) is $m_b/m_p = Q/f_i \gg Q$. Adding genomic DNA increases the concentration of specific sequence by $1 + Q$, and each uniformly dispersed sequence by $1 + Q/f_i$. If we perform this calculation using 85 ng of the 1.5-Mb chromosome and a genome size of 35 Mb (the fraction of the genome represented in the large-insert library), we obtain $Q = 2.52$ with 0.5 μg ml^{-1} (5 μg total) genomic DNA.

Pilot experiments were carried out with the trypanosome cosmid filters using denatured genomic DNA in the prehybridization solution. Since little was known about the distribution of repeated sequence within the *T.b.brucei* genome at that time, several concentrations of unlabeled genomic DNA were used and the results compared. Replica filters were

Table 9.1A[a]

Concentration of genomic DNA (μg/ml)	No. clones retrieved				No. expected
	Strong	Intermed	Weak	Total	
0.0	33	71	126	230	214
0.5	33	67	138	238	214
1.5	34	59	116	209	214
5.0	21	22	145	188	214

[a]The number of clones retrieved after prehybridization with genomic trypanosome DNA and probing with the MC7 chromosomal probe. The total given is the number of colonies which showed hybridization on replica filters hybridized in one tube.

Table 9.1B[a]

Concentration of genomic DNA (μg/ml)	No. clones retrieved Total	No. expected
0.0 (replica)	224	214
0.5 (filters)	228	214
0.5 (replica)	232	214
5.0 (filters)	177	214

[a]The number of clones retrieved from similar probings in which replica filters were subjected to different prehybridization conditions (see also Fig. 8.3). The number of clones expected is based on estimates of genome size and average cloned insert size (see text).

treated in one of two ways: (1) two filters were hybridized under the same conditions in order to check the reproducibility of results (Table 9.1A), or (2) each replica filter was prehybridized with a different amount of genomic DNA (sheared by passing through a 27G needle, denatured by heating to 95°C for 5 min), then incubated separately with an equal portion of the same probe separation (e.g., two probe preparations were mixed together, then split between two filters) (Table 9.1B). This provided information as to which type of signal was lost as the amount of genomic DNA in the prehybridization solution was increased. The experimental methods are summarized in Fig. 9.2.

When filters with 5000 immobilized cosmid clones were prehybridized with sheared herring sperm DNA but no trypanosome DNA prior to adding the radioactive chromosomal probe, there were 230 positive signals

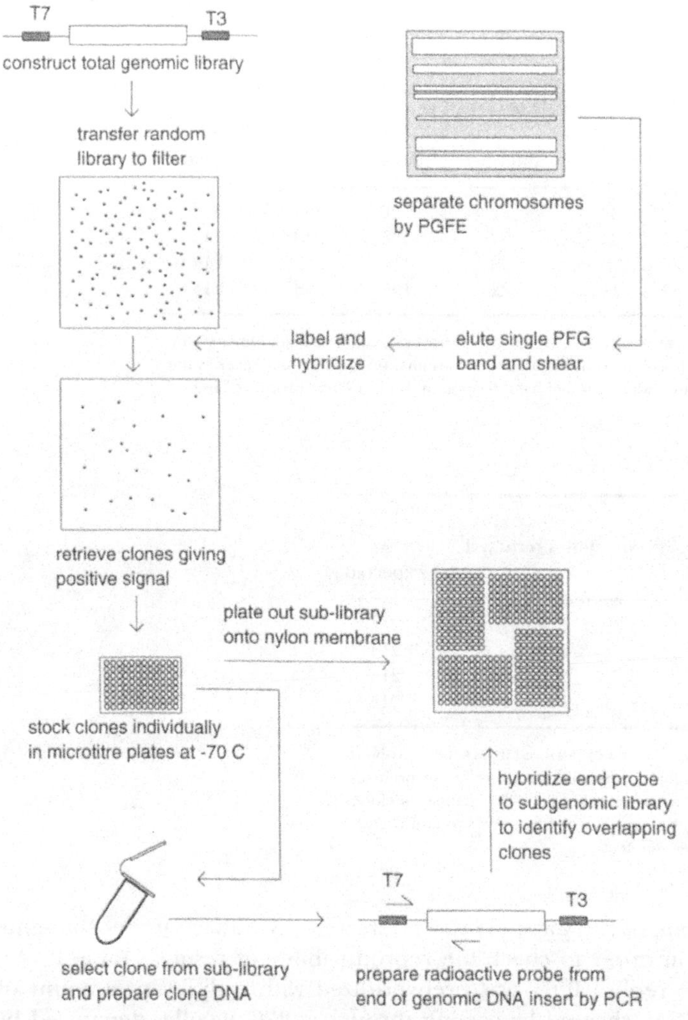

Figure 9.2 A method for the selection of a subset of cosmid clones, most of which are derived from one pair of largely homologous chromosomes, is summarized. In this case, the chromosome-specific library consists of frozen cultures in four microtitre plates.

which showed a considerable range of intensity. These signals were classed as strong, intermediate, or weak, in order to collate information as to which type of signal was affected by increasing the concentration of genomic DNA in the prehybridization. The weakest signals are just visible on the autoradiograph but are rendered less so by photographic reproduction. In this case, of the 230 positive signals, 33 were classed as very strong and 71 as intermediate (Table 9.1A). This result was reproducible, although the relative intensity of the signals may alter as this depends on the efficiency of cosmid DNA binding and of the radioactive labeling reaction.

The average cosmid insert length in this library was estimated to be 35 kb. If the chromosome is approximately 1.5 Mb in length, the expected number of positive colonies from a 5× haploid genome equivalent is 214. The number obtained is only 3.2 clones per genome equivalent more than the expected number. There is no evidence from this experiment to suggest that the probe selects a prohibitive number of clones which do not derive from that pulsed-field band.

The chromosomal probe was then hybridized to filters of 5000 clones after exactly 1 hr of prehybridization with 0.5 μg ml^{-1} ($Q = 2.53$) sheared total genomic MC7 DNA at 65°C. There were 238 positive signals of which 33 were classed as very strong and 67 were of intermediate strength (Table 9.1A). The prehybridization at this level has not reduced the number of hybridizations. After prehybridization of filters with 1.5 μg ml^{-1} ($Q = 7.5$) trypanosome DNA, there were 209 positive colonies with 34 very strong signals and 59 intermediate signals; again, this is an insignificant difference from that expected. The amount of trypanosome genomic DNA in the prehybridization solution was then increased substantially to 5 μg ml^{-1} ($Q = 25$). There was clearly some loss of signal: 188 positive colonies, including 21 very strong signals (Table 9.1).

Figure 9.3 shows the result of probing two replica filters after 1 hr prehybridization with 0.5 μg ml^{-1} (Filter A) and 5.0 μg ml^{-1} (Filter B) trypanosome genomic DNA. Comparison of these filters reveals that 12 strong signals present on Filter A are reduced in intensity on Filter B and there are only 19 intermediate strength signals on Filter B of which most are reduced in comparison to Filter A. Although the number of strong and intermediate signals has decreased and the number of weak signals has increased (these are not clearly reproduced in Fig. 9.3), comparing replica filters shows clearly that most of the signals lost at this high level of genomic DNA concentration were the weakest of those seen after prehybridization with 0.5 μg ml^{-1} genomic DNA. It is probable that the concentration of genomic DNA is sufficiently high to mask bone fide chromosome-specific clones containing single-copy DNA. Those clones which no longer gave an autoradiographic signal after prehybridization with 5 μg ml^{-1} genomic DNA but which gave a positive signal under other hybridization conditions were therefore included in the subgenomic library. Two of these were later tested and found to derive from that used to prepare the probe (see Section III.H).

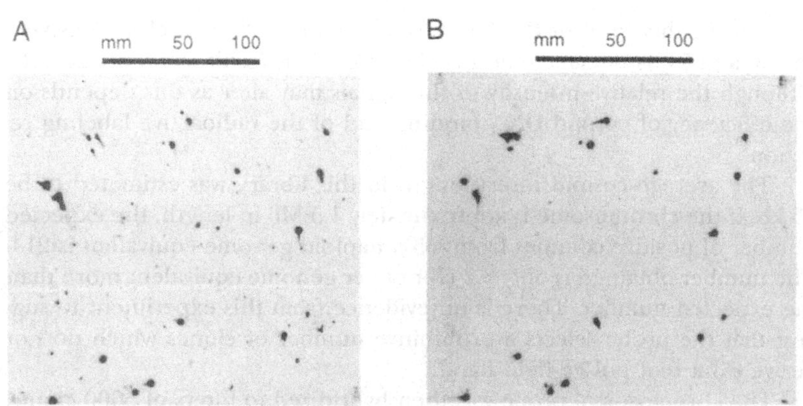

Figure 9.3 Two replica filters were prepared from a plate of approximately 5000 cosmid clones. Filter A was prehybridized for 1 hr with 0.5 μg ml^{-1} sheared MC7 genomic DNA before addition of the radioactive MC7 chromosomal probe; filter B was prehybridized with 10× the concentration of MC7 genomic DNA compared to filter A. The number of clones which gave a positive hybridization signal on filter B is considerably reduced compared to that on filter A (see Table 1).

The average number of autoradiographic signals obtained on hybridization with the 1.5-Mb probe is insignificantly different from that expected (see above). This result supports the observation, based on ethidium bromide fluorescence, that the selected pulsed-field gel band probably consists of a single chromosome and suggested a paucity of repeated sequence. Since these first experiments appeared so successful, pilot experiments involving the prehybridization of the probe with denatured genomic DNA prior to hybridization to the filter-bound DNA have not been carried out yet in our laboratory.

Subsequently, replica filters of the P1 library (1988 clones each) were probed with DNA prepared from a well-separated 1.2-Mb chromosome from trypanosome isolate TREU927/4 (an apparent homolog of the 1.5-Mb of MC7; PFGE separation not shown) after prehybridization with 1 μg ml^{-1} trypanosome genomic DNA ($Q = 4$). Figure 9.4 shows the result of hybridization to one filter: a total of 164 positive colonies were finally selected by comparison of this autoradiograph to several replica probings. The average insert size of this library is 65 kb, and therefore a 3.7× haploid library should yield 68 clones from the 1.2-Mb chromosome. The diffuseness of the fainter signals is more of a problem with high-density filters and rescreening is essential (see Section D.1). Many of the 164 clones were ruled out by a rescreen.

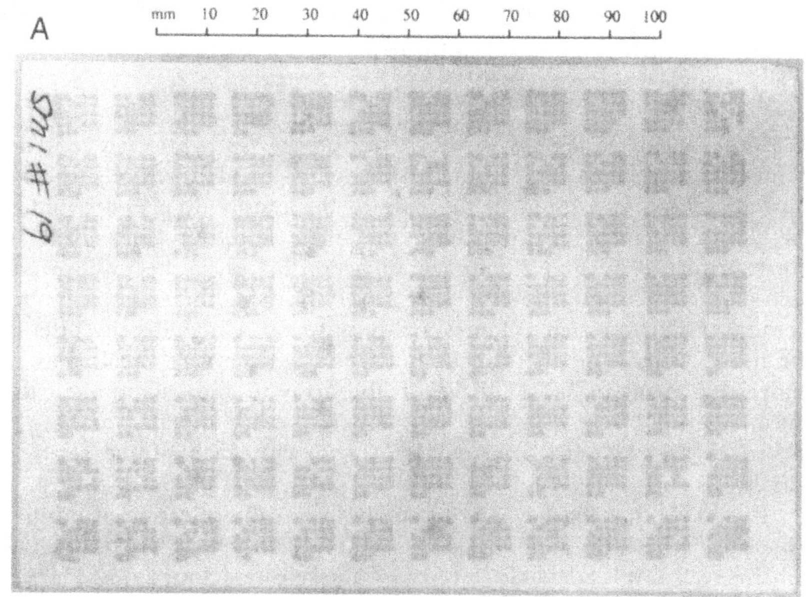

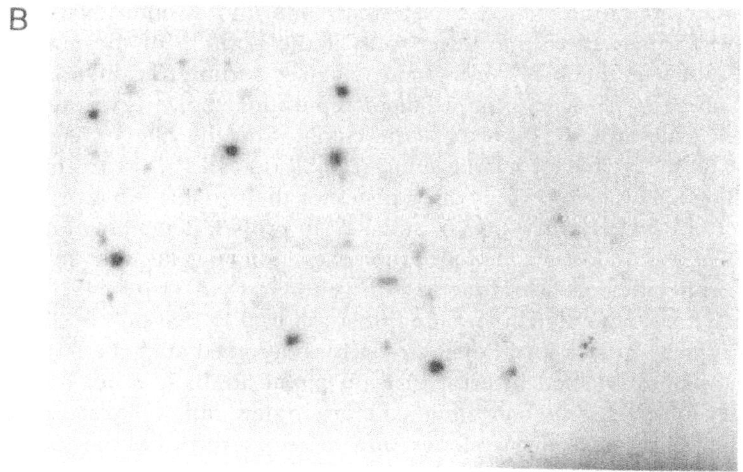

Figure 9.4 1988 clones of trypanosome DNA in bacteriophage P1 were robotically arrayed on a nylon filter. (A) The nylon filter with bacterial colonies after 16 hr growth and treatment of the filter to immobilize the DNA on the filter. The colonies were not wiped away. (B) An autoradiogram produced after prehybridization of the filter with 1 μg ml^{-1} genomic DNA followed by hybridization of a radioactive chromosomal probe to P1 clone DNA on the filter. This produced a range of signal strengths as seen on probing the cosmid library (compare Fig. 9.3).

D. Creation of Subgenomic Libraries

1. Retrieval of Chromosome-Specific Clones

Retrieval of the clones to which the probe has hybridized now depends on the exactitude of your markings, upon which you rely to align
autoradiograph and filter. In the case of the randomly plated cosmids,
the clones are retrieved from the master plate stored at 4 or −70°C. In
the case of the arrayed P1 library, the coordinates of the positive colonies are noted and the clones retrieved from a library copy in the
freezer. Deciding which colonies are positive on a high-density filter is
the more exacting of these techniques, and we use very small spots of
fluorescent paint placed in spaces in the array to allow us to align the
autoradiograph to a template. It may also be useful later to have a
record of the strength of signal associated with each of the selected
clones.

It is likely that there will be uncertainties in the selection of positive
clones using both the random plating method and the high-density array.
For this reason it is advisable to rescreen the selected clones at least once.
Prepare multiwell plates containing 200 μl L-broth, 10% glycerol, and the
appropriate antibiotic. Pick each selected clone into an individual well,
plus several alternatives into other wells if the position of the positive
colony is unclear. Incubate overnight at 37°C (see Section III.C.2), transfer
onto a filter on LB-agar using a 96-pin replicating tool, then store the
multiwell plate at −70°C. Incubate the colonies on the filters overnight
at 37°C, then prepare for hybridization (see Section III.C.1 for treatment
of the filter). This involves only three plates for the cosmids and two plates
for the P1s selected using the 1.5- and 1.2-Mb probes, depending on the
number of extra clones included. Probe these filters again with the gel-
eluted probe after prehybridization with genomic DNA (1 μg ml^{-1}) and
wash as before, always including the final wash in 0.1× SSC, 0.1% SDS at
65°C. Positive signals vary in their strength as observed after the original
probing. Those selected in error give no signal at all. It is not usually
necessary to recreate the subgenomic library in new multiwell plates without the extra nonhybridizing clones, only to keep a record of the screening results.

2. Storing the Chromosome-Specific Clones

The clones may be stored in the microtitre plates in LB medium with
10% glycerol and the appropriate antibiotic at −70°C, as described for
the P1 library in Section III.C.2. Replica libraries should be created. Many
copies of the subgenomic library on filters may be created for hybridization experiments using a manual replicating tool and clones subsequently

retrieved as required for DNA extraction from the multiwell plates in the freezer.

E. Confirmation of Chromosome Specificity

Rescreening of colony lifts of selected clones as described in the previous section confirms the hybridization of the chromosomal probe to that clone. This cannot provide definitive confirmation of the chromosome source of the clones. This must involve hybridization again, and the most convenient method to some extent depends upon the experiments to follow.

1. Hybridization to Pulsed-Field Gels

Given the low repetitive content of many small genomes, direct hybridization of an entire clone onto a PFGE separation of the chromosomes may show hybridization to one homologous pair only. To obtain sufficient hybridization signal it is usually necessary to isolate the insert from the vector by digesting the clone with *Not*I (Supercos) or *Not*I and *Sfi*I (p*Ad*10*sac*BII P1 vector), separating them by agarose gel electrophoresis and eluting the insert (or part of it, if there are internal restriction sites) by, for example, agarase digestion. The inclusion of competitive genomic DNA in the prehybridization fluid (1 μg ml^{-1}) will reduce the level of cross-hybridization to repeated sequences on other chromosomes. However, if there are other sites of hybridization, this gives no firm indication of the actual source of the clone. It is more informative to prepare probes from each end of the cosmid or P1 insert and then to hybridize these to separate Southern blots of PFGE separations. If both ends of such large inserts hybridize exclusively to the homologous pair, this is good evidence that the cloned insert originates from one of the pair. If one end-probe hybridizes to another chromosome and not to the chromosome pair of interest, the clone either originates from a different chromosome with some common sequence or is a chimeric clone. If an end-probe hybridizes to the chromosomes of interest and to another pair, its source will have to be investigated further. End-probes may be used in contig construction (see Section III.F), in which case each end-probe may be verified concurrently by PFGE probing. There are several methods for the preparation of end-probes, including inverse PCR, end-rescue (Chapter 7), linear amplification (Baxendale *et al.*, 1993), RNA transcription from T3, T7, and SP6 promoters (e.g., the Stratagene RNA transcription kit, although the SP6 promoter in the P1 vector used here is suboptimal), and a random primer technique (Wesley *et al.*, 1994). Here we describe the rapid preparation of end-probes by the polymerase chain reaction (PCR) from either end of the cosmid or P1 vectors, as devised by Wesley *et al.* The primer sequences are given in Section II.

Protocol 5: Preparation of End-Probes

(i) Extract DNA from a culture of the P1 or cosmid clone (10 ml of P1 culture or 1.5 ml of cosmid culture; see Chapter 7). (Once the reaction is working routinely, it is worth trying to use cells directly from the 96-well glycerol stocks: add 1 μl cells in broth to 11 μl H_2O, heat to 95°C immediately, and proceed as described in step (ii). This removes a time-consuming step in the procedure.)

(ii) Add 1 μl of P1 or cosmid DNA (20–50 ng) to 11 μl distilled H_2O, heat to 95°C for 5 min, centrifuge briefly, and chill on ice. Prepare two tubes for each end-probe preparation (one will act as a control). If you wish to prepare end-probes from both ends of a clone, prepare three tubes.

(iii) To each tube add 2.5 μl 10× PCR buffer, 4 μl 1.25 mM dNTPs, 2 units Taq polymerase, and 37.5 ng NS-2 primer to a total of 20 μl. Cover with mineral oil and put through one cycle of 95°C for 30 sec, 30°C for 4 min and 75°C for 4 min, incorporating a 4-min ramp between the last two temperatures.

(iv) Hold the temperature at 75°C. To two tubes add 50 ng of one specific vector primer (e.g., T3 to one and T7 to the other) and 25 ng NS-2 primer in a volume of 5 μl through the mineral oil. To the other (control) tube add 25 ng of NS-2 only in 5 μl. Amplify the DNA through 35 cycles of 95°C for 30 sec, 52°C for 1 min, and 75°C for 1 min.

(v) Electrophorese 5–7 μl of each PCR reaction through a 50-ml 1.2% NA agarose gel in 0.5× TBE. If there is no visible band in the control tube, it is very probable that a band in the tube containing NS-2 and vector primers is sequence derived from the end of the trypanosome DNA insert. Occasionally there are two bands, but in all cases we have tested so far the larger fragment contains within it the smaller fragment and results from a second random priming site further from the vector. If there is a complex pattern of bands, the required band may be excised from the gel and a small amount used in a reamplification reaction using the same primer pair (use an agarose tested for *in agaro* activity, e.g., Seakem, FMC). Alternatively, reamplify an aliquot (1 μl) from the first reaction using NS-2 plus a nested vector primer, i.e., closer to the cloning site than the first. Using NS-2 with the trypanosome DNA genomic clones, the end-probes are typically 0.4–2.5 kb in length.

(vi) Prepare radioactive probes using 25–50 ng of the PCR product directly in a random priming reaction and separate from unincorporated nucleotides as described in Section III.C.3.

Figure 9.5 shows the result of hybridizing an end-probe from a cosmid to a PFGE separation of the chromosomes.

2. Restriction Digests of Cosmid and P1 Clones

Separating restriction-digested clones on an agarose gel, Southern blotting and probing with the chromosomal probe provides more information about the cloned inserts than the use of end-probes, such as an estimate of insert size, the distribution of restriction sites, the number of hybridizing fragments, and the strength of hybridization to different fragments. Using a large insert library to cover quite small chromosomes (~1 Mb), this does not involve a prohibitive number of minipreps (70 P1 clones for the 1.2-Mb chromosome for 3–4× coverage, plus some false positives). If using cosmids or a considerably larger chromosome, the number of digests required should be considered. Cosmids have the advantage that DNA may be prepared using more rapid techniques, e.g., Wizard minipreps (Promega, Madison, WI), whereas these do not work well with P1s. If you plan to extract DNA from the clones for other rea-

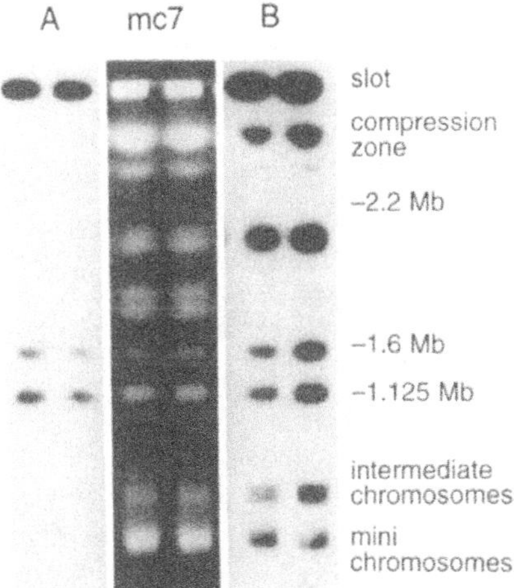

Figure 9.5 The central panel shows ethidium-stained chromosomes of trypanosome stock MC7 separated by PFGE (compare Fig. 9.1). This DNA was transferred to membrane and strips were probed with radioactive DNA probes: (A) The result of hybridizing a probe prepared from the end of a cosmid clone from the chromosome-specific library to the MC7 chromosome separation. The probe has hybridized to the chromosome used to select the subgenomic library and to its presumed homolog. (B) The result of probing the MC7 chromosome separation with the chromosomal probe prepared from DNA eluted from a pulsed-field gel. The probe has hybridized to the mini-, intermediate, and megabase chromosomes; however, hybridization to three of the larger chromosomes is very weak.

sons, for example to build contigs using restriction digest patterns (see Section III.F), then this is only an additional small screening step.

The extraction of cosmid or bacteriophage P1 DNA and their digestion by restriction enzymes are described by Wenzer and Herrmann (Chapter 7). Southern blotting is a standard technique (Sambrook *et al.*, 1989) and hybridization is carried out exactly as previously described for probing the colony lifts (see Section III.C.3).

Figure 9.6 shows a set of P1 clones containing inserts of trypanosome DNA restricted with *Bam*HI and *Eco*RI in a double digest. This gel was Southern blotted and the filter probed with the radioactive 1.2-Mb chromosomal DNA probe (Fig. 9.6). The radioactive probe should hybridize to all the fragments of a clone which derives from the correct chromosome (although, depending on how good the probe is, some of the small

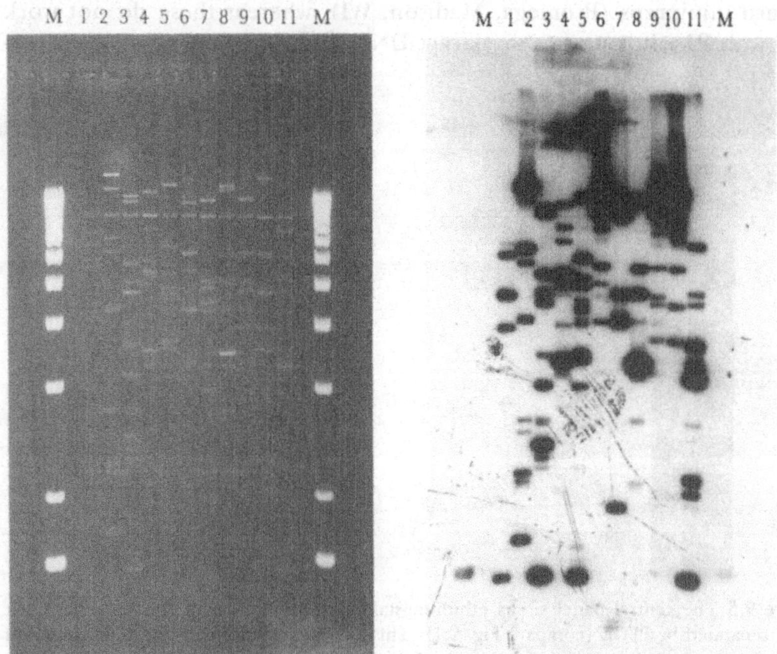

Figure 9.6 DNA was extracted from P1 colonies to which the chromosomal probe had hybridized, double-digested with *Bam*H1 and *Eco*RI, and fractionated on a 0.8% gel (left) in 0.5× TBE at 60 V for 20 hrs. The marker lane contains the 1-kb ladder (Gibco BRL); the sizes (in base pairs) in ascending order are: 1636, 2036, 3054, 4072, 5090, 6108, 7126, 8144, 9162, 10,180, 11,198, 12,216. This gel was Southern blotted and probed with the chromosomal probe (right).

single-copy fragments may give a very low signal). Note that the chromosomal probe has not hybridized to many bands in lane 1, suggesting that the original probing picked up a clone with some common sequence in a different environment on another chromosome. For example, we know that there is at least one gene array on this chromosome which also exists at another locus on a different homologous pair. This P1 clone was excluded from subsequent experiments on this chromosome. The strongly hybridizing bands indicate that there is repeated sequence, usually a tandemly repeated gene in the case of trypanosome DNA, and this correlates with some of the strongly hybridizing colonies in the first screen. Since the chromosomal probe has hybridized to all bands in these lanes, these clones are included in the chromosome-specific library. Clones selected in error from the master plates give no hybridization at all, although these should be screened out by the first rescreen of the selected clones on colony lifts. If the probe has hybridized to all except three or four small fragments in one lane, then this clone is included in the chromosome-specific library pending further investigation. Thus, this kind of screen can provide a lot of information and does not require a huge amount of work when studying small genomes.

F. Using the Subgenomic Library To Create Contigs and Select Markers

1. Creating Contigs

Methods for recognizing overlapping clones and constructing contigs are discussed in detail by Wenzel and Herrmann (Chapter 7) and by Hoheisel *et al.* (Chapter 11). If it is intended to reconstruct a chromosome from overlapping clones starting with known markers to that chromosome and selecting overlapping clones by walking, then there is only limited advantage in first selecting a whole chromosome-specific subset from the original library: one reason may be to have the convenience of an arrayed library in microtitre plates without having to array the entire genomic library. However, the main advantage lies in the fact that the creation of the subgenomic library allows the application of more global techniques of contig construction to a single chromosome or set of chromosomes: for example, the use of end-probes (Palazzolo *et al.*, 1991; Mizukami *et al.*, 1993; see Section III.E.1), restriction analysis (Coulson *et al.*, 1986; Olson *et al.*, 1986; Kohara *et al.*, 1987; see Section III.E.2), repeated sequence fingerprinting (Stallings *et al.*, 1990), or oligonucleotide fingerprinting (Craig *et al.*, 1990). Obviously, the size of genomic library probed with the chromosomal probe determines the size of the chromosome-specific library. It is necessary to take into account the expected number of clones and how they will be managed, the likely contribution of repeated sequences to false positives, and the coverage required for the contigging technique chosen. A 5–6× haploid coverage is a good compromise for chromosomes of this size range; gaps in a contig can be filled in by rescreening a larger genomic library with end-probes.

In the first example, end-probes from a clone which was positive on rescreening are used to probe the chromosome-specific library and the results are recorded. End-probes from a second clone to which the first did not hybridize are then used to probe the library and the results are recorded. The procedure is repeated until all colonies have given a positive signal and it is possible to work out which clones are overlapping. As discussed in Section III.E.1, if this method is chosen, the clones may be screened concurrently to check their chromosome of origin by including a strip of a Southern blot of a PFGE separation of the chromosomes in each of the probings (see Fig. 9.5). If contig construction is attempted by restriction analysis, then the clones may all be screened by hybridization before the analysis begins (see Section III.E.2) and those originating from other chromosomes omitted. This method can give the experimenter a lot of information about the clones, as discussed above, but for contig construction it requires almost 50% overlap between clones. Therefore a deep library of 5× haploid is usually required. However, the required overlap may be reduced by the use of repeated sequence fingerprinting using Alu probes, microsatellites (Stallings *et al.*, 1990) or, in the case of trypanosomes, RIME and INGI (S. E. Melville and C. S. Gerrard, unpublished).

2. Selecting Chromosome-Specific Markers from the Subgenomic Library

The selection of chromosome-specific markers is clearly simplified by the ability to elute the chromosome from the gel; there are many methods by which markers may be selected from clones of chromosomal DNA (for example, Green *et al.*, 1991; Bailey *et al.*, 1993; Rafalski *et al.*, Chapter 4). Most end-probes from cosmid or bacteriophage P1 clones of a genome of low repeated DNA content may conveniently serve as simple markers, the relative positions of which become known as contig construction proceeds. Direct sequencing from cosmids and P1 clones has been reported (Nurminsky and Hartl, 1993), allowing the creation of PCR-based sequence-tagged sites (STSs). Alternatively, the chromosomal DNA eluted from the PFG may be digested to completion with a restriction enzyme and cloned into a plasmid vector to create a chromosome-specific set of probes which are easily converted to STSs.

G. Probing Pulsed-Field Gels with the Chromosomal Probe

In genomes with little dispersed repetitive DNA, the chromosomal probe may be used in its entirety onto pulsed-field gels in order to detect regions of homology. This is especially useful in highly plastic genomes, such as trypanosomes, in which homologous chromosomes often vary considerably in size. Figure 9.5 shows that, in addition to the two homologs of 1.5 and 1.1 Mb, the chromosomal probe hybridizes to another separated chromosome of 2 Mb and to the compression zone (Fig. 9.5B). If the blot is prehybridized with dispersed repeats which are not of direct interest (the 70-bp repeats, Campbell *et al.*, 1984, and RIME and INGI), hybridization to the 2-Mb chromosome and the intermediate and minichromosomes is virtually absent (data not shown). It is not yet known why the probe hybridizes to the compression zone, but no end-probes tested thus far have done so.

H. Summary of Results

In the cosmid experiments decribed here, a total of 238 clones were selected to form the primary chromosome-specific library (plus 34 extra clones, selected as alternatives where the position was uncertain). On rescreening, the 1.5-Mb probe hybridized to 231. Of these, 40 clones were selected randomly; end-probes were prepared and hybridized to Southern blots of chromosome separations. Twenty-nine clones were clearly derived from the desired chromosome pair, 1 clone derived from another pair, 5 gave multiple hybridization from 1 end-probe, and 5 gave multiple hybridization from both. One of the clones was used to initiate a chromosome walk extending over 10 clones and all pairs of end-probes hybridized to the desired chromosome pair. However, three end-probes hybridized to all chromosomes, indicating the presence of a dispersed repetitive sequence.

The decision to prehybridize the high-density filters of the P1 library with 1 μg ml^{-1} TREU 927/4 genomic DNA was based on the trials with the MC7 cosmid library. However, although we expected only 68 true positives, 164 hybridizing clones were identified after the first screen, of which 123 were positive on rescreening. After probing blots of digested P1 DNAs, 81 true positives were selected and a contig was created across the 1.2-Mb chromosome (with two gaps and excluding telomeric and 50- to 100-kb subtelomeric regions; S. E. Melville and C. S. Gerrard, unpublished results); 75% of the false positives arose through hybridization to RIME and INGI sequences and a third, as yet uncharacterized, repetitive sequence. There are at least 40 copies of RIME and/or INGI in 800 kb of this chromosome, approximately five times the original estimates of Hasan *et al.*, (1984) and Kimmel *et al.*, (1987). It seems likely at this point that there are more copies of these repetitive sequences in the 1.2-Mb than in the 1.5-Mb chromosome, for they occur at 50% of the ends of inserts of the selected P1s but only 10% of the ends of the cosmids (they contain many *Sau*3AI sites). The reason for the larger (though manageable) number of false positives is not certain at this point; it may have been an experimental error, but the resulting observation of RIME and INGI copy frequency proved useful: we now routinely prehybridize filters with PCR products containing the entire RIME and INGI sequences, and also a stretch of 70-bp repeats (200 ng ml^{-1} each) for 1 hr before adding genomic DNA (1 μg ml^{-1}) for half an hour, followed by the radioactive probe.

Of the strongly hybridizing clones in the cosmid and the P1 library hybridizations, half contained a large section of the α-β-tubulin array which consists of approximately 10 copies of a 3.6-kb repeat, and no RIME or INGI sequences. Tubulin is one of many tandemly repeated genes described in *T.b.brucei* (Clayton, 1988). The remaining strongly hybridizing clones all derive from one region of the chromosome and contain no RIME or INGI sequences. The reason for the strong hybridization is not yet known.

With these caveats, the procedure works equally with both cosmids and P1s. However, the P1 clones are better able to accommodate the large tandem gene arrays common on trypanosomes, give more consistent growth on high-density filters, and have shown little cloning bias across the 1.2-Mb chromosome. They remain for us the vector of choice.

IV. Notes and Pitfalls

A. Low Levels of Hybridization

Some probe preparations give very low levels of hybridization: it is usually best to repeat the preparation of PFGE-eluted chromosomal DNA.

It is quite probable that insufficient DNA was eluted from the gel. The elution process is somewhat unpredictable, whichever method is chosen. Repeat, check agarase digestion carefully, and resuspend in less volume.

If the gel-eluted DNA contains contaminants, it may not be labeled efficiently in the random priming reaction. Try extracting the DNA solution with phenol and chloroform and reprecipitating with ethanol. Resuspend in half the volume. Alternatively, incubate the gel slice in NDS with 50 μg ml^{-1} proteinase K for several hours, then dialyse in 50 vol TE with 40 μg ml^{-1} phenylmethylsulfonyl fluoride (PMSF) for 1 hr at 50°C, followed by several dialyses against TE alone at 50°C (Sambrook *et al.*, 1989) before eluting the DNA from the agarose. The latter method is preferred in our laboratory, although it is usually not necessary. Finally, consider a different make of agarose.

B. The Gel-Eluted DNA May Be of an Inappropriate Size

If the DNA remains in very large fragments, excess hybridization of the probe to the colony lifts occurs. We have found that the DNA is sheared sufficiently during the extraction procedure but we note here that some experimenters who have reported the elution of YAC DNA from gels for use as probes have treated the DNA with restriction enzymes to ensure the fragments are of a suitable size (Bancroft *et al.*, 1992), while others have not (Holland *et al.*, 1993). Using genomic DNA digested with *Bam*HI and *Eco*RI before hybridizing to the Southern blots of digested P1 clones should reduce the formation of partially paired duplexes.

C. Some Chromosomes Are Less Effective Than Others

The chromosome which was chosen here to illustrate the technique is among the smaller of the trypanosome housekeeping chromosomes, can be very well separated from other chromosomes, and was thought likely to be a single haploid species. We have successfully prepared a probe from a chromosome of greater than 3.5 Mb (S. E. Melville, unpublished results) for the selection of chromosome-specific clones. However, not all chromosomes can be separated so well: in these cases it may be necessary to account for contamination from neighboring chromosomes in the planning of the experiments. Also, several bands in complex karyotypes may contain more than one species of chromosome. Although we have successfully selected clones using a PFGE band of 1 Mb containing at least three different chromosomes, we cannot distinguish the three chromosome-specific subsets contained in the selection. Nevertheless the number of clones involved in subsequent experiments is reduced.

D. Gel-Eluted Probes May Be Amplified

If the organisms are difficult to obtain and the supply of DNA plugs for PFGE is thereby limited, it is possible to amplify the DNA by one of two related techniques: (1) attaching linkers to restriction-digested chromosomal DNA, then amplifying by the polymerase chain reaction (PCR) using primers complementary to the linkers (Lüdecke *et al.*, 1989; Saunders *et al.*, 1989; K-L. Wan, personal communication) or (2) degenerate oligonucleotide-primed PCR which randomly amplifies all DNA (Telenius *et al.*, 1992a,b; J. P. Warner, personal communication). Either of these techniques ensures a constant source of DNA, although the amplification of the chromosomal DNA may be biased and we prefer to label unamplified DNA whenever possible.

E. High-Density Filters Present Particular Problems

Probing a well-spaced, randomly plated library proved easier than probing a high-density filter, at least partly because the close spacing made it more difficult to distinguish positive signals from each other and to determine the correct clone (compare Figs. 9.3 and 9.4). The signals were also more diffuse. Although not insurmountable, if dealing with a small genome it may be worth producing a set of less densely gridded filters specifically for chromosomal probings (i.e., 21 separate filters for the P1 library), using the 96-pin manual replicating tool.

However, the number of false positives from the P1 library varied more between replica filters than between cosmid replicas. Many of the extra colonies gave no signal on secondary screening, where we had washed off bacterial debris, indicating that the positive signal was not due to the DNA content of the clone or the vector. The main difference between the two libraries was the amount of bacterial debris remaining after filter preparation, and the number of false positives was greatest on the P1 filters prepared by the method of Olsen *et al.* (1993) on which the bacterial residue was most evident. This has caused no problems whatsoever when using less complex probes and shorter hybridization times. Our results indicate that it is preferable to remove all bacterial debris before using the chromosomal probe, since the faintest of signals are possible positives.

Because methods to remove colonies during processing can lead to less sharp hybridization signals (presumably due to smearing), this is best achieved by prehybridizing the filters in hybridization solution with no addition of DNA at 65°C for 30–60 min, then discarding the solution. Repeat if necessary. Finally, rinse the filters with hybridization solution before setting up the prehybridization with genomic DNA as described.

F. Use Your Knowledge about the Repeated DNA Content of the Genome

It can be very useful if something is known regarding the frequency and distribution of the repeated DNA in the genome under study. For example, cloned or amplified copies of the repeated DNA sequences may be included in the prehybridization solution (see Sections III.C.3 and III.H). We now always include PCR-amplified copies of RIME and INGI in prehybridizations. In addition, it was known that the longer tracts of another dominant repeat sequence in the trypanosome genome was commonly found in the subtelomeric regions and that it did not contain sites for the enzyme *Sau*3AI. Therefore, the use of this enzyme largely precluded the cloning of these sequences in the library (Van der Ploeg *et al.*, 1982), thus reducing the problems caused by hybridization despite the presence of the sequence in the probe. The ends of the chromosome under study will have to be cloned by other methods.

Acknowledgments

This work was supported by the U.K. Overseas Development Administration (S.E.M.) and the Medical Research Council (S.E.M. and C.S.G.). We are indebted to Professor A. Tait of the Wellcome Unit of Molecular Parasitology and Dr. C. M. R. Turner of the Department of Zoology, University of Glasgow, for the cloned stocks of field isolates STIB 247-L and TREU 927 and the cloned hybrid stocks derived from a laboratory cross, and to Du Pont–Merck Pharmaceutical Company for providing facilities for the preparation of the P1 library. We also thank A. Tait for his advice on karyotype analysis by pulsed field gel electrophoresis. The high density filters were created at the Sanger Centre, Hinxton, nr. Cambridge with the invaluable help of Andrew Dunham. The mcl7 gel-eluted chromosomal probe was prepared by Justin Sweetman and the photographs by Roger Williams and Philip Starling.

References

Bailey, D. M. D., Carter, N. P., de Vos, D., Leversha, M. A., Perryman, M. T., and Ferguson-Smith, M. A. (1993). Coincidence painting: A rapid method for cloning region-specific DNA sequences. *Nucleic Acids Res.* **21**(22), 5117–5123.

Bancroft, I., Westphal, L., Schmidt, R., and Dean, C. (1992). PFGE-resolved RFLP analysis and long-range restriction mapping of the DNA of *Arabidopsis thaliana* using whole YAC clones as probes. *Nucleic Acids Res.* **20**(23), 6201–6207.

Baxendale, S., MacDonald, M. E., Mott, R., Francis, F., Lin, C., Kirby, S. F., James, M., Zehetner, G., Hummerich, H., Valdes, J., Collins, F., Deaven, L., Gusella, J., Lehrach, H., and Bates, G. (1993). A cosmid contig and high resolution restriction map of the 2 megabase region containing the Huntington's disease gene. *Nat. Genet.* **4**, 181–186.

Beals, T. P., and Boothroyd, J. C. (1992). Sequence divergence among members of a trypanosome variant surface glycoprotein gene family. *J. Mol. Biol.* **225**, 973–983.

Borst, P., Van der Ploeg, L. H. T., Van Hoek, J. F., Tas, J., and James, J. (1982). On the DNA content and ploidy of Trypanosomes. *Mol. Biochem. Parasitol.* **6**, 13–23.

Britten, R. J., Graham, D. E., and Neufeld, B. R. (1974). Analysis of repeating DNA sequences by reassociation. *In* "Methods in Enzymology" (L. Grossman and K. Moldave, eds.), Vol. 29, pp. 363–406. Academic Press, New York.

Brun, R., and Schonenberger, M. (1979). Cultivation and *in vitro* cloning of procyclic culture forms of *Trypanosoma brucei* in semi-defined medium. *Acta Trop.* **36**, 289–292.

Budarf, M. L., McDermid, H. E., Sellinger, B., and Emanuel, B. S. (1991). Isolation and regional localization of 35 unique anonymous DNA markers for human chromosome 22. *Genomics* **10**, 996–1002.

Burke, D. T., Carle, G. F., and Olsen, M. V. (1987). Cloning of large segments of exogonous DNA into yeast by means of artificial chromosome vectors. *Science* **236**, 806–812.

Campbell, D. A., van Bree, M. P., and Boothroyd, J. C. (1984) The apparent limit of transposition and upstream barren region of a trypanosome VSG gene, Tandem 76- base pair repeats flanking $(TAA)_{90}$. *Nucleic Acids Res.* **12**, 2759–2774.

Choo, K. H., Earle, E., Vissel, B., and Filby, R. G. (1990). Identification of two distinct subfamilies of alpha satellite DNA that are highly specific for human chromosome 15. *Genomics* **7**, 143–151.

Clayton, C. E. (1988). The molecular biology of the Kinetoplastidae. *Genet. Eng.* **7**, 1–56.

Coulson, A., Sulston, J., Brenner, S., and Karn, J. (1986). Toward a physical map of the genome of the nematode *Caenorhabditis elegans. Proc. Nat. Acad. Sci. U.S.A.* **83**, 7826–7830.

Craig, A. G., Nižetić, D., Hoheisel, J. D., Zehetner, G., and Lehrach, H. (1990). Ordering of cosmid clones covering the *Herpes* simplex virus type I (HSV-I) genome: A test case for fingerprinting by hybridisation. *Nucleic Acids Res.* **18**(9), 2653–2660.

Evans, G. A., Lewis, K., and Rothenburg, B. E. (1989). High efficiency vectors for cosmid microcloning and genomic analysis. *Gene* **79**, 9–20.

Feinberg, A. P., and Vogelstein, B. (1983). A technique for radiolabelling DNA restriction endonuclease fragments to high specific activity. *Anal. Biochem.* **132**, 6–13.

Fuscoe, J. C., McNinch, J. S., Collins, C. C., and Van Dilla, M. A. (1989). Human chromosome-specific DNA libraries: Construction and purity analysis. *Cytogenet. Cell Genet.* **50**, 211–215.

Gottesdiener, K., Garcia-Anoveros, J., Lee, M.G.-S., and Van der Ploeg, L. H. T. (1990). Chromosome organisation of the protozoan *Trypanosoma brucei. Mol. Cell. Biol.* **10**(11), 6079–6083.

Green, E. D., Mohr, R. M., Idol, J. R., Jones, M., Buckingham, J. M., Deaven, L. L., Moyzis, R. K., and Olson, M. V. (1991). Systematic generation of sequence-tagged sites for physical mapping of human chromosomes: Application to the mapping of human chromosome 7 using yeast artificial chromosomes. *Genomics* **11**, 548–564.

Hanahan, D., and Meselson, M. (1980). Plasmid screening at high colony density. *Gene* **10**, 63–66.

Hasan, G., Turner, M. J., and Cordingley, J. S. (1984). Complete nucleotide sequence of an unusual mobile element from *Trypanosoma brucei. Cell (Cambridge, Mass.)* 37, 333–341.

Holland, J., Coffey, A. J., Giannelli, F., and Bentley, D. R. (1993). Vertical integration of cosmid and YAC resources for interval mapping on the X-chromosome. *Genomics* **15**, 297–304.

Kimmel, B. E., ole-MoiYoi, O. K., and Young, J. R. (1987). Ingi, a 5.2 Kb dispersed sequence element from *Trypanosoma brucei* that carries half of a smaller mobile element at either end and has homology with mammalian LINEs. *Mol. Cell. Biol.* **7**, 1465–1475.

Kohara, Y., Akiyama, K., and Isono, K. (1987). The physical map of the whole *E.coli* genome. *Cell (Cambridge, Mass.)* **50**, 495–508.

Lüdecke, H. J., Senger, G., Claussen, U., and Horsthemke, B. (1989). Cloning defined regions of the human genome by microdissection of banded chromosomes and enzymatic amplification. *Nature (London)* **338**, 348–350.

Milan, D., Yerle, M., Schmitz, A., Chaput, B., Vaiman, M., Frelat, G., and Gellin, J. (1993). A PCR-based method to amplify DNA with random primers: Determining the chromosomal content of porcine flow-karyotype peaks by chromosome painting. *Cytogenet. Cell Genet.* **62**, 139–141.

Miller, J. R., Dixon, S. C., Miller, N. G. A., Tucker, E. M., Hindkjær, J., and Thomsen, P. D. (1992). A chromosome 1-specific DNA library from the domestic pig *(Sus scrofa domestica)*. *Cytogenet. Cell Genet.* **61**, 128–131.

Mizukami, T., Chang, W. I., Garkavtsev, I., Kaplan, N., Lombardi, D., Matsumoto, T., Niwa, O., Kounosu, A., Yanagida, M., Marr, T. G., and Beach, D. (1993). A 13 kb resolution cosmid map of the 14 Mb fission yeast genome by non-random sequence-tagged site mapping. *Cell (Cambridge, Mass.)* **73**, 121–132.

Nižetić, D., Zehetner, G., Monaco, A. P., Gellen, L., Young, B. D., and Lehrach, H. (1991). Construction, arraying, and high-density screening of large insert libraries of human chromosomes X and 21: Their potential use as reference libraries. *Proc. Natl. Acad. Sci. U.S.A.* **88**, 3233-3237.

Nurminsky, D. I., and Hartl, D. L. (1993). Amplification of the ends of DNA fragments cloned in bacteriophage P1. *BioTechniques* **15**(2), 201–208.

Olsen, A. S., Combs, J., Garcia, E., Elliot, J., Amemiya, C., de Jong, P. J., and Threadgill, G. (1993). Automated production of high density cosmid and YAC colony filters using a robotic workstation. *BioTechniques* **14**, 116–123.

Olson, M. V., Dutchik, J. E., Graham, M. Y., Brodeur, G. M., Helms, C., Frank, M., MacCollin, M., Scheniman, R., and Frank, T. (1986). Random-clone strategy for genomic restriction mapping in yeast. *Proc. Natl. Acad. Sci. U.S.A.* **83**, 7826–7830.

Palazzolo, M. J., Sawyer, S. A., Martin, C. H., Smoller, D. A., and Hartl, D. L. (1991). Optimized strategies for sequence-tagged-site selection in genome mapping. *Proc. Natl. Acad. Sci. U.S.A.* **88**, 8034–8038.

Pierce, J. C., and Sternberg, N. (1993). Using the bacteriophage P1 system to clone high molecular weight genomic DNA. *In* "Methods in Enzymology" (R. Wu, ed.), Vol. 216, pp. 549–574. Academic Press, San Diego, CA.

Pierce, J. C., Sauer, B., and Sternberg, N. (1992). A positive selection vector for cloning high molecular weight DNA by the bacteriophage P1 system: Improved cloning efficiency. *Proc. Natl. Acad. Sci. U.S.A.* **89**, 2056–2060.

Pinkel, D., Landegent, J., Collins, C., Fuscoe, J., Segraves, R., Lucas, J., and Gray, J. (1988). Fluorescence *in situ* hybridization with human chromosome-specific libraries: Detection of trisomy 21 and translocations of chromosome 4. *Proc. Natl. Acad. Sci. U.S.A.* **85**, 9138–9142.

Saito, A., Abad, J. P., Wang, D., Ohki, M., Cantor, C. R., and Smith, C. L. (1991). Construction and characterization of a *Not1* linking library of human chromosome 21. *Genomics* **10**, 618–630.

Sambrook, J., Frisch, E. F., and Maniatis, T. (1989). "Molecular Cloning: A Laboratory Manual," 2nd ed. Cold Spring Harbor Lab. Press, Cold Spring Harbor, NY.

Saunders, R. D., Glover, D. M., Ashburner, M., Siden-Kiamos, I., Louis, C., Monstirioti, M., Savakis, C., and Kafatos, F. (1989). PCR amplification of DNA microdissected from a single polytene band: A comparison with conventional microcloning. *Nucleic Acids Res.* **17**, 9027–9037.

Sealey, P. G., Whittaker, P. A., and Souther, E. M. (1985). Removal of repeated sequences from hybridization probes. *Nucleic Acids Res.* **13**(6), 1905–1922.

Shepherd, N. S., Pfrogner, B. D., Coulby, J. N., Ackerman, S. L., Vaidyanathan, G., Sauer, R. H., Balkenhol, T. C., and Sternberg N. (1994). Preparation and screening of an arrayed human genomic library generated with the P1 cloning system. *Proc. Natl. Acad. Sci.* **91**, 2629–2633.

Stallings, R. L., Torney, D. C., Hildebrand, C. E., Longmire, J. L., Deaven, L. L., Jett, J. H., Doggett, N. A., and Moyzis, R. K. (1990). Physical mapping of human chromosomes by repetitive sequence fingerprinting. *Proc. Natl. Acad. Sci. U.S.A.* **87**, 6218–6222.

Sternberg, J., Turner, C. M. R., Wells, J. M., Ranford-Cartwright, L. C., Le Page, R. W. F., and Tait, A. (1989). Gene exchange in African trypanosomes: Frequency and allelic segregation. *Mol. Biochem. Parasitol.* **34**, 269–280.

Sternberg, N. (1990). Bacteriophage P1 cloning system for the isolation, amplification, and recovery of DNA fragments as large as 100 kilobase pairs. *Proc. Natl. Acad. Sci. U.S.A.* **89**, 103–107.

Telenius, H., Pelmear, A. H., Tunnacliffe, A., Carter, N. P., Behmel, A., Ferguson-Smith, M. A., Nordenskjöld, M., Pfragner, R., and Ponder, B. A. J. (1992a). Cytogenetic analysis by chromosome painting using DOP-PCR amplified flow-sorted chromosomes. *Genes, Chromosomes, Cancer* **4**, 257–263.

Telenius, H., Carter, N. P., Bebb, C. E., Nordenskjöld, M., Ponder, B. A. J., and Tunnacliffe, A. (1992b). Degenerate oligonucleotide-primed PCR: General amplification of target DNA by a single degenerate primer. *Genomics* **13**, 718–725.

Turner, C. M. R., Sternberg, J., Buchanan, N., Smith, E., Hide, G., and Tait, A. (1990). Evidence that the mechanism of gene exchange in *Trypanosoma brucei* involves meiosis and syngamy. *Parasitology* **101**, 377–386.

Van der Ploeg, L. H. T., Valerio, D., De Lange, T. , Bernards, A., Borst, P., and Grosveld, F. G. (1982). An analysis of cosmid clones of nuclear DNA from *Trypanosoma brucei* shows that the genes for variant surface glycoproteins are clustered in the genome. *Nucleic Acids Res.* **10**, 5905–5923.

Van der Ploeg, L. H. T., Cornelissen, A. W. C. A., Barry, J. D., and Borst, P. (1984). Chromosomes of Kinetoplastida. *EMBO J.* **3**(13), 3109–3115.

Van Dilla, M. A., and Deaven, L. (1990). Construction of gene libraries for each human chromosome. *Cytometry* **11**, 208–218.

Wesley, C. S., Myers, M. P., and Young, M. W. (1994). Rapid sequential walking from termini of cosmid, P1 and YAC inserts. *Nucleic Acids Res.* **22**(3), 538–539.

10

Analysis of the Dictyostelium discoideum Genome

Adam Kuspa and William F. Loomis

I. Introduction

A. Overview

The chromosomes of *Dictyostelium* are too small and too morphologically similar to allow the assignment of DNA markers to them by *in situ* hybridization techniques. At the same time, however, the 40-Mb genome is too large to allow direct restriction mapping of the chromosomes, as has been done with *Escherichia coli* and *Schizosaccharomyces pombe* (Smith *et al.*, 1987; Fan *et al.*, 1988). Thus, physical maps of the *Dictyostelium* genome have been built largely from the bottom up, clone by clone, marker by marker.

The *Dictyostelium* genome consists of six or seven chromosomes. The uncertainty in the chromosome number exists because of four sets of conflicting data. Seven kinetochores were clearly observed by Moens (1976) by electron microscopy of serial sections through the nucleus. In addition, while six linkage groups have been defined for some time (Newell, 1978; Loomis, 1987), a seventh linkage group has recently been marked by a single genetic locus (Darcy *et al.*, 1993). Cox and co-workers (1990) have obtained evidence for six chromosomes by separating intact chromosomes of the strain AX3 using pulsed-field gel electrophoresis. These workers identified five bands in the range of 5 to 10 Mb using genetically mapped gene-probes. Four of the bands could be correlated with a specific linkage group and one band was shown to consist of two equally sized chromosomes. Finally, we have nearly completed the physical maps of six chromosomes using a variety of methods and find no evidence for a seventh chromosome (A. Kuspa and W. F. Loomis, unpublished). These data indicate that there are six large chromosomes, and that the seventh chromosome, if it exists, must be small and relatively gene-poor.

By comparison to *S. pombe* chromosome size standards, Cox and co-workers (1990) estimated that the six chromosomes they observed for AX3 sum to 40 Mb, in agreement with previous estimates of genome size (reviewed by Kimmel and Firtel, 1982). In addition to the chromosomes, a linear 90-kb palindrome containing the rRNA genes is present in about 90 copies per nucleus (reviewed by Kimmel and Firtel, 1982).

B. Genetics and Physical Genome Analysis

Since high-frequency meiotic recombination has yet to be harnessed as a genetic tool in *Dictyostelium,* it has not been possible to use recombinational frequency between markers to generate high-resolution genetic maps. Thus, gene mapping is limited mainly to chromosome assignment using parasexual techniques (Newell, 1978; Loomis, 1987). Asexual haploid cells of *Dictyostelium* fuse at low frequency to generate stable diploid strains which can be induced to segregate haploid progeny. Random chromosomal assortment in the haploids provides a means of determining genetic linkage. In all, over 100 mutations have been assigned to one or another of the seven linkage groups (Newell *et al.,* 1993; Darcy *et al.,* 1993). At least 20 of these loci have been cloned to date. These clones provide the only link between the genetic map and the physical map. They have been used extensively to anchor physical maps to specific chromosomes (Kuspa *et al.,* 1992). A few genes in each linkage group have been ordered using the rare mitotic recombination events that occur in diploids generated by the fusion of marked haploid strains (Welker and Williams, 1982). However, too few of the loci ordered in this way have been cloned, so this information cannot be integrated with physical mapping data.

Parasexual analysis has also allowed the grouping of phenotypically similar mutations into complementation groups. For instance, Newell and Ross (1982) were able to place 32 "slugger" mutants into 10 complementation groups and estimate that no more than 12 such loci exist. Several estimates of this type have been made for different classes of mutations. They will become increasingly important as the genome is saturated by insertional mutagenesis (see below).

C. General Considerations

General aspects of the biology of *Dictyostelium* as well as methods for its experimental manipulation have been reviewed (Loomis, 1982; Spudich, 1987). The cultivation of *Dictyostelium* has been described recently by Sussman (1987), and a brief description of useful media is provided in the following section. Techniques for parasexual genetic analysis have also been reviewed extensively elsewhere (Newell, 1978; Loomis, 1987).

Most of the physical analyses have been carried out with derivatives of the axenic strain AX3 (Loomis, 1971), such as AX4 (Knecht *et al.*, 1986), HL328, and HL330 (Kuspa *et al.*, 1992). Differences in genome structure have been observed between AX3 (and its derivatives) and AX3's parent NC4 (Cox *et al.*, 1990; Kuspa *et al.*, 1992). The major difference is the presence in AX3 of an inverted duplication of the central region of chromosome 2, consisting of at least 0.5 Mb within each half of the duplication. AX2 (Watts and Ashworth, 1970), a different axenic derivative of NC4, does not have this duplication (Kuspa *et al.*, 1992). Thus, in order to make meaningful comparisons to previously established maps, the use of strains derived from AX3 is encouraged.

In this chapter we describe methods for studying the structure of the *Dictyostelium* genome, the distribution of gene families, and the function of genes. Most of the protocols and approaches used in analyzing the *Dictyostelium* genome were developed previously for use with other organisms. We will focus on those aspects of technique and analysis that are most specific to *Dictyostelium*. However, the general approaches presented should be transferable directly to the analysis of any organism with a relatively small haploid genome.

II. Materials

A. Growth Media for Dictyostelium

There are several simple media that will support growth of *Dictyostelium* for routine procedures. For review of the relevant conditions see Sussman (1987).

1. HL-5 Medium—For Routine Axenic Growth

	g/1000 ml
Oxoid bacteriological peptone (Unipath, England)	10
Yeast extract (Difco Laboratories, Detroit, MI)	5
Glucose	10
Na$_2$HPO$_4$	0.35
KH$_2$PO$_4$	0.35

Adjust pH to between 6.4 and 6.6 with H_3PO_4. Sterilize by autoclaving for 20 min and remove from the autoclave immediately to minimize caramelization.

2. SM Medium—For Growth in the Presence of Bacterial Lawns

g/1000 ml

Bacto-peptone (Difco Laboratories)	10
Yeast extract (Difco Laboratories)	1
Glucose	10
$MgSO_4$	1
K_2HPO_4	1.9
KH_2PO_4	0.6
Agar	20

Adjust pH to 6.0 to 6.4 with H_3PO_4. Sterilize by autoclaving. (SM liquid for growth of food bacteria is made with the same recipe, but without agar.)

3. FM Minimal Medium (Franke and Kessin, 1977)—For Selection of Uracil Prototrophs

FM can be conveniently made in 10-liter batches, which should supply four or five workers for 1–3 months. To make 10 liters, begin by dissolving 50 mmol of K_2HPO_4 in 9.5 liters of deionized water. Heat this buffer to 50–60°C and dissolve the amino acids listed below by occasionally adjusting the pH of the solution to 7.0.

Amino acids for FM	g/10 liters
Arginine (free base)	5.7
Asparagine	3.0
Cysteine	2.7
Glycine	9.0
Glutamic acid (free acid)	5.0
Histidine (free base)	2.2
Isoleucine	6.0
Leucine	9.0
Lysine (monohydrochloride)	9.1
Methionine	3.0
Phenylalanine	5.0
Proline	8.0
Threonine	5.0
Tryptophan	2.0
Valine	7.0

Add 100 ml of the 100× salts and trace elements stock solution (see below), 100 ml of the 100× vitamins stock solution (see below), 200 ml of a 50% glucose solution, and 100 ml of the 100× Pen/Strep stock solution. Adjust the pH with NaOH or H_3PO_4 to 6.5.

100× Vitamins	mg/1000 ml
Biotin	2.0
Cyanocobalamin	0.5
Folic acid	20.0
Lipoic acid(DL-6-8-thioctic acid)	40.0
Riboflavin	50.0
Thiamine·HCl	60.0

Adjust pH to 7.0

100× Salts and trace elements	g/1000 ml
NaOH	8
$NaHCO_3$	1.7
NH_4Cl	5.4
$CaCl_2 \cdot 2H_2O$	0.29
$FeCl_3 \cdot 6H_2O$	2.7
$MgCl_2 \cdot 6H_2O$	8.1
$Na_2EDTA \cdot 2H_2O$	0.48
H_3BO_3	0.11
$CoCl_2 \cdot 6H_2O$	0.017
$CuSO_4 \cdot 5H_2O$	0.015
$(NH_4)_6Mo_7O_{24} \cdot 4H_2O$	0.010
$MnCl_2 \cdot 4H_2O$	0.051
$ZnSO_4 \cdot 7H_2O$	0.230

Adjust pH to 7.0

100× Pen/Strep
10,000 U/ml pennicillin G
10 mg/ml streptomycin sulfate

Filter-sterilize the media through a 0.45-μm filter using cellulose prefilters.

B. Solutions

1. Nuclei Buffer: For the Isolation of Nuclei for the Preparation of Genomic DNA

Component (final concentration)	g/1000 ml
Tris·HCl (40 mM)	4.8
Sucrose (1.5%)	15.0
$EDTA^2Na \cdot 2H_2O$ (0.1 mM)	0.04
$MgCl_2 \cdot 6H_2O$ (6 mM)	1.2
KCl (40 mM)	3.0
Dithiothreitol (5 mM)	0.77
Nonionic detergent NP-40 (0.4%)	4.0 ml/1000 ml

Adjust to pH 7.8 and store at 4°C.

2. NDS Buffer: For the Preparation of High-Molecular-Weight DNA

Component (final concentration)	g/1000 ml
Tris·HCl (10 mM)	1.2
EDTA·2Na·2H$_2$O (0.5 M)	186
Sodium sarcosinate (1%)	10

Adjust pH to 8.0 and store at room temperature.

III. Preparation and Restriction Analysis of Genomic DNA

Cox *et al.* (1990) assigned certain genes to chromosomes by hybridization of gene-probes to chromosomes resolved on pulsed-field gels. The seven chromosomes of *Dictyostelium* average about 5.7 Mb in size, so the assignment of a gene to a specific chromosome does not provide much in the way of map resolution. Higher resolution can be obtained with large-scale restriction mapping (Kuspa *et al.*, 1992). With this type of map genes can be localized to ±100 kb, within a map that extends from 300 to 2000 kb.

Some restriction enzymes are better suited than others for analyzing the *Dictyostelium* genome because their distribution in the genome allows the construction of useful maps in the size range of 0.5–2.0 Mb. These enzymes can be ordered according to their usefulness as determined by the approximate average fragment size they generate when used on *Dictyostelium* DNA (given in parentheses below), their reliability of digesting *Dictyostelium* genomic DNA embedded in agarose, and their price per unit. This order, by decreasing usefulness, is: *Apa*I (400–600 kb), *Bgl*I (200–400 kb), *Nar*I (200–400 kb), *Sal*I (50–75 kb), *Sst*II (600–800 kb), *Sma*I 800–1000 kb), and *Eag*I (800–1000 kb). *Nar*I displays a marked site preference and usually gives one or two partial digestion products. This is not a problem for map construction, and the sites revealed by the partial products provide additional, informative landmarks for each map.

A. Preparation of High-Molecular-Weight Nuclear DNA

(1) Grow cells at 22°C in 1 liter HL-5 medium (see above) by shaking them at 200 rpm to 1–5 × 10^6 cells/ml.

(2) Harvest the culture in 500-ml centrifuge bottles at 1500 g and 4°C for 10 min. Decant as much of the media from the cell pellet as possible.

(3) Add 30–40 ml ice-cold nuclei buffer (see above). Keep on ice and gently disrupt the cells by stirring them into the buffer and then by forcing them up and down a 25-ml pipet (without foaming) for 1–3 min.

(4) Transfer the suspension to 40-ml tubes, and centrifuge at 10,000 *g* and 4°C for 10 min. Take care in decanting the supernatant since the nuclei do not form tight pellets.

(5) Resuspend the nuclear pellet in 10 ml nuclei buffer using gentle trituration, then fill the tubes with buffer and centrifuge as in step 4.

(6) While waiting for the centrifugation in step 5, dissolve 1 g low-melting-point agarose in 100 ml 125 mM EDTA (pH 8) by microwave-heating, and it allow to cool to 65–75°C.

(7) Decant the supernatant from step 5 and discard it. Resuspend the now white nuclear pellet in nuclei buffer to a calculated 10^9 nuclei/ml (about 1–5 ml total) using a pasteur pipet and gentle trituration.

(8) Warm up the nuclear suspension to room temperature with the heat of your hand, add 1.5 ml agarose per 1 ml of nuclear suspension with a 5- or 10-ml disposable plastic pipet, and mix the cells and agarose by gently, but quickly, pipeting up and down without producing bubbles. Quickly layer the suspension into an appropriate mold (see Chapter 1) such that the plugs are no thicker than 2 mm in at least one dimension. Allow them to solidify on ice and then cut into plugs.

(9) Put plugs into at least 2 vol (10 ml per 5 ml of the final plug volume) 1 mg/ml proteinase K in NDS buffer (see above) in a positive-seal 50-ml disposable polypropylene tube. Incubate at 50°C for 20–40 hr, mixing occasionally during the first few hours by inverting the tube several times.

(10) Rinse the plugs 5–10 times over a 6-hr period with 4 vol of 50 mM EDTA (to remove detergent and protease) at 50°C. The plugs can be stored at 4°C in 50 mM EDTA (pH 8.0) for at least 3 years.

Other protocols have been published detailing the preparation of high-molecular-weight DNA from whole *Dictyostelium* cells (Cox *et al.*, 1990; Birren and Lai, 1993). These protocols suggest starving the cells and preincubating the plugs with EDTA prior to the addition of detergent. These steps are unnecessary when preparing DNA from nuclei. One of the advantages of preparing DNA from nuclei is that the preparations contain less cellular debris, and should therefore be more susceptible to digestion with restriction enzymes (or less enzyme should be required). In addition, there is very little contaminating mitochondrial DNA in these samples.

B. Digestion of DNA Embedded in Agarose

(1) Cut plugs containing DNA prepared as described above into pieces of about 30 μl such that the dimensions fit the wells in the gel that you plan to use, and place each in a 2-ml straight-wall microcentrifuge

tube or the well of a 24-well plate. This corresponds to approximately 5 μg of genomic DNA.

(2) Rinse the plugs (by stationary incubation) with 100 μl of 0.5 mM phenylmethylsulfonyl fluoride (in TE buffer) at 37°C for >40 min.

(3) Rinse with 100 μl of the appropriate restriction enzyme buffer without enzyme at 37°C for >40 min. Remove buffer and replace with fresh buffer to repeat the rinse. Remove the final rinse buffer.

(4) Add 20 μl buffer and 20 U of enzyme. Incubate for 4–20 hr at 37°C. Digests can be stored for several days at 4°C by adding 5 μl 0.5 M EDTA to stop the reaction.

Standard pulsed-field gels and Southern blotting procedures are used to analyze the resulting restriction fragments. Of the enzymes listed above, only *Nar*I should give partial digestion products. If other enzymes give partial digestion, it is usually because the proteinase K was not inactivated, the EDTA was not rinsed out of the plugs, or not enough enzyme was used (40 U could be used but the glycerol from the enzyme storage buffer might exceed a concentration of 10% depending on the enzyme concentration provided by the supplier. Some problems may also be caused by the rDNA. The rDNA genes are carried on a 90-kb linear palindrome present in about 90 copies/nucleus. Since it represents only about 0.1% of the sequence complexity but 17% of the DNA content of the nucleus, false-positive hybridization signals that correspond to the rDNA bands are often observed with a number of different gene-probes. These signals are usually not a problem in analyzing the results of high-stringency hybridizations. However, problems can occur in low-stringency hydridization screens, especially when one is attempting to identify new members of a gene family.

Most single gene probes hybridize to unique fragments generated by single rare-cutting enzymes or by combinations of two such enzymes. The results allow one to construct a map of the rare sites relative to each other and the genetic loci in a region up to a megabase in length. Useful digests for this purpose are *Apa*I, *Bgl*I, *Sst*II, *Sma*I, *Nar*I, and all possible pairwise combinations.

IV. The Use of YACs in Genome Analysis

YAC clones have proven to be extremely useful for physical mapping (Burke *et al.*, 1987). By using a fivefold-redundant YAC library, high-resolution (gene placements of ± 20 kb) YAC contig maps have been generated for most of the *Dictyostelium* genome (Kuspa *et al.*, 1992; Kuspa and Loomis, 1996). YAC analysis has also been used to characterize the size and distribution of a number of gene families and repetitive elements, leading in some cases to the cloning of new members of established families. Finally, YACs have proven to be useful for isolating uncloned regions of previously characterized genes.

A. YAC Library Construction

YAC clones may be prepared by following any of the standard protocols (e.g., Burke and Olson, 1991). In about 3 months, one person should be able to carry out the DNA isolation procedure several times to accumulate enough DNA and produce a YAC library of several thousand stored clones. The general approach we have successfully used to construct YACs using the pYAC4 vector is described below (Kuspa *et al.*, 1992).

1. Preparation of High-Molecular-Weight DNA in Solution

Milligram quantities of high-molecular-weight DNA in solution can be obtained from 2 liters of *Dictyostelium* cells grown in suspension, by modification of a procedure described by Burke and Olson (1991).

(1) Nuclei are prepared from 2 liters of cells grown in HL-5, as described above, and resuspended to a calculated density of 2.5×10^9 nuclei ml in prelysis buffer [0.1 M Tris·HCl (pH 7.6), 15% sucrose (w/v), 10 mM EDTA].

(2) Slowly add 4 ml of this suspension to a 250-ml flask containing 7 ml of 22°C lysis buffer (0.4 M Tris·HCl, pH 9, 0.2 M EDTA, 3% sarkosyl) by dribbling it down the side of the flask over the course of 2–3 min while rocking the flask with a circular motion at 20–30 rpm. Care must be taken to avoid trapping unlysed cells in clumps within the viscous solution which results.

(3) Immediately immerse the lysed-cell suspension in a 65°C water bath and incubate for 15 min.

(4) Gently pour the entire 11-ml suspension onto a single sucrose "block" gradient. The gradient is made with 3 ml of 50% sucrose at the bottom of an SW27 tube (Beckman) which is overlayed with 13 ml of 25%, then 13 ml of 15% sucrose. In addition to sucrose the gradient contains 0.8 M NaCl, 20 mM Tris·HCl (pH 8), and 10 mM EDTA. The gradients are centrifuged at 26,000 rpm for 3 hr at 20°C in an SW27 Beckman rotor.

(5) Aspirate the top one-third of the gradient and collect the clear, viscous DNA from the bottom of the gradient with the open end of a 10-ml glass pipet.

(6) Concentrate the DNA in a collodion bag (Schleicher and Schuell; UH 100/1) with vacuum dialysis, against TE, to 1–2 mg/ml. The maximum DNA concentration that can be reached by this method while maintaining it in solution is about 2 mg/ml.

In this protocol, the amount of DNA loaded per gradient is an important parameter which, to ensure isolation of the DNA, must not exceed what is described above. When more material is layered on a single gradient, little separation of the DNA from other cellular components is observed.

2. YAC Library Construction

After analyzing small pilot digests, four 200-µl aliquots of high-molecular-weight DNA are digested with varying amounts of *Eco*RI (0.25 to 2.0 total enzyme units per aliquot) for about 30 min. The enzyme is carefully mixed with the DNA in enzyme buffer without magnesium ion. This is accomplished by stirring followed by a stationary incubation at 4°C for several hours. The reactions are then initiated by adding $MgCl_2$ to a concentration of 10 mM. The change in the viscosity of the DNA provides a good measure of the progress of the digestion, and can be used to determine the exact time of digestion. It is better to slightly overdigest the DNA since any small partial digestion products will be purified away from large fragments later. This optimal digestion is reached when the viscosity of the initially gelatinous DNA sample is still easily detectable by dragging some of the solution up on the wall of the test tube with a small plastic pipet tip. Ten percent of the digested samples is analyzed by pulsed-field gel electrophoresis to determine the DNA size distribution. Samples that show a decreased average size compared to the no-enzyme control, but which have maintained an average size great enough to allow fragments of the desired size to be cloned, are used for ligation to vector arms. The no-enzyme control should show no change in size distribution relative to an untreated control. If the size distribution does change, the DNA sample is probably contaminated with other endonucleases.

Properly digested DNA is mixed with an equal weight of *Eco*RI/*Bam*HI-digested and calf intestinal phosphatase-treated pYAC4 vector arms in 1× ligation buffer (Maniatis *et al.*, 1982) at a total DNA concentration of 0.5–1.0 mg/ml. The viscous DNA mixture is stirred slowly for several minutes and allowed to sit on ice for 4–6 hr prior to the addition of T4 DNA ligase. After 16 hr of ligation at 14°C, large ligation products are purified by sucrose gradient sedimentation (Burke *et al.*, 1987). Fractions containing DNA larger than 75 kb are pooled, concentrated, and dialyzed against TE by collodion vacuum dialysis as described above. Yeast strain AB1380 is transformed by spheroplasting (Burgers and Percival, 1987), with the size-fractionated ligation products (at about 1 mg/ml), and transformants are selected on yeast minimal (YM) media plates lacking uracil. This procedure should give approximately 10^6 transformants per microgram of plasmid DNA, and approximately 300 transformants per microgram of YAC ligation products can be expected. Primary transformants are picked to YM plates lacking uracil and tryptophan to test for the presence of both vector arms on each clone. Individual clones can then be grown and stored, and high-molecular-weight DNA samples can be prepared from them for analysis of the YACs (Chapter 2).

In general the average insert size in the YAC library obtained will be less than the average DNA size in the final ligation mixture. This is most likely due to bias in the transformation step favoring smaller clones. We were also able to estimate directly the total number of chimeric clones in our YAC library because 20% of the DNA in *Dictyostelium* nuclei is present in about 100 copies of an extrachromosomal palindrome of 90 kb. Since the largest *Eco*RI fragment that can be generated from this palindrome is 85 kb, and the YAC arms add only another 9.5 kb, any YAC greater than 95 kb that carries these sequences must be a chimera in which a portion of the palindrome was attached to an independent genomic fragment within the same YAC. Thus, we used the 90-kb palindrome to probe the YAC library for palindrome-containing clones >100 kb. About 1% of the YACs in our library were found to have inserts derived in part from the 90-kb palindromic DNA. Since the palindromic DNA makes up 20% of the total nuclear DNA, we expect the frequency of all chimeras to be about five-times the number of events involving the palindromic DNA, or about 5%. This assumes that most chimeras form during the ligation step prior to transformation of the YACs into yeast cells, and not by recombination in yeast. It is possible that we avoided producing a large percentage of chimeric clones because of the care we took to fully mix the vector arms with the genomic DNA prior to the addition of ligase. This should decrease the probability of two genomic fragments ligating together by making the theoretical vector/insert ratio of 40:1 (moles of vector arms: moles of genomic fragments) an experimental reality. We may have also avoided cloning a high percentage of chimeras by using a small amount of ligation DNA in each transformation (<1 μg of DNA per 10^6 cells). The actual reasons for the low percentage of chimeras in our library are not known since the above suppositions are based on the uncontrolled experiment of a single library construction. However, the extra work required to carefully mix the vector and insert DNA, and to keep the amount of DNA per transformation low, is minimal and should be considered by anyone attempting to construct a YAC library.

B. YAC Contig Construction

The most robust methods for contig construction using *Dictyostelium* YAC clones are based on the use of random hybridization probes. Small groups of linked YACs can be identified easily by hybridization of the YAC set with single-copy probes. We have found that this is best done by probing Southern blots of pulsed-field gels since it is required that the YACs be separated from the bulk of the yeast chromosomes to avoid high background caused by cross-hybridization of the *Dictyostelium* probes to their yeast homologs. It is possible to run at least 348 samples on one large 24 × 16-cm gel by using 12 29-well combs spaced 2 cm apart and by running the gel for 4 hr under standard conditions used to separate yeast chromosomes (Chapter 1). The YACs grouped in this way can be aligned roughly by mapping the positions of shared rare restriction sites. To do this, the YACs are digested with *Apa*I, *Bgl*I, or *Nar*I, and then Southerns of the resulting fragments resolved on pulsed-field gels are sequentially probed with a YAC end-specific probe (usually a fragment from one of the YAC vector arms) followed by the gene-probe used to define the group. Alignment of the YACs is made relative to the rare sites, and YAC orientation is determined by whether the YAC end-specific probe identifies a YAC fragment that is the same size or a different size than that recognized by the gene-probe. Partial digestion of the YAC DNA can also be used to determine the position of every site for a single frequent cutting restriction enzyme, and will refine the alignment of the YACs (Burke *et al.*, 1987; Kuspa *et al.*, 1992). The enzymes most useful for the partial digestion of *Dictyostelium* YACs include *Bam*HI, *Xho*I, and *Sal*I. The recognition sites for these enzymes occur every 10–50 kb in the genome.

There is an unequal nucleotide composition in the coding and noncoding regions of the *Dictyostelium* genome, where intergenic regions and introns are usually >90% A + T, while coding regions vary from 60 to 70% A + T. As a consequence, random genomic clones are not a good source of probes for grouping YACs since many have such a high A+T content that they do not hybridize specifically. Our experience is that less than 30% of random genomic fragments give specific hybridization signals. For the same reason, probes generated from the ends of YACs will not hybridize uniquely. Thus, for *Dictyostelium* YAC contigs have been constructed by a random-probe strategy which relies on the chance distribution of probes for extension at the ends of contigs.

YAC contigs can be assigned to specific chromosomes when they encompass one or more genes previously mapped to a chromosome. The contigs can be aligned to the large-scale restriction map of the region by determining the positions of as many of the relevant rare restriction sites as required for unambiguous positioning. As the YAC contigs become large, they eventually can be used to link disparate pieces of mapping information. They can be used, in essence, as linking clones connecting restriction maps or RFLP-marked fragments (see below). However, extended maps based on YAC contigs are subject to cloning anomalies, so methods of independent verification of proposed linkages are needed. This is where comparison with long-range restriction maps is essential. Identification of landmark restriction sites, obtained directly from the analysis of genomic DNA, in the proper location within the YAC contigs serves to weed out anomolous YAC clones and false linkages. This points out the general principle that, performed in isolation, YAC contig construction is not likely to yield an accurate map. Rather, accurate physical mapping requires the comparison of data from several independent sources. For *Dictyostelium*, we have used linkage group assignments, long-range restriction mapping, and RFLP analysis (see below) to increase the accuracy of the physical map. When YAC contig construction is performed within such a multipronged approach, cloning anomalies are revealed as trivial abberations within the interlocking data sets.

C. Characterization of Gene Families Using YACs

YACs represent an efficient mapping tool that can provide a unique advantage when analyzing gene families or repetitive elements. A single hybridization of a gene-family probe to the ordered YACs will identify all of the regions of the genome that contain at least one member of the family. The exact number of related genes present at each locus can be determined by detailed restriction mapping (by Southern hybridization) of the relevant YAC clones. Using this approach we have mapped 13 myosin genes and 18 actin genes (Titus *et al.*, 1994), the genes of two tRNA gene families, and the members of five different repetitive element families.

In addition to mapping the family members, the YACs from each locus also provide an enriched source for cloning any newly discovered gene. For example, they can be used as substrates for amplification of individual members of a gene family by polymerase chain reaction (PCR). This can be especially valuable when the design of PCR oligonucleotides can be guided only by sequence information from one or a few cloned members of the family, and amplification of an uncloned member is desired. The DNA substrate for the PCR reaction can be highly purified by isolating the YAC DNA as a gel slice from a pulsed-field gel made from low-melting-point agarose. Varying amounts of molten agarose can then be added directly to the PCR reaction (Titus *et al.*, 1994). Direct subcloning from purified YAC DNA has also proven to be extremely useful for isolating genomic clones.

D. Subcloning *Dictyostelium* Genes from YACs

For cloning genes from YACs it is best to choose the smallest YAC that still contains all of the desired sequences. This reduces the complexity of the sublibrary source DNA, and allows for easier separation of the YAC from the endogenous yeast chromosomes. The enzymes to be used for subcloning can be determined from standard genomic Southern analysis. Also, it is best to make sure that the YAC of choice has the appropriate genomic fragment by Southern analysis, comparing the YAC DNA (a standard agarose block of YAC DNA cut with frequently cutting restriction enzymes will work) to genomic DNA [2 μg of low-molecular-weight DNA (Nellen *et al.*, 1987)].

1. Isolation and Restriction Digestion of YAC DNA

The following procedure can be used to obtain about 1 μg of pure YAC DNA which can be cloned into the appropriate bacterial vector. The resulting YAC "minilibrary" can then be screened for the desired clones.

(1) Make at least 10 blocks (1.2 × 3 × 12 mm or equivalent volume— about 70 μl per block) of high-molecular-weight DNA from the appropriate YAC strain (see Chapter 2).

(2) Run 10 blocks of DNA on a pulsed-field gel, such as CHEF, optimizing the separation to fit the YAC size. If you are isolating a YAC of 50–150 kb, one sample of a control yeast DNA should be run alongside the YAC DNA for size comparison. In some DNA preparations degraded chromosomal DNA running at about 100 kb on pulsed-field gels appears similar to the YAC band.

(3) Stain entire gel with ethidium bromide (1 μg/ml in H_2O) for 30 min, and destain for at least 30 min. With the gel on a sheet of plastic wrap on a UV light box (312 nm) use a ruler and razor blade to cut the gel just below the YAC band. Turn the UV light off and proceed to cut a gel strip 2 to 3-mm wide containing the YAC DNA. Put this strip in a 15-ml plastic screw cap test tube (store at 4°C in the dark if you want to stop here). Turn the UV light on to confirm that the YAC band was excised.

(4) Incubate the YAC gel strip in the 15-ml tube with 10 ml restriction enzyme buffer, lacking enzyme, with gentle agitation at 37°C for >1 hr. Carefully pour off the buffer and repeat by adding another 10 ml of buffer. Carefully decant the second wash.

(5) Add 1 ml of restriction enzyme buffer containing 100–200 U of restriction enzyme, invert several times to mix, and incubate the tube on its side without agitation at 37°C for 12–16 hr (a minimum of 4 hr is required for most enzymes to diffuse completely into the strip).

(6) To collect the digested fragments from the gel strip, lay pieces of DE81 paper (Whatman) directly on the strip, and sandwich the DE81/gel strip between two previously solidified agarose slabs (containing 1 µg/ml ethidium bromide) in a standard submarine gel box. To monitor the migration of the YAC fragments out of the slice, load DNA standards that are similar in size to the YAC fragment(s) desired in a well somewhere on the gel. Following the DNA standards provides an indirect way to monitor progress since you cannot visualize the DNA in the strip. Electrophoresis is continued until the fragment size desired would be expected to have migrated the width of the gel strip and onto the DE81 paper.

(7) For every 2–3 cm^2 of DE81 paper, vortex the paper with 0.5 ml of 1 *M* NaCl in TE and 10 µg yeast tRNA in a 1.5-ml microcentrifuge tube. Leave at room temperature for at least 30 min. Collect the eluate and put it in a fresh tube. Extract the eluate twice with an equal volume of phenol/chloroform and once with chloroform/isoamyl alcohol. Avoid the paper fragments at the interface when recovering the supernatants. Add 1 ml of ethanol, vortex, and spin at 15,000 rpm for 15 min. Pour off the ethanol supernatant, and rinse the pellet with 70% ethanol.

(8) At this point you should have about 1 µg of digested YAC DNA (with <5% yeast DNA contamination) for cloning into the desired vector.

An alternative procedure for the isolation of DNA from agarose involves the use of the enzyme agarase, which eliminates the need for organic extractions (see Chapter 1).

V. Restriction Enzyme-Mediated Integration (REMI)-RFLP Analysis

Introducing the restriction enzyme used to linearize a transforming plasmid into cells along with the plasmid DNA dramatically increases the frequency of integration into genomic restriction sites recognized by the specific enzyme (Schiestl and Petes, 1991; Kuspa and Loomis, 1992). Since a wide range of different restriction enzymes can be used, this technique has been named Restriction Enzyme-Mediated Integration (REMI). REMI can be used to generate strains carrying a single copy of a plasmid designed to generate restriction fragment length polymorphisms, or REMI-RFLPs (Kuspa and Loomis, 1994). The plasmid used, DIV6 (Fig. 10.1),

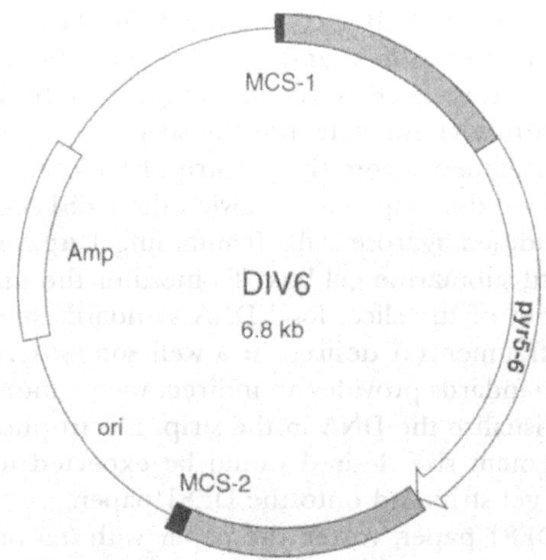

MCS-1 (useful sites): ApaI, EagI, SfiI, SstII, NotI.

MCS-2 (useful sites): XbaI, BamHI, AvaI, SmaI, SstI.

Figure 10.1 The *Dictyostelium* integrating vector, DIV6. A schematic representation of a vector used for the generation of REMI-RFLPs is shown. The thin line represents bacterial sequences pGEM5Zf(+) (Promega; Madison, WI), and the shaded areas and arrow represent *Dictyostelium* sequences. For details of its construction see Kuspa *et al.* (1992). Only the unique and useful restriction sites in each multiple cloning site are shown.

carries restriction sites corresponding to many of the restriction enzymes that cleave *Dictyostelium* genomic DNA infrequently. Thus, in each REMI-RFLP strain, for any given rare-cutting enzyme, a single restriction fragment is altered in size, which can be detected easily on Southern blots of pulsed-field gels (Fig. 10.2). Any gene-probe that identifies an altered fragment can be mapped unambiguously within that fragment.

Mapping genes relative to one another by REMI-RFLP allows the establishment of long-range linkage that is not possible in *Dictyostelium* by other physical mapping techniques. The genes linked in this way can then be ordered, sometimes by fortuitious multiple insertions within the same region, or more reliably by restriction mapping and YAC contig analysis.

A. Construction of REMI-RFLP Strains

The *Dictyostelium* *pyr5-6* gene encodes a bifunctional uracil biosynthetic enzyme (Jacquet *et al.*, 1988). Starting with a strain in which the *pyr5-6* gene has been inactivated, transformants can be selected for integration

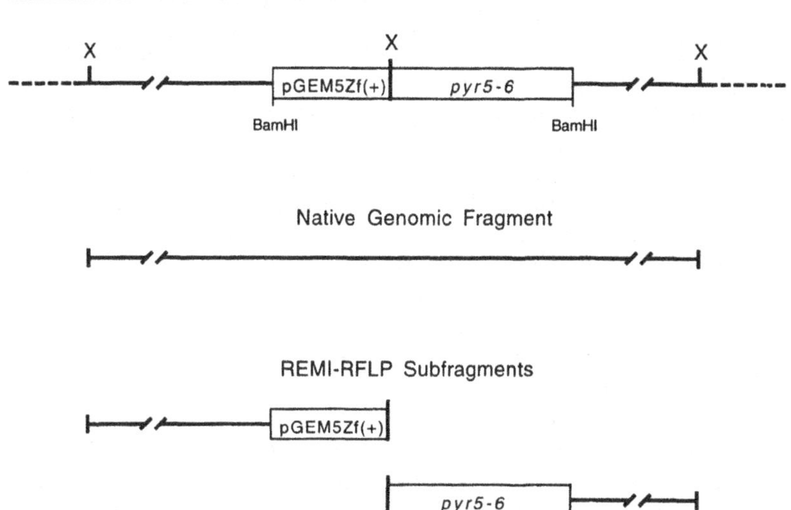

Figure 10.2 REMI-RFLP mapping strategy. *Bam*HI REMI is used to integrate a DIV6 vector (open boxes) into the *Dictyostelium* genome (horizontal line) at random *Bam*HI restriction sites. The top line shows a schematic map of such an insertion. Digestion of the genomic DNA of such insertion strains with restriction enzyme "X" liberates a native genomic fragment in untransformed strains and strains that did not acquire a DIV6 within this specific fragment. Two smaller subfragments are liberated from the DNA of strains that did acquire a DIV6 integration within the fragment.

of vector DNA containing the *pyr5-6* gene by requiring growth in media deficient in uracil (Kalpaxis *et al.*, 1990). A single copy of the *pyr5-6* gene is sufficient for uracil prototrophy. A convenient aspect of this selection is that the prerequisite strain, a *pyr5-6*⁻ mutant, can be generated from any strain able to grow axenically by selecting for growth in 5-fluoro-orotic acid (Kalpaxis *et al.*, 1990).

B. REMI Transformation Protocol

pyr5-6⁻ mutant cells to be transformed are grown axenically in HL-5 medium containing 20 μg/ml uracil, and must be harvested between 1 and 4×10^6 cells/ml for efficient transformation. The optimum growth stage at the time of harvesting should be determined for any untested strain. In the following protocol, the electroporation conditions are adapted from Howard *et al.* (1988), and the REMI methodology has been described previously (Kuspa and Loomis, 1992). When the following steps are carried out, we routinely recover REMI transformants at a frequency of greater than 10^{-5} (transformants per input cell).

(1) Chill the medium containing exponentially growing cells (25 to 50 ml for a typical experiment) on ice for 15 min with occasional swirling.

(2) Spin the cells down in a 50-ml tissue culture tube by centrifuging at 1500–2500 g at 4°C for 4 min.

(3) Cool the electroporation cuvettes (0.4-cm gap width) and large sterile glass test tubes (16 × 125 mm) on ice for at least 5 min.

(4) Decant the growth media and leave the 50-ml tube upside-down for a minute. Carefully aspirate the remaining liquid from the walls of the tube, especially near the pellet and around the rim and edges of the tube.

(5) Gently resuspend the pellet by tritutration with a pasteur pipet in ice-cold electroporation buffer [10 mM NaPO$_4$ (pH 6.1), 50 mM sucrose] at 10^7 cells per milliliter.

(6) Distribute 0.8-ml aliquots of cells to the cold glass tubes, and add 50 μg of *Bam*HI-digested DIV6 (at 1 μg/ml in TE) and 200 U of *Bam*HI restriction enzyme (20 μl in the storage buffer as supplied by the manufacturer) and mix by briefly swirling the tube. If 0.2-cm-gap curvettes are used, the volume should be reduced to 0.4 ml.

(7) Put the DNA/cell mix in a cold cuvette and electroporate at 2.5 kV/cm. Other electrical parameters vary with the particular electroporation device. Most manufacturers can provide the optimal conditions for *Dictyostelium*. It is probably best to adjust the conditions such that the time constant is 0.5 to 1.1 msec. The transformation efficiency decreases after the cells have been in electroporation buffer for more than 30 min. Approximately 12–20 aliquots from the same batch of cells can be processed in this time.

(8) Immediately after each electroporation, distribute the cells from one cuvette into four standard petri plates, each containing 10 ml of FM media (see below). Control samples consisting of cells electroporated with no DNA can be distributed to plates containing FM media as well as plates containing FM with uracil added to 20 μg/ml. The control sample lacking DNA tests the uracil selection in FM, while the plates containing FM with added uracil tests for the viability of the cells (which should be >95%). Incubate the cells at 22°C in a humid chamber.

(9) Colonies will appear after leaving the plates undisturbed for 6–8 days. At this time replace the medium with fresh FM and incubate for 6–8 more days. After 12–16 days the background of untransformed cells will be negligible and the transformant colonies will be about 1–3 mm in diameter, each containing at least 10^4 cells.

(10) At this time collect the transformants by directing a stream of medium over the surface of each plate from a pipet. The cells from each plate may be kept separate or pooled, diluted appropriately, and spread on SM agar plates with a few drops of a saturated suspension of *Klebsiella*

aerogenes (grown in SM liquid medium; see above). Since the plating efficiency varies, diluting the supension of primary transformants in a series of four 10-fold dilutions into liquid SM medium and spreading 0.1, 0.2, and 0.4 ml of the suspension from the three highest dilutions onto separate plates ensures that there will be adequate numbers of clones on one or more plates. Transformants appear as barely visible plaques in about 3 days, and grow to form larger plaques in the next few days. There may be a number of small plaques that appear after 4–6 days on the bacterial plates that result from untransformed cells that have survived the FM selection.

(11) With a sterile loop, pick a 1- to 2-mm-diameter ball of cells from the edge of large plaques (3–6 mm in diameter) to tubes containing 2 ml HL-5 (without uracil but containing 100 units/ml penicillin G and 100 μg/ml streptomycin sulfate to suppress growth of bacteria). Each transformant should grow to $>10^6$ cells/ml after 2 days of shaking at 22°C. Isolates that take longer than 3 days to grow to high titer in 2 ml of HL-5 are probably not stable transformants.

(12) Individual cultures are expanded to 1 liter in groups of eight and grown to $1–4 \times 10^6$ cells/ml. The cells from 25 ml of each culture are then pelleted, resuspended in 2 ml of HL-5, mixed with 0.2 ml of fresh DMSO, and frozen in two 1-ml aliquots at -70°C. High-molecular-weight DNA is prepared from the remainder of the cultures as described above.

C. REMI-RFLP Analysis

When the DIV6 plasmid integrates via its *Bam*HI site into chromosomal DNA, the bacterial sequences of DIV6 are on one side, and the *pyr5-6* gene is on the other side, with most of the restriction sites useful for RFLP analysis in between (Fig. 10.2; Kuspa and Loomis, 1994). Digestion of DNA from a REMI-RFLP strain with an appropriate enzyme liberates two subfragments of the native genomic fragment bounded by the sites of that enzyme. Each subfragment can be identified by hybridization with either the bacterial sequences (pGEM-5Zf) or a *pyr5-6* probe.

To carry out a large-scale analysis, DNA samples are prepared from a set of 100 to 200 REMI-RFLP strains, and digested with an appropriate restriction enzyme as described above. The resulting fragments are resolved by pulsed-field gel electrophoresis under conditions designed to separate 10 to 2000 kb. To facilitate hybridization with a large number of probes several sets of blots can be prepared. A minority (~10%) of the REMI-RFLP strains will have one of several possible anomalies revealed upon Southern blot hybridization with the vector-specific probes. Some strains will have two DIV6 insertions. Some strains acquire only one part of the DIV6 plasmid or have none of the plasmid, indicating that they were never truely transformed. Still other strains may have uninterpretable patterns of many fragments with a signal strength 10–20 times more than the other strains as if DIV6 sequences have been amplified. However, most of the strains (>90%) will contain a single DIV6 insert and will give the expected hybridization pattern with the vector-specific probes.

When the Southern blots are hybridized with a gene-probe, the native fragment will be seen in the great majority of the REMI-RFLP strains. In a few strains the fragment will be smaller, and will corresponded to one of the fragments that carried a portion of the DIV6 vector. The presence of a RFLP is confirmed in such samples by the absence of the native fragment. Thus, several independent alterations in the hybridization pattern indicate that the fragment carrying the specific gene is marked by an insertion in a REMI-RFLP strain. By measuring the sizes of the modified fragments the insertion sites can be positioned relative to one end of the native fragment.

Genes can be mapped to a specific large restriction fragment when they identify a RFLP fragment in a REMI-RFLP strain which displayed a RFLP fragment for a previously mapped gene. Many regions will be defined by a single insertion site allowing only a simple grouping of genes within subfragments. However, where two or more insertions occur in the same fragment, genes can be localized to a smaller segment defined by the flanking insertion sites. This type of relational mapping provides a resolution of a few hundred kilobasepairs. When combined with other physical mapping data obtained from large-scale restriction mapping and YAC contigs, the REMI-RFLP data add significantly to the resolution and reliability of the maps.

The probability that a particular gene-probe will identify a RFLP depends on the extent to which the REMI integration events have sampled all portions of the genome, in the set of strains used. Consider the use of *Apa* I in such an analysis. The average *Apa* I fragment in the *Dictyostelium* genome is 630 kb (unpublished observations). Assuming random integration by REMI, about 180 simple DIV6 insertion events would be required to sample >95% of the *Apa* I fragments in the genome. Naturally occuring RFLPs also appear in this type of analysis and should be helpful in identifying additional unique fragments not otherwise tagged by an insertion.

YAC contigs can be aligned with the RFLP fragment by determining the positions within the contigs of the restriction sites used in the RFLP analysis by standard restriction mapping techniques. For many of the contigs, the position of such sites will have been determined previously. As mentioned above, those contigs that span these restriction sites can be used to join otherwise unrelated fragments identified by REMI-RFLP, just as one might use linking clones.

More of the genome can be sampled with the same set of REMI-RFLP strains by analyzing fragments generated by a second rare-cutting restriction enzyme. Although there would be extensive overlap in coverage between the sets of fragments generated by the two enzymes, the second enzyme should sample new regions outside of those sampled by the first enzyme.

VI. Random Insertional Mutagenesis Using REMI

Advances during the past 10 years in the molecular genetic techniques for *Dictyostelium discoideum* have allowed detailed explorations into many aspects of its growth and development. These studies have benefited tremendously from an ability to isolate new genes by insertional mutagenesis. The large number of genes now potentially available to mutational analysis should expand our knowledge of the function of a significant proportion of the genome.

A method for generating insertional mutants by REMI and recovering the genomic DNA flanking the insertion sites has been devised (Kuspa *et al.*, 1992). An integrating vector is linearized with a specific restriction enzyme and introduced by electroporation into the cells along with the same restriction enzyme. Addition of restriction enzymes not only determines the sites of integration but also stimulates the rate of vector integration 20- to 60-fold in *Dictyostelium* such that transformants can be recovered at a frequency of 4×10^{-5} or more (Kuspa and Loomis, 1992, and unpublished data).

For a restriction enzyme to mediate integration, its recognition site must correspond to the site used to linearize vector. For instance, *Sau*3AI (which recognizes GATC) will stimulate integration of vector DNA linearized with *Bam*HI (which recognizes GGATCC), but *Eco*RI (which recognizes GAATTC) will not. The majority of the integration sites in *Sau*3AI REMI transformants are *Sau*3AI sites even when the plasmid is linearized with *Bam*HI. Thus, the mediating enzyme determines the site of integration. Of the enzymes that have been tested to date, *Bam*HI, *Sau*3AI, *Dpn*II, *Eco*RI, *Aha*III, *Bgl*II, *Cla*I, *Not*I, *Xba*I, and *Pst*I have been shown to stimulate integration. Thus, the broad range of enzymes available for REMI, and the apparently random nature of the integration they stimulate, should provide the distribution of sites necessary for the inactivation of every nonessential gene in *Dictyostelium*. *Dpn*II has proven to be extremely useful since its sites are present in coding regions about every 200 bp and it is less expensive than its isoschizomer, *Sau*3AI.

A. REMI Transformation and Mutant Screening

(1) Linearize a *pyr5-6*-containing vector with the enzyme that will be used for REMI except that *Bam*HI should be used to linearize the vector for *Sau*3AI or *Dpn*II REMI, and *Eco*RI should be used to linearize the vector for *Aha*III REMI. Purify the linearized vector by standard phenol extraction and ethanol precipitation methods.

(2) Follow the REMI protocol described above for REMI-RFLP strain construction, using a *pyr5-6* mutant strain. For each 0.8-ml aliquot cells, use 40 μg vector DNA and 100–200 U of restriction enzyme. Twelve to 20 aliqouts can be processed conveniently in 1 hr and will yield 2000 to 4000 transformants.

(3) Plate the cells in FM medium and select for uracil prototrophs. After 1 week, colonies of transformants will appear at a frequency of 10^{-5} to 6×10^{-5}.

(4) Collect cells from each plate separately and plate for clonal growth in a lawn of *K. aerogenes* on SM plates as described above. As the cells exhaust the bacterial food supply within each plaque, they begin to develop into many distinct multicellular aggregates, the majority of which continue through development to form mature fruiting bodies. After 5 to 6 days there are multiple structures in each plaque that can be visually inspected under a dissecting microscope. About 0.3–1% of the clones will display some easily visible developmental aberration.

(5) Cells from clones of interest are picked from the edges and streaked on fresh SM plates spread with *K. aerogenes* to test that the aberrations are hereditarily stable. It is unlikely that two independent mutants with the same morphological phenotype will arise from the population of 2×10^6 cells in the original selection plate, so only one representative of each distinct phenotype should be picked. Similar mutant plaques are likely to be sibling transformants, harboring the vector in the same site. In fact, the occurence of several clones that show identical morphological defects gives added assurance that an insertional mutant of this type had arisen in the original selection plate.

Once a set of mutants is obtained, the next step is to isolate the genomic sequences surrounding the insertion site such that the affected gene can be characterized. The plasmid and flanking sequences can be isolated from each strain by plasmid rescue, which is carried out by restriction digestion of genomic DNA followed by selection of recircularized plasmid in *E. coli*. An origin of replication and the ampicilin gene in the plasmid makes this a rapid and efficient process. The critical requirement step is to have sufficient DNA flanking the insertion site to permit selection of a 10- to 15-kb fragment containing *Dictyostelium* DNA. The following procedure can be used as a guide.

B. Cloning Sequences Flanking Insertion Sites

(1) Isolate genomic DNA from the insertion strains and purify on CsCl gradients (Nellen *et al.*, 1987). Digest 0.5 μg of purified DNA with a restriction enzyme that does not cut the integrating vector. The enzyme will cut in flanking *Dictyostelium* sequences. Since there is an upper limit on the size of plasmids that are easily isolated in *E. coli*, it is best to use an enzyme that will produce a vector-containing fragment of less than 15 kb. (This is predetermined by Southern analysis of the mutant strain's DNA, using the REMI vector as a probe.) Purify the DNA by phenol extraction and ethanol precipitation, and dissolve it in 20 μl of sterile water.

(2) To the DNA solution, add 0.48 ml ligase buffer (66 mM Tris·HCl, 5 mM MgCl$_2$, 5 mM dithiothreitol, 1 mM ATP, pH 7.5), and 5 U of T4 DNA ligase. Incubate at 12–15°C for at least 12 hr. The low concentration of the DNA is designed to favor circularization of monomolecular fragments over concatemerization. Precipitate the ligation products by mixing in 10 μl of 5 N NaCl, followed by 1 ml of ethanol at room temperature. Vortex, and centrifuge at 12,500 g for 10 min. Rinse the DNA precipitate twice with 1 ml 70% ethanol followed by brief centrifugation (2 min), removing as much of the wash ethanol at each step as possible. It is necessary to reduce the salt concentration as much as possible for the high-voltage electroporation of the bacteria that follows. Dissolve the pellet in 40 μl of sterile water.

(3) Add several different volumes in the range of 1–6 μl of the ligation products to a 40-μl aliquot of electrocompetent *E. coli* SURE cells (Stratagene), and electroporate with a gene pulser (Bio-Rad), or other suitable device, according to the manufacturer's suggested protocol. Plate out the entire population from each electroporation on bacterial plates containing 75 μg/ml carbenicilin and incubate at 37°C. Expect 10–100 bacterial transformants after a 16-hr incubation.

If the same restriction enzyme is used to both linearize the vector and stimulate its integration, the recovery system can be tested by excising the vector alone from the genome with that enzyme, then following the procedure just described above. A large number of bacterial transformants should be recovered in this test (as much as 100-times more than for the actual experiment) because the efficiency of several of the cloning steps varies inversely with the size of the fragment. These control transformants should contain just the original plasmid.

C. Recapitulating the Mutant Phenotype with Cloned DNA

After obtaining a plasmid carrying the regions flanking the insertion site, it is possible to verify that the original mutant phenotype is a direct consequence of the insertion event. The procedure entails using homologous recombination to regenerate the lesion in wild-type test cells. The plasmid DNA cloned in the above procedure is linearized with the same enzyme used to cut it out of the mutant genome. This generates a disruption fragment with the original REMI vector flanked by genomic DNA derived from the locus of integration. *Dictyostelium* cells that are $pyr5$-6^- are then transformed with the linearized fragment without any added restriction enzyme. Homologous recombination reestablishes the vector at the original insertion site in 20–100% of the transformants, depending on the amount of flanking DNA in a given clone. Absolute correspondence between the genomic structure seen in the original mutant and the original mutant phenotype proves that the insertion caused the phenotype. The linearized clone can also be used to disrupt the gene in a variety of genetic backgrounds.

One important difference between REMI and chemical mutagenesis is the frequency of obtaining morphological mutants. With some types of chemical mutagenesis about seven mutational events per genome can be achieved, and about 1 out of 40 survivors is a mutant (Loomis, 1987). Developmental mutants are more rare among REMI transformants, occuring at a frequency of about 1 in 300 transformed strains (Kuspa and Loomis, 1992). Since the insertions in these strains can only affect a single locus at a time, they would not be expected to generate morphological mutants at the same rate as chemical mutagenesis. Attempts to tag all 300 developmental genes by REMI will require generating and screening about 5×10^5 transformants. The same number of transformants will have to be screened to have a 95% chance of recovering an insertion in any one specific gene. Thus, it is quite possible that the developmental genes of *Dictyostelium* will be saturated with tags during the coming years. Only those genes that are present in multiple redundant copies and those that are necessary for the growth are inaccessible to REMI mutagenesis.

References

Birren, B., and Lai, E. (1993). "Pulsed field gel electrophoresis: A practical Guide." Academic Press, San Diego, CA.

Burgers, P., and Percival, K. (1987). Transformation of yeast spheroplasts without cell fusion. *Anal. Biochem.* **161,** 391–397.

Burke, D., and Olson, M. (1991). Preparation of clone libraries in yeast artificial chromosome vectors. *In* "Methods in Enzymology" (C. Guthrie and G. R. Fink, eds.), Vol. 194, pp. 251–270. Academic Press, San Diego, CA.

Burke, D., Carle, G., and Olson, M. (1987). Cloning of large segments of exogenous DNA into yeast by means of artificial chromosome vectors. *Science* **236,** 806–812.

Cox, E. C., Vocke, C. D., Walter, S., Gregg, K. Y., and Bain, E. S. (1990). Electrophoretic karyotype for *Dictyostelium discoideum. Proc. Natl. Acad. Sci. U.S.A.* **87,** 8247–8251.

Darcy, P. K., Wilczynska, Z., and Fisher, P. R. (1993). Phototaxis genes on linkage group V in *Dictyostelium discoideum. FEMS Microbiol. Lett.* **111,** 123–127.

Fan, J., Chikashige, Y., Smith, C. L., Niwa, O., Yanagida, M., and Cantor, C. R. (1988). Construction of a *Not*I restriction map of the fission yeast *Schizosaccharomyces pombe* genome. *Nucleic Acids Res.* **17,** 2801–2818.

Franke, J., and Kessin, R. (1977). A defined minimal medium for axenic strains of *Dictyostelium discoideum. Proc. Natl. Acad. Sci. U.S.A.* **74,** 2157–2161.

Howard, P., Ahern, K., and Firtel, R. A. (1988). Establishment of a transient expression system for *Dictyostelium discoideum. Nucleic Acids Res.* **16,** 2613–2623.

Jacquet, M., Gilbaud, R., and Garreau, H. (1988). Sequence analysis of the DdPYR5-6 gene coding for UMP synthase in Dictyostelium and comparison with orotate phosophoribosyl transferases and OMP decarboxylases. *Mol. Gen. Genet.* **211,** 441–445.

Kalpaxis, D., Werner, H., Boy Marcotte, E., Jacquet, M., and Dingermann, T. (1990). Positive selection for Dictyostelium mutants lacking uridine monophosphate synthase activity based on resistance to 5-fluoro-orotic acid. *Dev. Genet. (Amsterdam)* **11,** 396–402.

Kimmel, A. R., and Firtel, R. A. (1982). The organization and expression of the *Dictyostelium* genome. *In* "The Development of *Dictyostelium discodieum*" (W. F. Loomis, ed.), pp. 234–334. Academic Press, New York.

Knecht, D. A., Cohen, S. M., Loomis, W. F., and Lodish, H. F. (1986). Developmental regulation of *Dictyostelium discoideum.* actin gene fusions carried on low-copy and high-copy transformation vectors. *Mol. Cell. Biol.* **6,** 3973–3983.

Kuspa, A., and Loomis, W. F. (1996). Ordered yeast artificial chromosome clones representing the *Dictyostelium discoideum* genome. *Proc. Natl. Acad. Sci. U.S.A.,* in press.

Kuspa, A., Maghakian, D., Bergesch, P., and Loomis, W. F. (1992). Physical mapping of genes to specific chromosome in *Dictyostelium discoideum. Genomics* **13,** 49–61.

Kuspa, A., and Loomis, W. F. (1992). Tagging developmental genes in *Dictyostelium* by restriction enzyme-mediated integration of plasmid DNA. *Proc. Natl. Acad. Sci. U.S.A.* **89,** 8803–8807.

Kuspa, A., and Loomis, W. F. (1994). REMI-RFLP mapping in the *Dictyostelium* genome. *Genetics* **138,** 665–674.

Loomis, W. (1971). Sensitivity of *Dictyostelium discoideum.* to nucleic acid analogues. *Exp. Cell. Res.* **64,** 484–486.

Loomis, W. F., ed. (1982). "The Development of *Dictyostelium discoideum.*" Academic Press, New York.

Loomis, W. F. (1987). Genetic tools for *Dictyostelium discoideum. Methods Cell Biol.* **28,** 31–65.

Maniatis, T., Frisch, E., and Sambrook, J. (1982). "Molecular Cloning." Cold Spring Harbor Lab. Press, Cold Spring Harbor, NY.

Moens, P. B. (1976). Spindle and kinetochore morphology of *Dictyostelium discoideum. J. Cell Biol.* **68,** 113–122.

Nellen, W., Datta, S., Reymond, C., Sivertsen, A., Mann, S., Crowley, T., and Firtel, R. A. (1987). Molecular biology in Dictyostelium: Tools and applications. *Methods Cell Biol.* **28,** 67–100.

Newell, P. (1978). Genetics of the cellular slime molds. *Annu. Rev. Genet.* **12**, 69–93.

Newell, P., and Ross, F. (1982). Genetic analysis of the slug stage of *Dictyostelium discoideum*. *J. Gen. Microbiol.* **128**, 1639–1652.

Newell, P. C., Williams, K. L., Kuspa, A., and Loomis, W. F. (1993). Genetic map of *Dictyostelium*. *In* "Genetic Maps: Locus Maps of Complex Genomes," (S. J. O'Brien, ed.), 6th ed., pp. 3.1–3.10. Cold Spring Harbor Laboratory Press, Cold Spring Harbor, NY.

Schiestl, R. H., and Petes, T. D. (1991). Integration of DNA fragments by illegitimate recombination in *Saccharomyces cerevisiae*. *Proc. Natl. Acad. Sci. U.S.A.* **88**, 7585–7589.

Smith, C. L., Econome, J. G., Schutt, A., Klco, S., and Cantor, C. R. (1987). A physical map of the *Escherichia coli* K12 genome. *Science* **236**, 1448–1453.

Spudich, J. A., ed. (1987). *Dictyostelium discoideum:* Molecular approaches to cell biology. *Methods Cell Biol.* **28**, 1–516.

Sussman, M. (1987). Cultivation of *Dictyostelium*. *Methods Cell Biol.* **28**, 9–29.

Titus, M., Kuspa, A., and Loomis, W. F. (1994). The myosin family of *Dictyostelium:* A YAC-based approach to identifying members of a gene family. *Proc. Natl. Acad. Sci. U.S.A.* **91**, 9446–9450.

Watts, D. J., and Ashworth, J. M. (1970). Growth of myxamoebae of the cellular slime mould *Dictyostelium discoideum* in axenic culture. *Biochem. J.* **119**, 171–174.

Welker, D., and Williams, K. (1982). A genetic map of *Dictyostelium discoideum* based on mitotic recombination. *Genetics* **102**, 691–710.

11

Integrated Genome Mapping by Hybridization Techniques

Jörg D. Hoheisel,[1] Elmar Maier, Richard Mott, and Hans Lehrach

I. Introduction

Developments in the application of hybridization techniques have brought a detailed analysis of large genomic regions or indeed entire genomes well within the range of experimentation (for reviews, see Hoheisel and Lehrach, 1993; Hoheisel, 1994). Hybridization permits a parallel examination of large clone numbers, since the degree of handling per individual clone is small (Fig. 11.1). Integrated genome mapping (Lehrach et al., 1990) utilizes different libraries in parallel, so that information produced on one level of DNA handling (e.g., radiation hybrid cells, YAC clones, bacteriophage P1 and cosmid libraries, genomic λ and plasmid libraries, cDNA and exon-trap clones) will instantly assist the analysis on another. In the experimental stage, immediate integration and correlation of the data are already possible, because basically any piece of nucleic acid can be used as both a probe and a target. This feature also allows the selection of experimental arrangements that yield maximal gain in information. In principle, genomic analyses by the hybridization of a defined set of short oligonucleotides would be the most effective technique. It combines the benefits of a hybridization technique with the additional advantage that the amount of information obtained per experiment is independent of the genome size and that, in comparison to hybridiza-

<hr>

[1]To whom correspondence should be addressed. Fax: [+49] (6221) 42-4682.

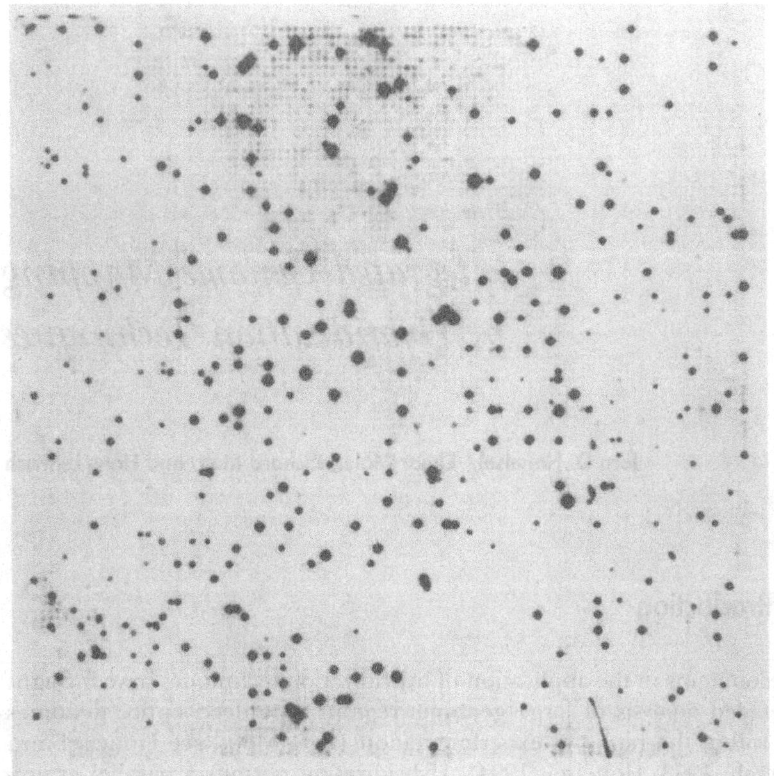

Figure 11.1 High-density cosmid filter. The DNA of 9216 (96×96) *E. coli* colonies was attached to the filter *in situ*. Hybridization was carried out with a pool of probes made from 12 cosmid inserts and autoradiographed.

tions of longer fragments, there are far smaller effects from repeat sequences. Technically, however, oligonucleotide hybridization is demanding and thus for smaller projects not necessarily the optimal technique. Hybridization mapping of mammalian genomes with unique probes utilizing cosmid and bacteriophage P1 libraries does not seem practical due to the large number of experiments necessary to accumulate the required amount of information. However, the technical simplicity makes unique probes the favorite tool for mapping projects on relatively short genomic regions of up to 30 or 40 million base pairs (Mbp). This is true especially for genomes that do not contain as many repetitive elements as mammalian DNA, since this vastly simplifies the isolation of unique marker molecules.

Here, we describe the basic techniques and strategies that were applied successfully to the mapping analysis of several genomes, in particular the completed mapping of the fission yeast *Schizosaccharomyces pombe* (Maier *et al.*, 1992; Hoheisel *et al.*, 1993), and work done on the fruit fly *Drosophila melanogaster* (Hoheisel *et al.*, 1991, 1994). The methodology does not require any exceptional experimental skills nor expensive equipment and, with some additions, such studies can be carried out in laboratories equipped for standard molecular-genetic work. The necessary computational tools are uncomplicated and were tested rigorously in practical applications. Thus even for relatively small groups, the means to analyze moderately sized genomes are at hand.

II. Materials

A. Solutions and Reagents

1. Stock Solutions

(1) 1 *M* Ammonium sulfate: 12.33% (w/w) ammonium sulfate, 87.67% (w/w) H_2O.

(2) 50 mg/ml Ampicillin: 4.76% (w/w) ampicillin (anhydrous), 12.84% (w/w) 1 *M* NaOH, 82.40% (w/w) H_2O; pH should be about 8.5.

(3) 1 *M* Dithiothreitol (DTT): 13.3% (w/w) dithiothreitol, 86.7% (w/w) H_2O; store at $-20°C$.

(4) 0.5 *M* EDTA, pH 8: 16.9% (w/w) EDTA (ethylendiaminetetraacetic acid disodium salt dihydrate), 1.9% (w/w) NaOH pellets, 81.2% (w/w) H_2O.

(5) 40% Glucose: 40% (w/w) glucose (water-free), 60% (w/w) H_2O; autoclave.

(6) 3 *M* KCl: 19.77% (w/w) KCl, 80.23% (w/w) H_2O.

(7) 1 *M* $MgCl_2$: 18.95% (w/w) $MgCl_2$ ($6 \cdot H_2O$), 81.05% (w/w) H_2O.

(8) 1 *M* Na-citrate (pH 5.8): 133.3 g Na_3-citrate, 46.5 ml 1 *M* citric acid, H_2O to 500 ml.

(9) 5 *M* NaCl: 24.64% (w/w) NaCl, 75.36% (w/w) H_2O.

(10) 1 *M* NaOH: 4 g NaOH pellets, 100 g H_2O.

(11) 1 *M* Na-phosphate (pH 7.2): 6.68% (w/w) Na_2HPO_4, 0.52% (w/w) H_3PO_4, 92.80% (w/w) H_2O.

(12) 2 *M* Sorbitol: 32.56% (w/w) D($-$)-sorbitol, 67.44% (w/w) H_2O (dissolves slowly); autoclave.

(13) 1 *M* Tris-HCl (pH 9.0): 9.70% (w/w) Trizma base (Sigma), 2.75% (w/w) Trizma hydrochloride (Sigma), 87.55% (w/w) H_2O.

2. Media and Buffers

(1) 2YT Growth medium/agar: 1.6% (w/w) bacto tryptone (Difco), 1% (w/w) bacto yeast extract (Difco), 0.5% (w/w) NaCl, 96.9% (w/w) H_2O; autoclave. For 2YT agar add 1.5% (w/w) agar before autoclaving.

(2) 5× PCR amplification buffer [250 mM Tris-HCl (pH 9.0), 75 mM ammonium sulfate, 35 mM $MgCl_2$, 250 mM KCl, 0.85 mg/ml BSA]: 2.5 ml 1 M Tris-HCl (pH 9.0), 0.75 ml 1 M ammonium sulfate, 0.35 ml 1 M $MgCl_2$, 0.083 ml 3 M KCl, 0.85 ml 10 mg/ml BSA, 5.647 ml H_2O. Store in aliquots at $-20°C$.

(3) 10× H.M.F.M. freezing medium (1× concentration: 36 mM K_2HPO_4, 13.2 mM KH_2PO_4, 0.4 mM $MgSO_4$, 1.7 mM Na_3-citrate, 6.8 mM $(NH_4)_2SO_4$, 4.4% (v/v) glycerol): 0.76 g $MgSO_4$ (7·H_2O), 4.50 g Na_3-citrate (2·H_2O), 9.00 g $(NH_4)_2SO_4$, 440.00 g glycerol; add water to 800 ml and autoclave; 18 g KH_2PO_4, 47 g K_2HPO_4; add water to 200 ml and autoclave. Mix both solutions to make up final solution.

(4) AHC selective medium/agar: 3.35 g yeast nitrogen base, 5 g casamino acids, 5 mg adenine hemisulfate, 475 g H_2O; autoclave, cool to about 65°C and add 25 ml 40% sterile glucose. For agar add 15 g agar before autoclaving.

(5) Filter denaturing solution (0.5 M NaOH, 1.5 M NaCl): 1.87% (w/w) NaOH-pellets, 8.15% (w/w) NaCl, 89.98% (w/w) H_2O.

(6) Filter neutralization buffer [1 M Tris-HCl (pH 7.6), 1.5 M NaCl]: 1322.0 g Trizma hydrochloride (Sigma), 194.0 g Trizma base (Sigma), 876.6 g NaCl, 9000 g H_2O.

(7) Filter processing buffer [50 mM Tris-HCl (pH 8.5), 50 mM EDTA, 100 mM NaCl, 1% (v/v) sodium sarcosyl, 0.25 mg/ml proteinase K or pronase]: 186.1 g EDTA, 58.4 g NaCl, 315 g (300 ml) 30% Na-sarkosyl, 33.3 g Trizma base (Sigma), 35.3 g Trizma hydrochloride (Sigma), 9620 g H_2O; add 0.5 g pronase or proteinase K to 1800 ml buffer and use 600 ml per filter. The solution can be reused.

(8) Synthetic dextrose medium (2% glucose, 0.7% yeast nitrogen base without amino acids, 1.4% casamino acids, 100 μg/ml adenine hemisulfate, 55 μg/ml tyrosine): 50 ml 40% glucose, 7 g yeast nitrogen base without amino acids (Difco), 14 g casamino acids (Difco), 0.1 g adenine hemisulfate, 0.055 g tyrosine, 950 ml H_2O; autoclave.

(9) Hybridization buffer [0.5 M sodium phosphate (pH 7.2), 7% SDS, 1 mM EDTA]: 6.95% (w/w) SDS, 0.20% (w/w) 0.5 M EDTA, 49.75% (w/w) 1 M Na-phosphate (pH 7.2), 43.10% (w/w) H_2O.

(10) SCE (1 M sorbitol, 0.1 M sodium citrate, 10 mM EDTA, 10 mM dithiothreitol): 25 ml 2 M sorbitol, 5 ml 1 M Na-citrate (pH 5.8), 1 ml 0.5 M EDTA, 19 ml H_2O.

(11) YPD medium/agar: 5 g yeast extract (Difco), 10 g bacto-peptone (Difco), 475 g H_2O; autoclave, cool to about 65°C, and add 25 ml 40% sterile glucose. For agar add 15 g agar before autoclaving.

B. Special Materials

Listed below are some nonstandard materials used for the techniques described under *Procedures* (further information can be requested from the authors).

(1) 384-well dishes (four interleaving grids of 96 wells of normal spacing in standard format dishes; maximal volume per well 70 μl; Cat.No. X5001) and 384-pin replicators (Cat.No. X5050) can be purchased from Genetix Ltd., UK.

(2) 96-well replicators are available from Sigma–Aldrich Techware (Cat.No. R2508); information about the design of 12-prong wheels for clone picking can be requested from the authors.

(3) The cosmid vectors of the Lawrist series have been described by de Jong *et al.* (1989; contact Pieter J. de Jong, Human Genetics Department, Roswell Park Cancer Institute, Elm & Carlton Streets, Buffalo, NY 14263, e-mail pieter@dejong.med.buffalo.edu). The average insert size was found to be 37 kb with DNA from various sources.

(4) Hybond N+ filter membranes are sold by Amersham Life Science (Cat.No. RPN 2222B).

(5) Novozym for the conversion of yeast cells to spheroplasts is available from Sigma (Cat.No. L 3768).

(6) The program package for the handling and analyses of hybridization results for genome mapping which is referred to in this manuscript was described by Mott *et al.* (1993) and is available from the EMBL software server Netserv@EMBL-Heidelberg.DE. The software is written in C and runs on a SUN SPARCstation I, II, and IPX, running SUNOS 4.1.1 and the OpenWindows window manager, version 3.

III. Procedures

A. Tools for Hybridization Analyses

1. Creation of Libraries for Complete Genome Coverage

Five is a mystic number for scientists generating genomic libraries, because with a statistical probability of 99% every fragment of a cloned genome will be present at least once in a library of fivefold coverage

(Clarke and Carbon, 1976). While such clone coverage is generally satisfactory for the identification and isolation of specific DNA pieces by library screening, it is inadequate for the continuous mapping of large genomic areas. Not only will there be a few real gaps, but more importantly there will be regions of clone representation that is insufficient for unambiguous map generation. The phenomenon is intensified further on account of variations in the cloning efficiency, an effect that becomes more manifest as the insert sizes get smaller. To obtain a sufficient and homogeneous DNA coverage three strategies can be followed, with the best results to be expected when all three are applied simultaneously.

First and most importantly, a coverage well above five genome equivalents should be sought. Calculations on the basis of the *S. pombe* mapping experiments indicated that a 10-fold representation was the minimum necessary for the formation of a continuous and unambiguous YAC map of the genome (Maier *et al.*, 1992). Higher degrees of redundancy proved to be extremely helpful for the analysis, and were actually essential for resolving problematic areas at the cosmid and P1 level (Hoheisel *et al.*, 1993). Library construction and data acquisition by hybridization from larger libraries is comparably far less work-intensive and time-consuming than a subsequent investigation of unresolved regions. In short, the bigger the library the better.

Second, different techniques should be used for the fragmentation of the insert DNA in an attempt to avoid a method-induced bias in representation. Partial enzymatic digestion is prone to representational variations due to site preferences or the local absence of sites in certain regions. To compensate for the former as much as possible two enzymatic techniques were applied in actual experiments (Hoheisel *et al.*, 1991; Larin *et al.*, 1993). Either DNA was partially cut with limited amounts of restriction enzyme alone (*Mbo*I for cosmid/P1 clones; *Eco*RI for YACs) or the digestion was carried out by a combined reaction of the restriction enzyme and its related methylase (Dam methylase in case of *Mbo*I; Hoheisel *et al.*, 1989). The latter method allows for easy control of the frequency of cleavage by the ratio of the two antagonistic enzymes and permits the generation of representative *E. coli*-based clone libraries even from very limited amounts of DNA, such as flow-sorted chromosomal material (Nižetić *et al.*, 1991) or gel-purified YAC-DNA (Whittaker *et al.*, 1993). Alternatively to the enzymatic cleavage, genomic DNA was sheared mechanically, enzymatically treated to produce blunt ends, ligated to appropriate adapters, and subsequently size-selected on a gel, before being inserted into a vector (Hoheisel *et al.*, 1991; Ajioka *et al.*, 1991). Although less biased, more material is needed for this procedure.

On three cosmid libraries, made by the above methods from *D. melanogaster* (Hoheisel *et al.*, 1991), the hybridization of 107 single-copy markers yielded on average 1.40 clone per genome equivalent, the standard deviation being 0.62. Taking into account remaining biases, i.e., that euchromatic regions are most likely better represented in the libraries than heterochromatin due to a better clonability, the above value comes fairly close to the ideal of 1.

The third mechanism for coming closer to uniform coverage is a combined use of different cloning systems. Due to differences, like host (e.g., yeast for YACs; *E. coli* for cosmids and P1) or copy number (single copy: YACs, P1; several copies: cosmids), for instance, a given DNA fragment is reasonably likely to be cloned in at least one system. Biases may be reduced further by a substitution of biological components of an experimental set-up with physicochemical procedures, such as the replacement of the packaging of P1 phages for bacteria transfection (Pierce *et al.*, 1992) by electroporation (Ioannou *et al.*, 1994) or the complete avoidance of biological hosts by using PCR amplification of the DNA.

Apart from the improvement in representation, any mapping analysis is considerably simplified by an exploitation of a multitude of vectors on account of their different insert sizes. "Short" insert clones, of which a relatively high percentage lies in between repeats, can be superior in an initial phase of mapping unique genomic regions, for example, and are advantageous as probes for the same reason. Long DNA fragments on the other hand are often better suited for spanning problematic regions. In the mapping of *S. pombe*, for example, unambiguous cosmid contigs could not be built across the centromeres. Placing cosmid clones within particular centromeric regions, however, was requisite for the identification of P1 clones that specifically cover each of the three centromeres. For their length, YAC clones are widely used for final gap closure, as in the *Caenorhabditus elegans* and *Drosophila* mapping projects (Coulson *et al.*, 1991; Merriam *et al.*, 1991). Concerning YAC mapping, however, one has to be cautious about a preoccupation with creating very large clones. Dependent on the length of the DNA segment under investigation, YACs that are too large can rather be a disadvantage for their lack of accessibility to a detailed analysis of specific regions. Also, any significant increase in average clone length is usually offset by a reduction in transformation efficiency. Unless a genome is of a very substantial size, the clone length is not that critical a point for sufficient coverage; a mere 250 YACs of 200 kb each, nowadays a rather low average size, would cover 50 Mbp once, for example. Detailed protocols for the generation of representative libraries in the various vector systems can be found in the literature (e.g., cosmids: Hoheisel *et al.*, 1991; P1 phage: Smoller *et al.*, 1991; YACs: Larin *et al.*, 1993) and are given elsewhere in this issue (Chapters 7 and 9).

2. Generation of High-Density Clone Filters

One major advantage of hybridization analyses is the capability of parallel examination of large clone numbers, ranging from hundreds to multiple tens of thousands. However, while filter lifts from libraries plated in a random pattern are sufficient for the identification of clones in a few hybridization experiments, a great deal of information has to be gathered for an unambiguous characterization (fingerprinting) of every clone, as is required for map assembly. Although filters can be reused, many copies of identical hybridization targets are a prerequisite for such elaborate analyses, because a multiplication in productivity by the simultaneous use of different probes is imperative. An arrangement of the clones in ordered grids very much simplifies signal registration and clone identification. It is essential for an automated image analysis, which is obligatory for very large projects (mapping the human genome, for instance), and for a quantification of signal intensities. These three advantageous features—accessibility of large clone numbers, the availability of many identical copies, and a presentation in an ordered format—are combined in the scheme of high-density reference filters (Lehrach *et al.*, 1990).

Primary transformants are plated randomly and individual colonies are picked into multiwell dishes for growth and storage (Protocol 1). From these multiwell dishes, minute amounts of the cultures are transferred by a pin-gadget onto membranes of 22×22 cm in interleaving patterns (e.g., Fig. 11.1). The filters are put onto agar plates and incubated for colony growth. The DNA is attached to the filter by two steps of denaturation and neutralization, after which cellular protein is removed with proteinase K (Protocol 2). The proteinase K buffer can be reused at least six times (Fig. 11.2). The processing of yeast colony filters includes an additional step of converting the cells to spheroplasts prior to the above procedure. *Escherichia coli* filters were also digested with the much cheaper pronase instead of proteinase K, exhibiting the same durability in repeated hybridization (>30). More recently, a quicker technique has been developed by David J. Munroe (Massachusetts Institute of Technology, Cambridge, MA; personal communication), binding the DNA of freshly grown colonies by autoclaving the filters for 90 sec, followed by UV irradiation and drying. *E. coli* libraries that were treated this way could be hybridized repeatedly under the conditions described below, producing signal intensities similar to filters which had been processed as described in Protocol 2.

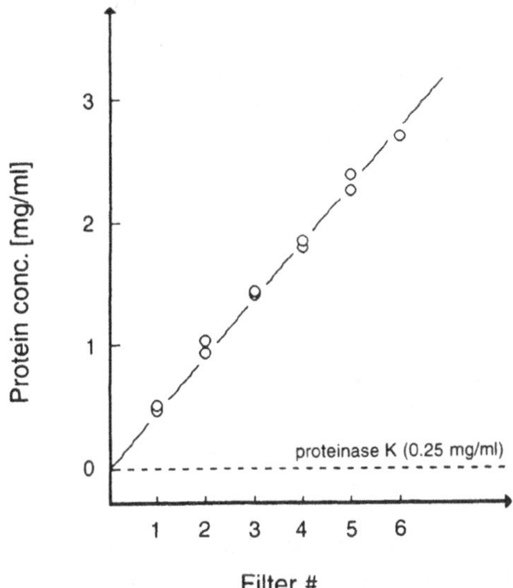

Figure 11.2 Reuse of the proteinase K buffer during filter processing. As determined by measuring the absorption at 260 and 280 nm, per *E. coli* filter of 22×22 cm about 270 mg protein is released into the buffer in six subsequent incubations.

a. Protocol 1: Multiwell Dish Storage of Clone Libraries

E. coli:

(1) Grow bacteria on selective agar plates made of 2× YT agar and antibiotic (e.g., 100 μg/ml ampicillin or 30 μg/ml kanamycin) at 37°C overnight.

(2) Fill multiwell dishes with about a 0.4-fold well volume (384-well dish: 30 μl; standard 96-well dish: 80 μl) of 2× YT growth medium supplemented with 1/10 vol of 10× H.M.F.M freezing medium and the relevant antibiotic.

(3) Pick one individual colony into each well. Wrap up stacks of dishes with Saran Wrap (Genetic Research Instrumentation) and grow overnight at 37°C without shaking.

(4) Freeze and store the plates at −70°C.

Yeast:

(1) Grow colonies for about 3 days at 30°C on inverted selective agar plates lacking uracil and tryptophan (AHC agar).

(2) Pick individual colonies into separate wells of standard 96-well dishes filled with 100 μl YPD medium.

(3) Wrap in Saran Wrap and incubate at 30°C for 1 to 2 days.

(4) Add 100 μl YPD medium plus 40% glycerol to the wells and mix. Transfer half the volume to a fresh 96-well dish, or split in four 384-well dishes.

(5) Freeze and store at −70°C.

b. Protocol 2: Filter Processing

Processing of E. coli filters:

(1) Prior to the spotting of *E. coli* bacteria, wet the Hybond N^+ filters (22×22 cm) in 2YT medium; keep on absorbent paper pads (23×23 cm; Whatman 3MM) soaked in 2YT medium during spotting.

(2) Grow colony filter overnight on agar plates containing the respective antibiotic; subsequently store the filter plates at 4°C for up to 1 week. Grow P1 clones overnight at 30°C, transfer the filter to a fresh agar plate supplemented with 1 m*M* IPTG, and grow at 37°C for another 6 to 8 hr for an induction of the copy number.

(3) Soak two sheets of 3MM-paper in filter denaturing solution.

(4) Transfer filter to first 3MM-sheet, colony side up, and leave for 4 min at room temperature.

(5) Move filter onto second sheet and put both on a glass plate that is positioned above a steaming waterbath (>90°C); keep in the steam atmosphere of the covered bath for 4 min.

(6) Put filter onto a third sheet of 3MM-paper soaked in filter neutralizing buffer for more than 5 min.

(7) Prewarm filter processing buffer to 37°C and incubate each filter separately in 500–600 ml for 30 min *without* shaking; the buffer can be reused at least six times.

(8) Blot the filter on 3MM-paper and dry at room temperature for more than 36 hr.

(9) UV-crosslink the DNA to the filter (254 nm, 1200 μJ).

Processing of YAC filters:

(1) Premoisten filter in synthetic dextrose medium containing 0.02% tryptophan and 0.25% calcium propionate [the latter to act as an antifungal agent (Ross *et al.*, 1992)].

(2) Spot cells under a sterile air flow; grow colonies on agar plates of the above medium for 2 to 3 days at 30°C.

(3) Transfer filter to 3MM paper soaked in SCE and 4 mg/ml Novozym (Sigma); cells are converted to spheroplasts by overnight treatment at 37°C under these conditions.

(4) Expose filter to denaturing solution for 15 min as above. Dry at room temperature for 5 to 10 min on fresh 3MM-paper, and neutralize by floating on filter neutralizing solution for 5 min.

(5) Float filter on 0.1 M Tris-HCl, pH 7.6, 0.15 M NaCl (filter neutralizing solution, diluted 1:10) for 5 min and subsequently digest each filter separately in 500 ml of the same buffer containing 0.25 mg/ml proteinase K for 30 to 60 min at 37°C.

(6) Blot the filter on 3MM-paper, air-dry at room temperature for more than 36 hr, and UV-crosslink the DNA.

3. The Necessity of Robotics?

Many investigators seem to be reluctant to deal with "large" clone numbers, for one because they think that elaborate and costly technical investments are required for their management. One needs to put the numbers in perspective, though. A single multiwell dish of 384 wells containing individual cosmid clones already represents about 14-fold coverage of a 1-Mbp DNA region, a value that more than doubles with a P1 phage library. A *20 genome equivalent* cosmid library of the *S. pombe* yeast genome, for instance, requires 7680 clones which fit into only 20 384-well dishes. Picking such numbers of clones does not necessitate the use of robotics, either. With simple mechanical tools (e.g., picking wheels of 12 prongs that roll into standard 96-well plates) a single person picks more than 4000 clones each working day. Duplicators of 96 or 384 pins allow for an easy and quick clone transfer from dish to dish and dish to membrane. Manual gridding onto filter membranes can be accomplished rather effectively. A clone density four times that of a 384-well plate (one interleaving pattern in each dimension; Fig. 11.1) can be administered with reasonable effort and spots 9216 clones on filters of 22×22 cm. Such big filters are not even necessary for many studies. Bacteriophage P1 filters of the size shown in Fig. 11.5 were actually used in the mapping of *S. pombe* and contain a 17-fold coverage of the yeast genome. The creation of a robotic infrastructure consisting of machines for clone picking, PCR amplification, filter spotting, and automated image analysis on a scale reported by Meier-Ewert *et al.* (1993) is only practical if very large numbers are being dealt with, as for studies on very large (e.g., mammalian) genomes and/or for an organization in the form of a resource center that provides services to many different projects. The value of such centers and their benefit to a wider group of laboratories has clearly been demonstrated by the impact of organizations such as the Human Polymorphism Study Center (C.E.P.H.) or the ICRF Reference Library System, but they are not required for an *individual* mapping project of moderate size, although extremely useful if at hand.

B. Hybridization Techniques

1. Probe Generation

Most of our technical expertise was compiled with radioactive label, mainly ^{32}P. Still, some experiments were carried out with nonradioactive labeling methods. Apart from the specific system-inherent characteristics, matching results were obtained and similar basic problems occurred.

For probe generation, the random hexamer priming procedure (Feinberg and Vogelstein, 1983) seems to be the most efficient labeling technique. One has to keep in mind that even for mapping moderate-sized genomes, hundreds of probes have to be generated. At least during an initial phase, the random priming procedure keeps the manipulative effort small and thus permits rapid data accumulation. Since the whole DNA is used as a probe, repetitive elements contained in the DNA are initially identified and localized as a contig break flanking a unique region. A physical map, particularly if generated as a preparation for a subsequent sequence analysis, should preferably indicate the locations of repeats rather than try to avoid them, even more so in view of their potential molecular significance, for example, the role of trimer repeats in the manifestation of some genetically determined diseases. If necessary, different experimental procedures (e.g., Protocol 3) can be used in a second phase to bridge any gaps resulting from the repeats.

Plasmid or cosmid DNA isolated by standard alkaline lysis procedures are adequate as templates of a random priming reaction, as are PCR-products and gel-purified YACs. For P1 DNA better purification is required, since the yield is comparatively low, even after induction to increase the copy number, and a quick preparation usually contains relatively more chromosomal DNA. The colabeling of the vector in addition to the insert is much less of an experimental problem than it seems to be at first sight, for one because often there is no cross-hybridization between different vector systems, for example, the cosmids of the Lawrist-series and YAC vector pYAC4 (Hoheisel *et al.*, 1993). But also, hybridization of a probe back to the same library type, the worst-case scenario, is quite satisfactory. An average recombinant Lawrist cosmid, for instance, contains about 5.5 kb vector and 37 kb insert DNA. In a hybridization to a cosmid library the probe from the vector portion is diluted, because it binds to all clones present, while the insert DNA binds to only a relatively small number of clones. In the *S. pombe* project, probes were usually hybridized to filters containing 3000 cosmid clones, representing the genome about eight-fold. Under those conditions, theoretically a 60-bp nonrepetitive portion of a cosmid insert should generate a signal as intense as the 5500-bp vector DNA. In practice, hybridization of substantially longer stretches was required for clear identification of positive signals due to experimental variations such as the deviation in signal intensities across a filter and the occurrence of background signal due to *E. coli* chromosomal DNA in the probe and unspecific binding effects. Rather than the ideal average of 16 clones (eightfold coverage of either clone end), about 12 clones were actually identified per hybridization (Hoheisel *et al.*, 1993). Still, this means that overlaps of less than 10% of the insert length were identified.

From the above number it follows that hybridizations of pooled clones to a library of the same vector type demand a better probe, avoiding the vector contamination. This can be achieved by a polymerase extension from known vector sequences located directly adjacent to the insert termini, either by RNA polymerase reactions (Sambrook *et al.*, 1989), if the promoter sequences are present, or by an extension of oligonucleotide primers that specifically anneal to the vector-DNA (Protocol 3). By cycling the latter reaction in a pseudo-PCR incubation (Saluz and Jost, 1989) high probe concentrations can be generated. Incorporation of a label at the 5' end of the oligomer primer further reduces nonspecific background. Probe produced from mixtures of up to 84 cosmid clones has been hybridized successfully. However, smaller pools are in fact more practical, because of the lower probability of labeling a repeat element. Pool hybridizations are most effective, and worth the additional manipulations involved, when the probes are arranged in pooling schemes which allow each hybridization signal be related to a particular probe (e.g., two-dimensional matrices, Fig. 11.3; Evans and Lewis, 1989). Repeat sequences in a probe pool, however, make it more complicated or even impossible to discern the relation between the observed signals and the individual probes responsible, and thus transform this and all intersecting pools into multilocus probes rather than a group of unique probes. The pool size for optimal efficiency mapping should be deduced for each project from the number of experiments saved by increased pool size and the expected or observed frequency of repeat sequences in the clone library.

a. Protocol 3: Generation of End-Specific Insert-Probe by Primer Extension

(1) Per sample (cosmids made with a Lawrist-series vector), use 3.5 pmol of the 21mer d(TAGGGAGACCGGAAGCTTAGG) or the 30mer d(CATACACATACGATTTAGGTGACACTATAG) in separate reactions. Both primer molecules have a T_m of 60°C, calculated with the program described by Rychlik and Rhoads (1989).

(2) 5' label the oligomer with 25 μCi [γ-^{32}P]ATP (5000 Ci/mmol) and 5 units T4 polynucleotide kinase for 1 to 2 hr at room temperature.

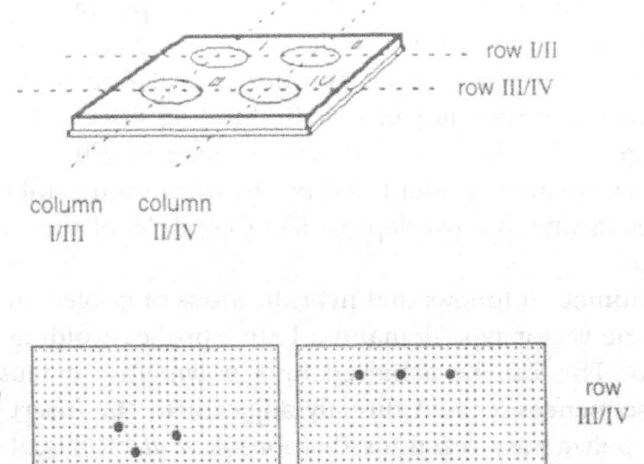

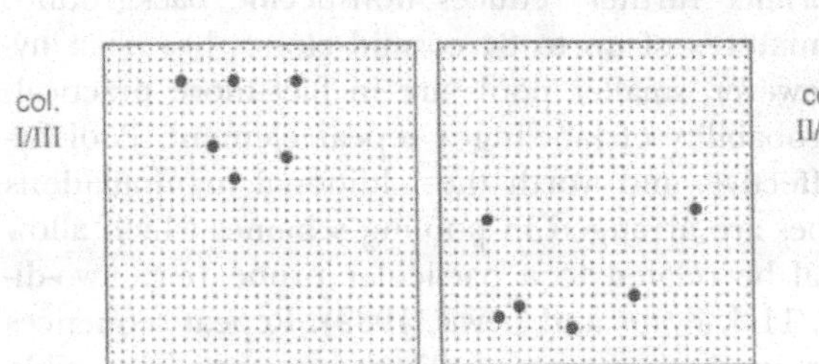

Figure 11.3 Clone arrangement in a two-dimensional matrix. Each clone (I to IV; represented by a schematic microdish well) is present once in two different pools. While the number of hybridization experiments only increases with double the square root of the number of clones (e.g., 24 for 144 clones), each hybridization signal is still interrelated to the corresponding probe. More-dimensional schemes are possible.

(3) Purify oligomer by phenol and chloroform/isoamylalcohol (24:1) extractions, add 0.5 µl 5 *M* NaCl, and precipitate with 3.5 vol ethanol at −70°C for 1 hr.

(4) Spin for 40 min at 14 krpm and take the DNA up in 13 µl water; use 0.5 µl as a control.

(5) Mix 12.5 µl with 10 µl 5× PCR amplification buffer, 3.3 µl of all four nucleotides (5 m*M* each), 1.5 µl AmpliTaq DNA polymerase (8 units; Perkin–Elmer Cetus), and 23 µl cosmid DNA; use 50 to 200 ng per cosmid, and up to 48 cosmids per sample.

(6) Carry out primer extension with 60 cycles of 2 min 92°C, 2 min 42°C, and 2 min 73°C. Check the efficiency of extension on an acrylamide gel.

2. Hybridization

For all probes longer than about 200 nucleotides, the procedure listed in Protocol 4 was used for hybridization. However, homologous DNA segments as short as 25 bp were found responsible for a specific signal under these conditions, with hybridization carried out at 50°C and washing at 30 to 50°C. Hybridizations to YAC filters are most reliable with fragments larger than 800 bp. Up to 15 filters of 22×22 cm have been hybridized simultaneously. In our hands, agitation of the hybridization liquid was found unnecessary to assure probe hybridization between the filter layers. In boxes, enough volume has to be added to just submerge the filter(s). When performed in bags, a uniform distribution of the liquid within each bag is important to avoid differences in the background intensities across the filters. This can be achieved by pressing the bags between perspex plates, for instance. At a total probe concentration above about 10^6 cpm/ ml increasing nonspecific binding occurs. Therefore, in experiments with complex probes, an extreme example being probe pools made from total mRNA preparations (Gress *et al.*, 1992), hybridization in larger volumes of lower probe concentration for longer periods is advantageous to reduce the degree of nonspecific binding. On cosmid filters a much higher degree of nonspecific hybridization to clones that contained only vector concatemers was observed. Particularly at high probe concentrations this signal can become as intense as those from specific hybridization (Fig. 11.4). Following the procedure given in Protocol 4, YAC filters could be used 3 or 4 times, while *E. coli* filters were hybridized more than 30 times without major reduction of the signal intensities. The lifespan of YAC-filters extends, when no stripping is carried out between probings. However, this should only be done in an initial mapping phase, since overlaps can easily be missed.

a. Protocol 4: Clone Hybridization

(1) Prehybridize library filters in hybridization buffer supplemented 0.1 mg/ml yeast tRNA at 65°C for 2 hr to overnight.

(2) Hybridize in the same buffer at 65°C overnight, the probe concentration being 0.2 to 0.5 Mcpm/ml.

(3) Briefly rinse at room temperature with 40 mM sodium phosphate, pH 7.2, 0.1% SDS.

(4) Add the same buffer of room temperature and wash the filters once rocking slowly in a waterbath of 65°C for 10 min (YACs) or 15 to 30

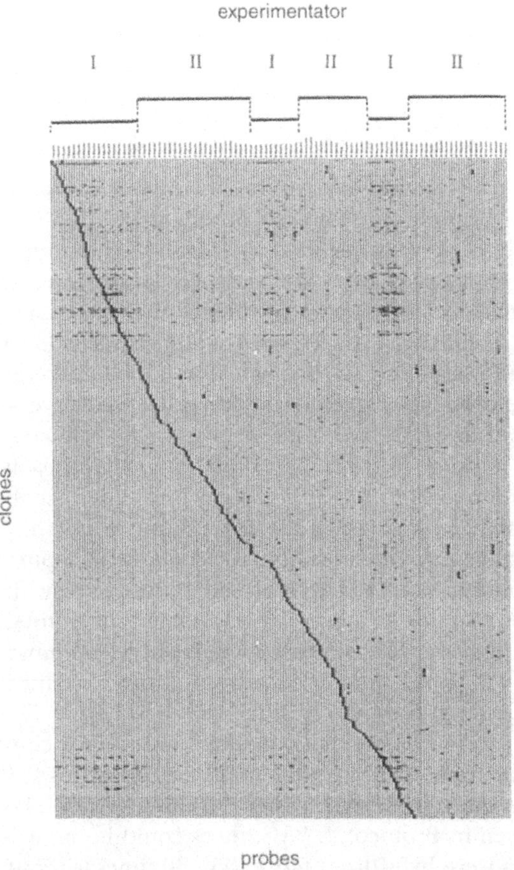

Figure 11.4 Quality control of *S. pombe* hybridization data. The vertical lines represent probes, listed in order of their entry date into the database. Clones correspond to rows. Positive hybridization of a probe to a clone is indicated by a short horizontal line. The signal-diagonal is artificial, since the computer program shuffled the clones trying to arrange them into contigs. Unspecific signals produced by experimentator I on a distinct subpopulation of the clones (predominantly vector concatemers) by using high-radioactivity concentrations are clearly identifiable at top and bottom.

min *(E. coli);* only filters hybridized with identical probe are washed together.

(5) Drain the filters of excess liquid and expose film 2 hr to overnight at −70°C using intensifying screen.

(6) To strip a probe off pour 1 to 2 liters of 5 mM sodium phosphate, pH 7.2, 0.1% SDS at room temperature into a box containing up to 20

filters and keep in a waterbath of about 90°C for 30 min, after which the procedure is repeated.

(7) Between hybridizations, keep filters at prehybridization conditions for up to 3 days or store them air-dried at room temperature.

3. Signal Scoring

Hybridization results on autoradiographs can be read manually or scored with digitizing equipment as used for reading sequencing gels, with slight adaptations to the software; the results can also be recorded by commercially available scanning or CCD camera equipment. Alternatively, the signal intensities of radioactivity can be detected directly, for example, by phosphor imaging (Molecular Dynamics). As with the clone handling, the access to such means is very helpful, but even large projects do not necessitate *full* automation. Already simple tools very much assist in the scoring process (Fig. 11.5). A manual analysis has the intrinsic advantage that all the raw hybridization results are examined in detail, which in case of the *S. pombe* project, led to several important observations, such as the occurrence of well-to-well contaminations in the clone libraries. Subsequently, appropriate software tools could be developed to deal with the recognized problem. By using the grids shown in Fig. 11.5, identifying the clones is uncomplicated and reliable, even at a high clone density and despite shrinking of the filter material and other variations caused by repeated hybridization. A fully automated image analysis is sensible for very large clone and hybridization numbers, and actually essential for hybridizations with very short oligomers, but requires a rather sophisticated experimental design and software (R. Mott and S. Meier-Ewert, personal communication).

C. Map Construction

1. Mapping Strategy

For an efficient analysis of large genomic areas it is advantageous to use entirely anonymous clones as probes. In such a case, no information prior to the start of the project is necessary. The ordering process will not rely on the incidence of other mapping data, although any information that is available represents a bonus. Two basic configurations are possible: either the probes are made out of the clone library that is analyzed or they originate from a different source. Depending on the experimental structure, different strategies are required.

The former approach allows a directed probe selection process by "sampling without replacement" and has been applied successfully in the

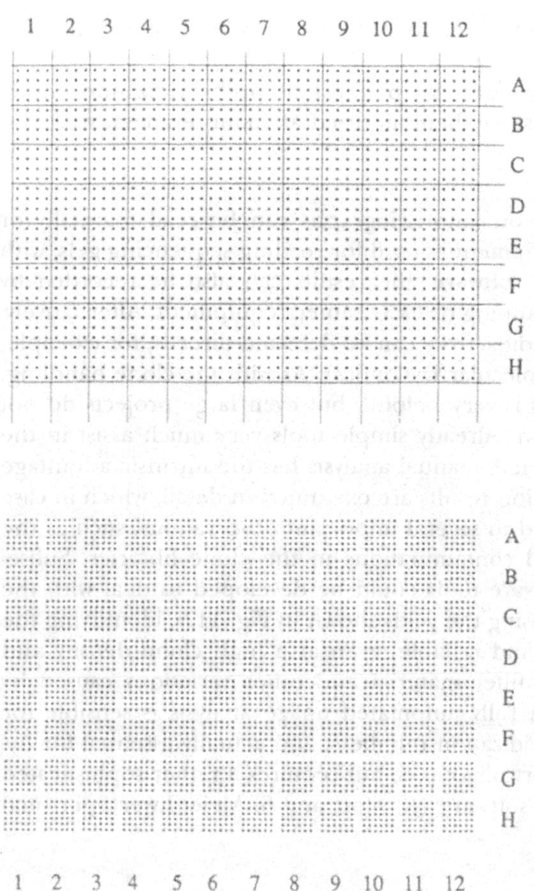

Figure 11.5 Signal score sheets; enlarge by factor 1.66 to original size. On a filter the size of a multiwell dish the clones from 4/16 (top) or 9/36 (bottom) 384/96-well dishes are spotted according to these patterns and probe hybridized. By overlaying the resulting autoradiograph with a transparent to which the score sheet has been copied, the microdish location and thus the "name" of each positive clone can be determined.

cosmid/P1-mapping of the *S. pombe* genome (Hoheisel *et al.*, 1993; Mizukami *et al.*, 1993). Clone DNA is isolated and hybridized back to the entire library. For the following round of hybridizations, the probes are picked at random from the ever-decreasing number of library clones that were not positive in any prior experiment. Genetic markers can be used as if they were part of the clone library. The process is repeated until all clones have been hit. By this strategy, the probes, although anonymous (apart from included markers), are relatively evenly spaced throughout the genome. Also, a redundant analysis of existing contigs is avoided. The efficiency of the iterative selection procedure can be improved further, if ordered DNA fragments of lower resolution already exist. By hybridizing a representation of the low-resolution map to the yet unordered library (e.g., YACs or *Not*I fragments to cosmids or P1 phages) the clones are subdivided. Rather than picking from the unhit portion at random, probes can be isolated from different subdivisions for each new cycle of hybridization experiments (Fig. 11.6), avoiding redundant probings.

Once all clones have been hit by at least one probe and, thus, are arranged in contigs, the terminal clones of most neighboring contigs should overlap, but these overlaps cannot be detected for the lack of probe in the overlapping region. Thus, in a second phase, the contig end clones are used as probes in order to define the overlaps. All remaining gaps occur because the relevant piece of DNA is not present in the library or the overlap is too short to be identified; a third reason for interruptions is long stretches of repeat sequences that cannot be bridged unambiguously.

In a project in which clones from one library are used as probes on another, for example, the ordering of YACs by cosmid hybridization (Maier *et al.*, 1992), the strategy described above does not apply. Probes should be picked randomly until about 75% of the (YAC) clones have been mapped. In terms of efficiency, at this point the degree of redundant information produced for the already existing contigs usually begins to outweigh the simplicity of the probe selection, and a change of strategy becomes advantageous. A converse hybridization of the existing (YAC) clone contigs to the (cosmid) *probe* library will find all the (cosmid) probes which underlie the contigs, thereby indirectly identifying the (cosmid) probes that belong to the unmapped 25% of the (YAC) clone library. By definition, hybridization of these probes in turn will provide information specifically on the yet unmapped areas. One reason for gaps that cannot be bridged in this fashion could be an insufficient coverage of the probe library rather than of the clone library. In such a case, probes isolated from the ends of the relevant clones themselves should circumvent the problem.

2. Contig Assembly

Since there are far fewer probes than library clones, ordering the probes rather than the clones is a more efficient procedure. Moreover, a comparison of two probes is based on the information from the relatively large number of positive clones (e.g., in a library of 10-fold genome cov-

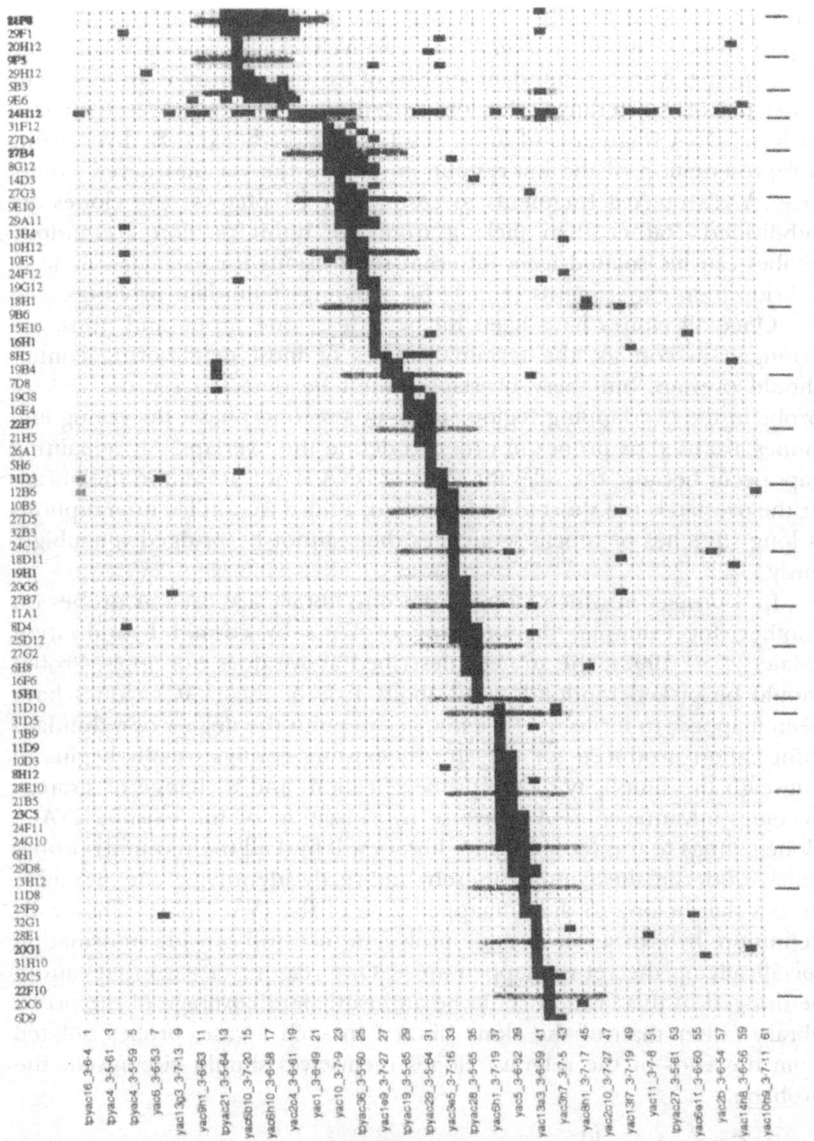

Figure 11.6 Probe selection during the cosmid mapping of *S. pombe*. Clones that had not been hit in earlier cosmid-to-cosmid hybridizations were sorted on the basis of the results of hybridizations with already ordered YAC clones. More evenly spaced probe-clones (marked by shaded horizontal lines) could so be selected from different subdivisions of the cosmid library for the next batch of hybridization experiments. YAC probes are represented by columns, cosmid clones by rows. Positive hybridization is indicated by a black square.

erage, statistically 10 clones per probe). A clone–clone comparison, on the other hand, relies on less data, as the result of the small number of probes binding to each particular clone. Hence, ordering probes is more reliable. As a consequence, a map is defined as a sequence of probes, to which the clones subsequently are fitted according to the very same hybridization results that are already the foundation of the probe map. Such a map is user-friendly, since it is short and easy to edit and modify.

For the mapping of *S. pombe*, ordering algorithms and programs were developed along this line, one based on distance measurements, the other applying heuristic rules to clean the data into a consistent set (Mott *et al.*, 1993). The former algorithm was used more extensively, although both methods produced nearly identical results. The distance between each possible pair of probes is calculated from the percentage of clones hybridizing only to one probe but not to both (Fig. 11.7). If two probes are identical, for example, the numerator (number of clones the probes do *not* share), and thus the whole ratio, becomes zero. Unrelated probes have no clones in common; numerator and denominator (sum of the clone numbers found by probe A and probe B) are identical. The ratio therefore equals one, meaning that the probes are one clone length (or more) apart. In the intermediate case depicted in Fig. 11.7, each probe binds to three clones, with only one clone found by both probes. The calculated distance ratio is $(2+2)/(3+3)=0.66$.

From such distances, the shortest possible linear succession of all probes is calculated by a simulated annealing algorithm, and the probes are ordered in contigs. A contig break occurs when the distance to the nearest probe(s) exceeds a variable threshold (Fig. 11.8). All potentially inconsistent hybridizations, for example, linking a clone in one contig to a probe in another, are listed. In the initial analysis, a reasonable threshold usually is one that is roughly equivalent to a two-clone link between a pair of probes. Signals due to experimental noise are largely avoided, but not too rigorous a selection is attached to the contig building. At later stages, a higher value should be applied for confirmation. Some of the initial contigs will break up, but by then many potential, though erroneous, links are eliminated, for the probes in question map elsewhere, making the identification of real links easier. As with the accuracy of DNA sequencing, this value is the experimentor's choice. The more clones define a probe connection the less likely it is a random event and the better the quality of the eventual map.

Intentionally, mapping data other than the anonymous hybridization results are not *directly* integrated into the probe-order process. An independent generation of the same probe order in different map types is the best assurance for an accurate map. An independent contig building and

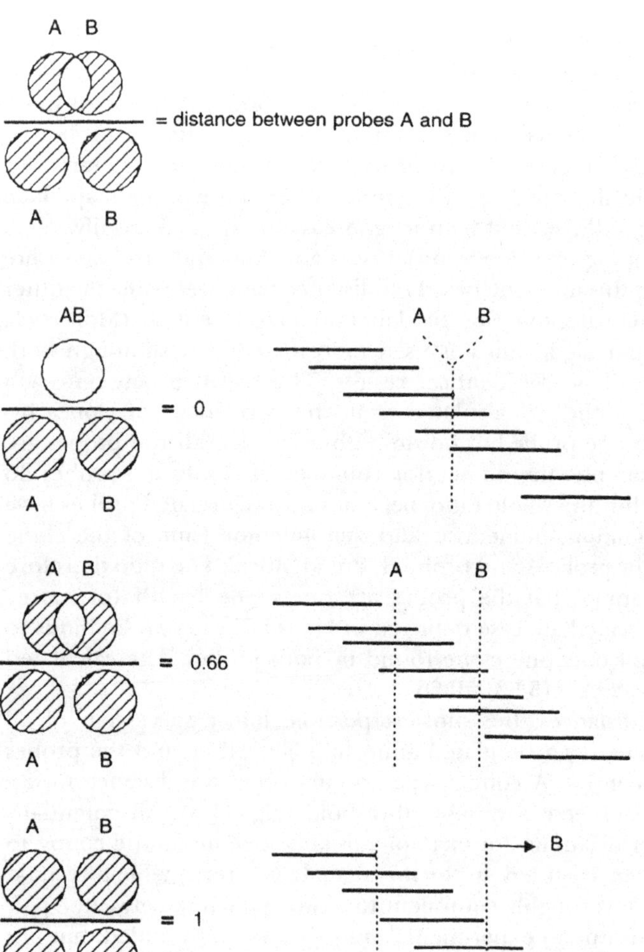

Figure 11.7 Algorithm for probe distance calculation. The distance (in average clone length) between a pair of probes is described by the ratio of the number of clones they do *not* have in common and the sum of the clone numbers found by probe A and probe B.

a *subsequent* comparison of the various types of mapping information does not only highlight areas of inconsistency, but also assists in the design of experiments aimed at resolving the conflicts. Based on the independent mapping information (usually of lower resolution) the emerging contigs

#	clone	1	2	3	4	5	6	7	8	9	10	11	12	13	14	15	16	17	18	19	20	21	22	23	24	25	26
																			distances								
1	14f2	0	62	82	.	.	.	.	.	.	.	.	.	.	.	.	.	.	.	.	.	.	.	.	.	.	.
2	15g9	62	0	52	64	70	.	.	.	.	.	.	.	.	.	.	.	.	.	.	.	.	.	.	.	.	.
3	57b9	82	52	0	21	41	.	.	.	.	.	.	.	.	.	.	.	.	.	.	.	.	.	.	.	.	.
4	56f9	.	64	21	0	38	.	.	.	.	.	.	.	.	.	.	.	.	.	.	.	.	.	.	.	.	.
5	56g7	.	70	41	38	0	.	.	.	.	.	.	.	.	.	.	.	.	.	.	.	.	.	.	.	.	.
6	8c9	.	.	.	.	.	0	67	80	90	.	.	.	.	.	.	.	.	.	.	.	.	.	.	.	.	.
7	56g3	.	.	.	.	.	67	0	63	79	.	.	.	.	.	.	.	.	.	.	.	.	.	.	.	.	.
8	17a6	.	.	.	.	.	80	63	0	70	.	.	.	.	.	.	.	.	.	.	.	.	.	.	.	.	.
9	5a7	.	.	.	.	.	90	79	70	0	47	50	67	74	.	.	.	.	.	.	.	.	89	.	.	.	.
10	57b6	.	.	.	.	.	.	.	.	47	0	20	45	61	78	.	.	.	.	.	.	90	87	.	.	.	.
11	56e2	.	.	.	.	.	.	.	.	50	20	0	47	57	65	.	.	.	.	.	.	.	.	.	.	.	.
12	56e9	.	.	.	.	.	.	.	.	67	45	47	0	27	36	.	.	.	.	.	.	.	.	.	.	.	.
13	57a4	.	.	.	.	.	.	.	.	74	61	57	27	0	20	.	.	.	.	.	.	.	.	.	.	.	.
14	8h6	.	.	.	.	.	.	.	.	.	78	65	36	20	0	.	.	.	.	.	.	.	.	.	.	.	.
15	57a11	.	.	.	.	.	.	.	.	.	.	.	.	.	.	0	64	.	.	.	.	.	.	.	.	.	.
16	56f2	.	.	.	.	.	.	.	.	.	.	.	.	.	.	64	0	85	88	.	.	.	.	.	.	.	.
17	56e11	.	.	.	.	.	.	.	.	.	.	.	.	.	.	.	85	0	30	.	.	.	.	.	.	.	.
18	57a2	.	.	.	.	.	.	.	.	.	.	.	.	.	.	.	88	30	0	.	.	.	.	.	.	.	.
19	2g12	.	.	.	.	.	.	.	.	.	.	.	.	.	.	.	.	.	.	0	42	72	81	83	90	.	.
20	11g5	.	.	.	.	.	.	.	.	.	.	.	.	.	.	.	.	.	.	42	0	52	71	72	82	83	.
21	15g10	.	.	.	.	.	.	.	.	.	90	.	.	.	.	.	.	.	.	72	52	0	57	58	68	72	87
22	56e8	.	.	.	.	.	.	.	.	89	87	.	.	.	.	.	.	.	.	81	71	57	0	8	27	35	63
23	56f12	.	.	.	.	.	.	.	.	.	.	.	.	.	.	.	.	.	.	83	72	58	8	0	19	28	62
24	7e8	.	.	.	.	.	.	.	.	.	.	.	.	.	.	.	.	.	.	90	82	68	27	19	0	16	57
25	56e12	.	.	.	.	.	.	.	.	.	.	.	.	.	.	.	.	.	.	.	83	72	35	28	16	0	51
26	57b8	.	.	.	.	.	.	.	.	.	.	.	.	.	.	.	.	.	.	.	.	87	63	62	57	51	0

1 2 3 4 5 6 7 8 9 10 11 12 13 14 15 16 17 18 19 20 21 22 23 24 25 26

Figure 11.8 Generation of probe contigs. For each possible pair of probes a distance was determined. The probes were ordered into four contigs using a simulated annealing algorithm. The maximal acceptable distance was set to 90% of the average clone length. Larger values were ignored and are printed as dots only. Numbers away from the main diagonal (connecting probes 9, 10 to 21, 22) indicate cross-hybridization events between different probe contigs due to repetitive sequences.

are sorted, checked, and oriented. If after this process two adjacent contigs are linked, this link is more likely to be genuine and calls for further examination.

3. Monitoring the Progress

For libraries of a clone coverage better than six genome equivalents, the progress in hybridization mapping can be described rather accurately. Relatively simple equations permit inspection of the map development and indicate critical stages when, for example, a change of strategy would be beneficial (for equations, see Grigoriev, 1993). The quality of a map can be anticipated and some important parameters can be determined: for example, the number of real gaps in a map approximately equals the number of contigs consisting of a single probe, while the number of gaps caused by nondetected overlaps is described by the number of probe contigs that contain more than one probe. In addition, the course of a project can be predicted. Using an optimal strategy, there are four distinct situations: (1) at an average probe density equivalent to 0.5 probe per clone *on a onefold genome coverage* the number of real contig breaks reaches a maximum; (2) the same is true for the number of contigs at 1 probe per clone; (3) at a density of 1.5 the number of nondetected overlaps peaks; and (4) when 3 probes are hybridized per clone interval, the number of real gaps in the map is near zero, apart from unclonable areas and, obviously, breaks between contigs on different chromosomes. Checking these theoretical figures against an actual experiment (the mapping of *S. pombe;* Grigoriev, 1993) proved their value for the evaluation of the progress of an experiment. Decisions can be made about whether and when the strategy should be adapted for a more efficient mapping.

D. Data Handling

1. Software Tools for Genome Mapping

One of the major problems faced in contig assembly are the management and visualization of very large data sets, the efficient selection of probes to minimize the number of experiments, and the resolution of contradictions in the experimental data. The software developed by Mott *et al.* (1993) addresses the needs and difficulties which were faced in the mapping analyses of the *S. pombe* genome (Maier *et al.*, 1992; Hoheisel *et al.*, 1993). It consists of several subprograms (for data management, optional selection and filtering processes, various types of display, and the ordering algorithms), some of which complement or supplement each other. The compartmental structure of the package allows for a modification or addition of programs.

There are many matters to be dealt with, which in part are rather banal but nevertheless important for the realization of a clone map: for example, during a project quite often different data formats are created (filter coordinates, microdish well coordinates, optical density values) and entered to the database via different input devices (camera system, digitizer, keyboard); these input formats are merged into a single canonical format. Data differences such as missing values (due to hybridizations to only a part of the clones) or variable spotting order on the clone filters have to be accounted for. For the "sampling without replacement" selection, a list of the yet unhit clones has to be produced. Mapping is simplified if the data obtained on different clone libraries can be combined. In the actual analysis, the importance of an identification of potential artefacts (e.g., Fig. 11.4) cannot be overestimated. Furthermore, since the ordering processes are based on the assumption that each probe is unique, an implementation of an optional filtering of the data detects probable clone contaminations in the library, causing false connections, or clones containing repeat sequences and excludes them from further calculations. Using different mapping algorithms on the same data provides a check that the produced contigs are legitimate.

An informative visualization of the data is imperative for map generation, since even with the help of sophisticated computational tools, human interaction is still required for checking, interpreting, and correcting the results. A display of the data as a matrix with clones as rows and probes as columns (e.g., Figs. 11.4 and 11.6) is extremely helpful for such work. A hybridization event is shown as a printed line at the intersection of clone and probe. Data can either be summarized on a single sheet or spread over as much area as is needed for a detailed visual analysis. This sort of data representation allows for an immediate identification of inconsistencies in the data, since all information is visible, including the results which do not fit to the current probe and clone order. Different data sets can be printed side by side, with, for example, the clone order only based on one sort of probe hybridization. This way, the congruence between different probe types can be checked.

2. Communication Concepts

Hybridization techniques require two experimental samples, probe and target; these often overlap or are exchangeable. Hence, hybridization analyses produce redundant information, since for instance the probes as well as the clones define a physical map. Probes can therefore be used as a description of a genomic area enabling its isolation anywhere. In principle, this would serve to free laboratories from the distribution of biological samples in favor of an exchange of written or electronic information. However, the clone libraries themselves are an integral portion of the map information. A mere transfer of information would lead to a duplication of work in many laboratories, whose avoidance is one major reason for communication. The importance of a continuous existence and distribution of clone libraries for analyses is best documented by the impact the distribution of the human Megabase YAC-library (Bellanne-Chantelot *et al.*, 1992) has had on the Human Genome Project.

IV. Conclusions

Hybridization techniques provide the means for an efficient physical mapping. High-coverage clone libraries are generated and stored. By the concept described above, whole libraries become accessible, with the clones nevertheless individually identifiable *and easily retrievable*. Work can be done in many places simultaneously on the same material and redundancies in library generation and characterization are much avoided. Due to the common nature of the information obtained, hybridization results are instantly correlated and easily interrelated. Rather than following a top-down or bottom-up approach, mapping preferably should be started in parallel on different levels of analysis; information generated on one level immediately serves the advance on another level. Particularly, a combination of long-range (e.g., YAC) and short-distance (e.g., cosmid) mapping considerably accelerates an ordering process; especially, the gap closing in both systems is simplified. In an apparent extension to the mere physical mapping, cDNA and exonic libraries can be used as probe sources, thus directly combining the ordering of the genomic DNA with the localization of transcribed sequences. By a simultaneous hybridization to the genomic and back to the transcriptional libraries, such experiments would additionally produce results on sequence homologies between transcribed sequences.

Acknowledgment

The contributions of Andrei Grigoriev are gratefully acknowledged.

References

Ajioka, J. W., Smoller, D. A., Jones, R. W., Carulli, J. P., Vellek, A. E. C., Garza, D., Link, A. J., Duncan, I. W., and Hartl, D. L. (1991). *Drosophila* genome project: One-hit coverage in yeast artificial chromosomes. *Chromosoma* **100**, 495–509.

Bellanne-Chantelot, C., Lacroix, B., Ougen, P., Billault, A., Beaufils, S., Bertrand, S., Georges, I., Glibert, F., Gros, I., Lucotte, G., JSusini. L., Codani, J.-J., Gesnouin, P., Pook, S., Vaysseix, G., Lu-Kuo, J., Ried, T., Ward, D., Chumakov, I., Le Paslier, D., Barillot, E. and Cohen, D. (1992). Mapping the whole human genome by fingerprinting yeast artificial chromosomes. *Cell (Cambridge, Mass.)* **70**, 1059–1068.

Clarke, L., and Carbon, J. (1976). A colony bank containing synthetic ColE1 hybrid plasmids representative of the entire *E. coli* genome. *Cell (Cambridge, Mass.)* **9**, 91–99.

Coulson, A., Kozono, Y., Lutterbach, B., Shownkeen, R., Sulston, J., and Waterston, R. (1991). YACs and the *C. elegans* genome. *BioEssays* **13**, 413–417.

de Jong, P. J., Chen, C., and Garnes, J. (1989). Application of PCR for the construction of vectors and the isolation of probes. *In* "Polymerase Chain Reaction" (H. A. Erlich, R. A. Gibbs, and H. H. Kazazian, eds.), pp. 205–210. Cold Spring Harbor Lab. Press, Cold Spring Harbor, NY.

Evans, G. A., and Lewis, K. A. (1989). Physical mapping of complex genomes by cosmid multiplex analysis. *Proc. Natl. Acad. Sci. U.S.A.* **86**, 5030–5034.

Feinberg, A. P., and Vogelstein, B. (1983). A technique for radiolabeling DNA restriction endonuclease fragments to high specific activity. *Anal. Biochem.* **132**, 6–13.

Gress, T. M., Hoheisel, J. D., Lennon, G. G., Zehetner, G., and Lehrach, H. (1992). Hybridization fingerprinting of high density cDNA-library arrays with cDNA pools derived from whole tissues. *Mamm. Genome* **3**, 609–619.

Grigoriev, A. V. (1993). Theoretical predictions and experimental observations of genomic mapping by anchoring random clones. *Genomics* **15**, 311–316.

Hoheisel, J. D. (1994). Application of hybridization techniques to genome mapping and sequencing. *Trends Genet.* **10**, 79–83.

Hoheisel, J. D., and Lehrach, H. (1993). Use of reference libraries and hybridization fingerprinting for relational genome analysis. *FEBS Lett.* **325**, 118–122.

Hoheisel, J. D., Nizetic, D., and Lehrach, H. (1989). Control of partial digestion combining the enzymes Dam methylase and MboI. *Nucleic Acids Res.* **17**, 4571–4582.

Hoheisel, J. D., Lennon, G. G., Zehetner, G., and Lehrach, H. (1991). Use of high coverage reference libraries of *Drosophila melanogaster* for relational data analysis; a step towards mapping and sequencing of the genome. *J. Mol. Biol.* **220**, 903–914.

Hoheisel, J. D., Maier, E., Mott, R., McCarthy, L., Grigoriev, A. V., Schalkwyk, L. C., Nižetić, D., Francis, F., and Lehrach, H. (1993). High-resolution cosmid and P1 maps spanning the 14-Mbp genome of the fission yeast *Schizosaccharomyces pombe*. *Cell (Cambridge, Mass.)* **73**, 109–120.

Hoheisel, J. D., Ross, M. T., Zehetner, G., and Lehrach, H. (1994). Relational genome analysis using reference libraries and hybridization fingerprinting. *J. Biotechnol.* **35**, 121–134.

Ioannou, P. A., Amemiya, C. T., Garnes, J., Kroisel, P. M., Shizuya, H., Chen, C., Batzer, M. A., and de Jong, P. J. (1994). A new bacteriophage P1-derived vector, for the propagation of large human DNA fragments. *Nat. Genet.* **6**, 84–89.

Larin, Z., Monaco, A. P., Meier-Ewert, S., and Lehrach, H. (1993). Construction and characterization of yeast artificial chromosome libraries from the mouse genome. *In* "Methods in Enzymology" Vol. 255, pp. 623–637. Academic Press, San Diego, CA.

Lehrach, H., Drmanac, R., Hoheisel, J. D., Larin, Z., Lennon, G., Monaco, A. P., Nižetić, D., Zehetner, G., and Poustka, A. (1990). Hybridization fingerprinting in genome mapping and sequencing. *In* "Genome Analysis" (K. E. Davies and S. M. Tilghman, eds.), Vol. 1, pp. 39–81. Cold Spring Harbor Lab. Press, Cold Spring Harbor, NY.

Maier, E., Hoheisel, J. D., McCarthy, L., Mott, R., Grigoriev, A. V., Monaco, A. P., Larin, Z., and Lehrach, H. (1992). Yeast artificial chromosome clones completely spanning the genome of *Schizosaccharomyces pombe*. *Nat. Genet.* **1**, 273–277.

Meier-Ewert, S., Maier, E., Ahmadi, A. R., Curtis, J., and Lehrach, H. (1993). An automated approach to generating expressed sequence catalogues. *Nature (London)* **361**, 375–376.

Merriam, J., Ashburner, M., Hartl, D. L., and Kafatos, F. C. (1991). Toward cloning and mapping the genome of *Drosophila*. *Science* **254**, 221–225.

Mizukami, T., Chang, W. I., Garkavtsev, I., Kaplan, N., Lombardi, D., Matsumoto, T., Niwa, O., Kounosu, A., Yanagida, M., Marr, T. G., and Beach, D. (1993). A 13 kb resolution cosmid map of the 14 Mb fission yeast genome by nonrandom sequence-tagged site mapping. *Cell (Cambridge, Mass.)* **73**, 121–132.

Mott, R., Grigoriev, A., Maier, E., Hoheisel, J. D., and Lehrach, H. (1993). Algorithms and software tools for ordering clone libraries; application to the mapping of the genome of *Schizosaccharomyces pombe*. *Nucleic Acids Res.* **21**, 1965–1974.

Nizetic, D., Zehetner, G., Monaco, A. P., Gellen, L., Young, B. D., and Lehrach, H. (1991). Construction, arraying and high density screening of large insert libraries of human chromosomes X and 21: Their potential use as reference libraries. *Proc. Natl. Acad. Sci. U.S.A.* **88**, 3233–3237.

Pierce, J. C., Sauer, B., and Sternberg, N. (1992). A positive selection vector for cloning high molecular weight DNA by the bacteriophage P1 system: Improved cloning efficiency. *Proc. Natl. Acad. Sci. U.S.A.* **89**, 2056–2060.

Ross, M. T., Hoheisel, J. D., Monaco, A. P., Larin, Z., Zehetner, G., and Lehrach, H. (1992). High-density gridded YAC filters: their potential as genome mapping tools. *In* "Techniques for the Analysis of Complex Genomes" (R. Anand, ed.), pp. 137–154. Academic Press, San Diego, CA.

Rychlik, W., and Rhoads, R. E. (1989). A computer program for choosing optimal oligonucleotides for filter hybridization, sequencing and in vitro amplification of DNA. *Nucleic Acids Res.* **17,** 8543–8551.

Saluz, H., and Jost, J.-P. (1989). A simple high-resolution procedure to study DNA methylation and in vivo DNA-protein interactions on a single-copy gene level in higher eukaryotes. *Proc. Natl. Acad. Sci. U.S.A.* **86,** 2602–2606.

Sambrook, J., Fritsch, E. F., and Maniatis, T. (1989). "Molecular Cloning: A Laboratory Manual," Cold Spring Harbor Lab. Press, Cold Spring Harbor, NY.

Smoller, D. A., Petrov, D., and Hartl, D. L. (1991). Characterization of bacteriophage P1 library containing inserts of *Drosophila* DNA of 75-100 kilobase pairs. *Chromosoma* **100,** 487–494.

Whittaker, P. A., Mathrubutham, M., and Wood, L. (1993). Construction of phage sublibraries from nanogram quantities of YAC DNA purified by preparative PFGE. *Trends Genet.* **9,** 195–196.

Index

AFLP, *see* Amplified fragment length
  polymorphism
Agarose
  elution of DNA from gels, 264–265, 307
  embedding for high-molecular weight
      DNA preparation
    bacteria, 13–14, 16
    Dictyostelium, 299
    erythrocytes, 234–236
    protoplasts, 64, 68
    protozoan parasites, 143–144
  enzymatic reactions in DNA–agarose
      plugs, 18–19, 65–66, 69–71, 187–189
  selection for pulsed-field gel
      electrophoresis, 13
Allele-specific amplification, polymorphism
  assay, 109
Allele-specific ligation, polymorphism assay,
  109–110
Amplified fragment length polymorphism
  comparison to other marker systems, 78
  DNA marker system in plants, 76
    advantages, 122, 128
    DNA preparation, 115–116, 121
    DNA sequencing, 122–123
    gel electrophoresis, 120–121
    materials, 79–80, 113–114
    polymerase chain reaction, 117, 119–
      120
    preamplification, 116–117
    precautions, 121–122
    principle, 111–112
  multiplex ratio, 79, 110–111
ASA, *see* Allele-specific amplification
ASL, *see* Allele-specific ligation

Bacteriophage λ, *see* Cosmid cloning

CAPS, *see* Cleavable amplified polymorphic
  sequences
Chromosome
  determination of number and size in
      bacteria, 171–172
  Dictyostelium characteristics, 293–294
  linearity in bacteria, 219
  preparation from bacteria
    cell growth, 14, 16
    cell lysis, 14, 16
    DNA concentration determination in
        agarose, 17–18
    embedding cells in agarose, 13–14, 16
    yield, 17
  preparation from fungi for karyotyping,
      30–37
  preparation from yeast
    cell growth, 14–15
    cell lysis, 14–15
    DNA concentration determination in
        agarose, 17–18
    embedding cells in agarose, 13–14
    yield, 17
  size standards for electrophoresis, 38
Chromosome walking, *see* Cosmid cloning
Cleavable amplified polymorphic sequences
  comparison to other marker systems, 78
  DNA marker system in plants, 76, 87–88
  materials, 79–80
Cloning, *see* Cosmid cloning; P1 artificial
  chromosome library; Yeast artificial
  chromosome